MW01626293

RANCHLANDS

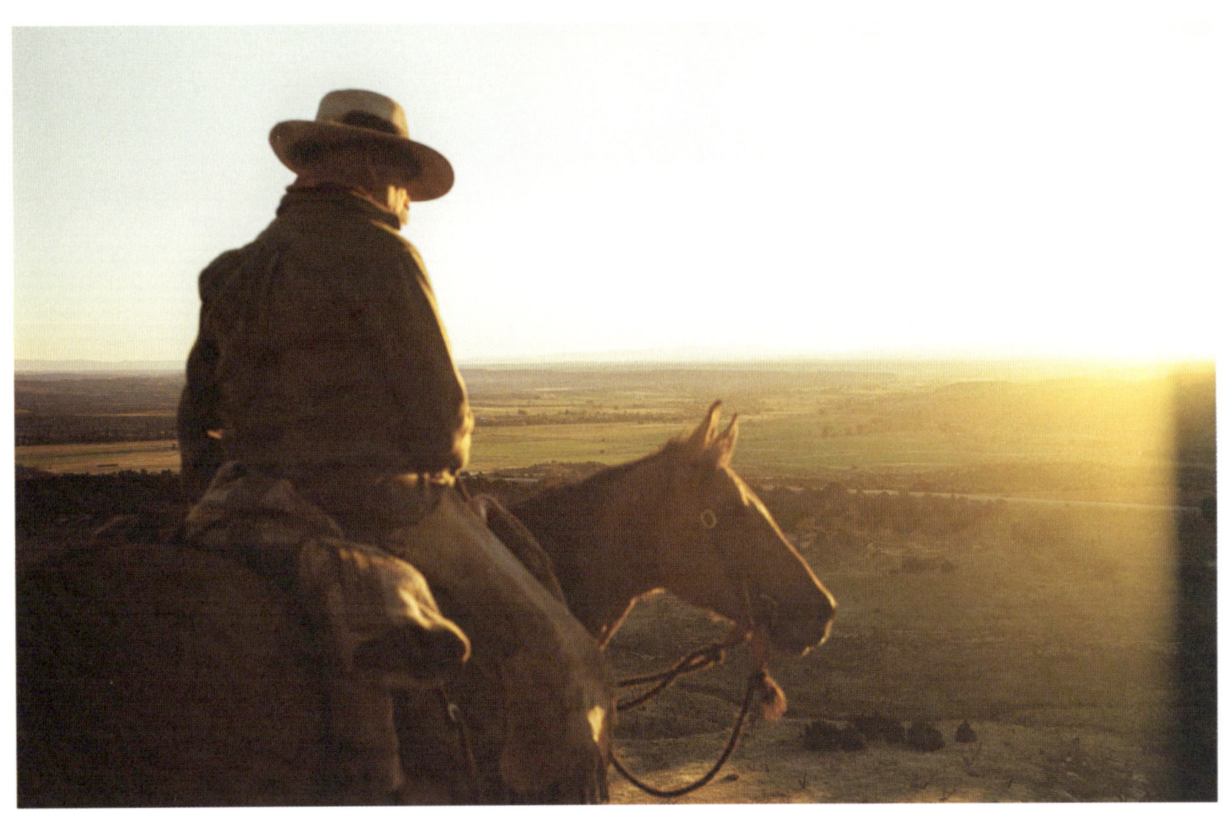

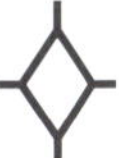

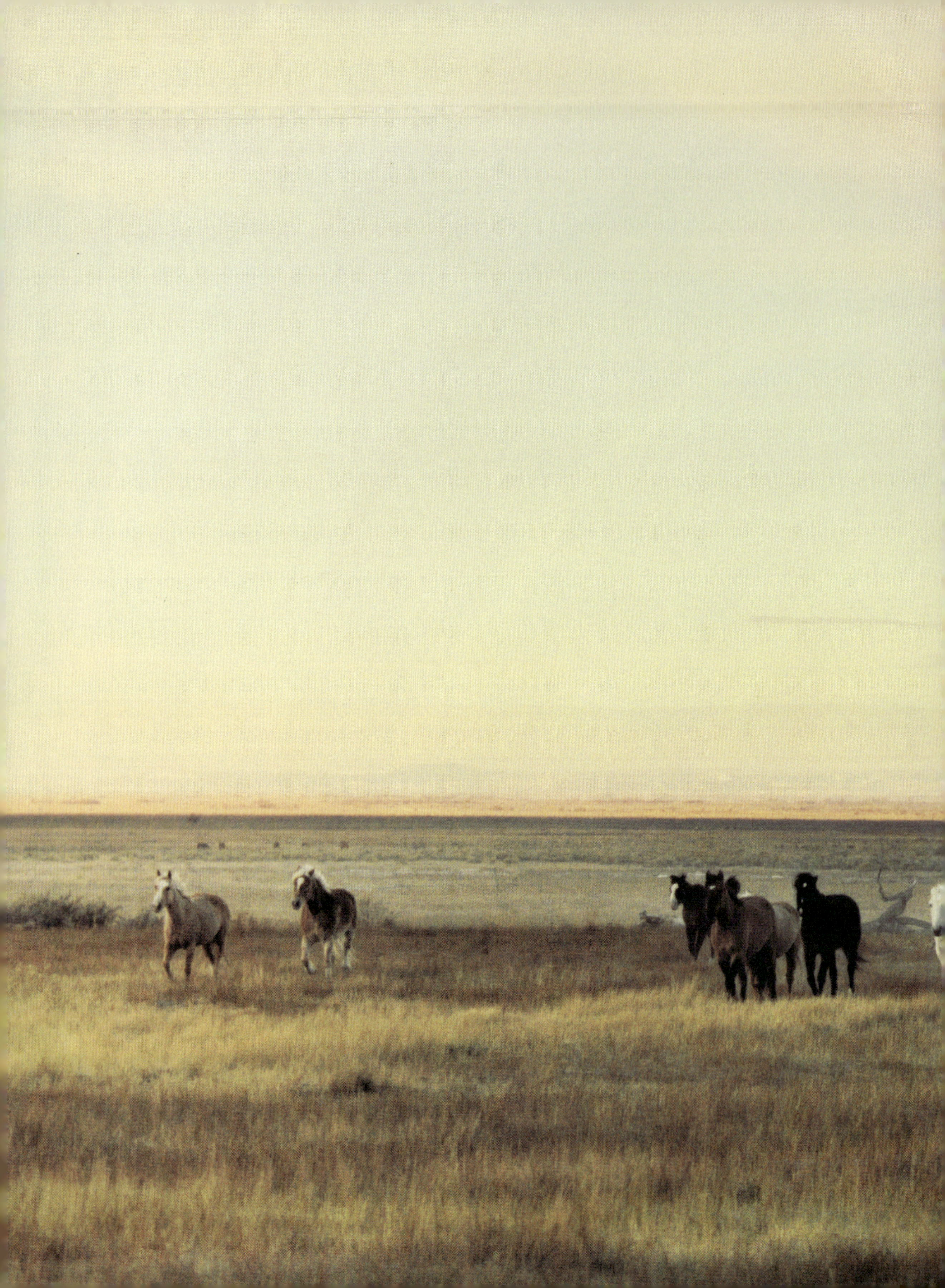

RANCHLANDS

from

THE LAND

for

THE LAND

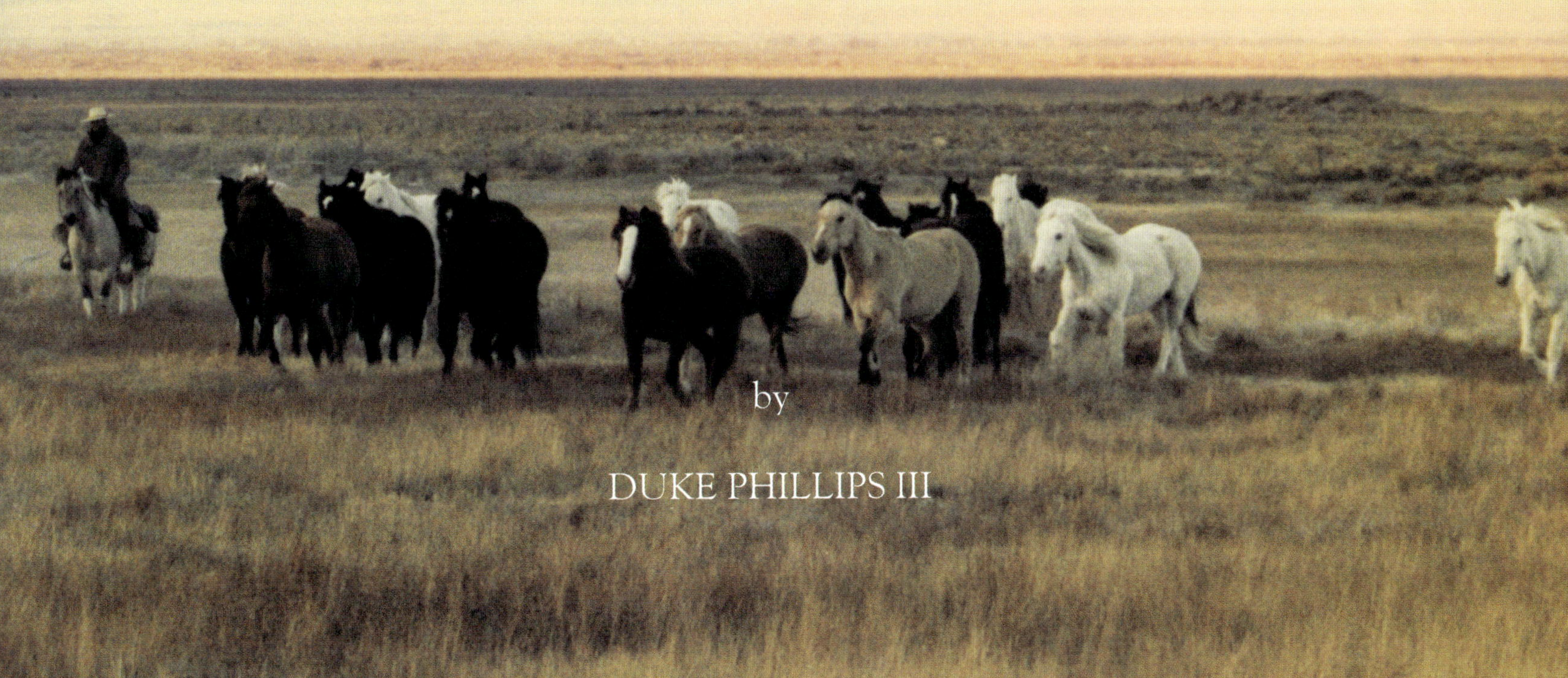

by

DUKE PHILLIPS III

RIZZOLI
NEW YORK

New York Paris London Milan

CONTENTS

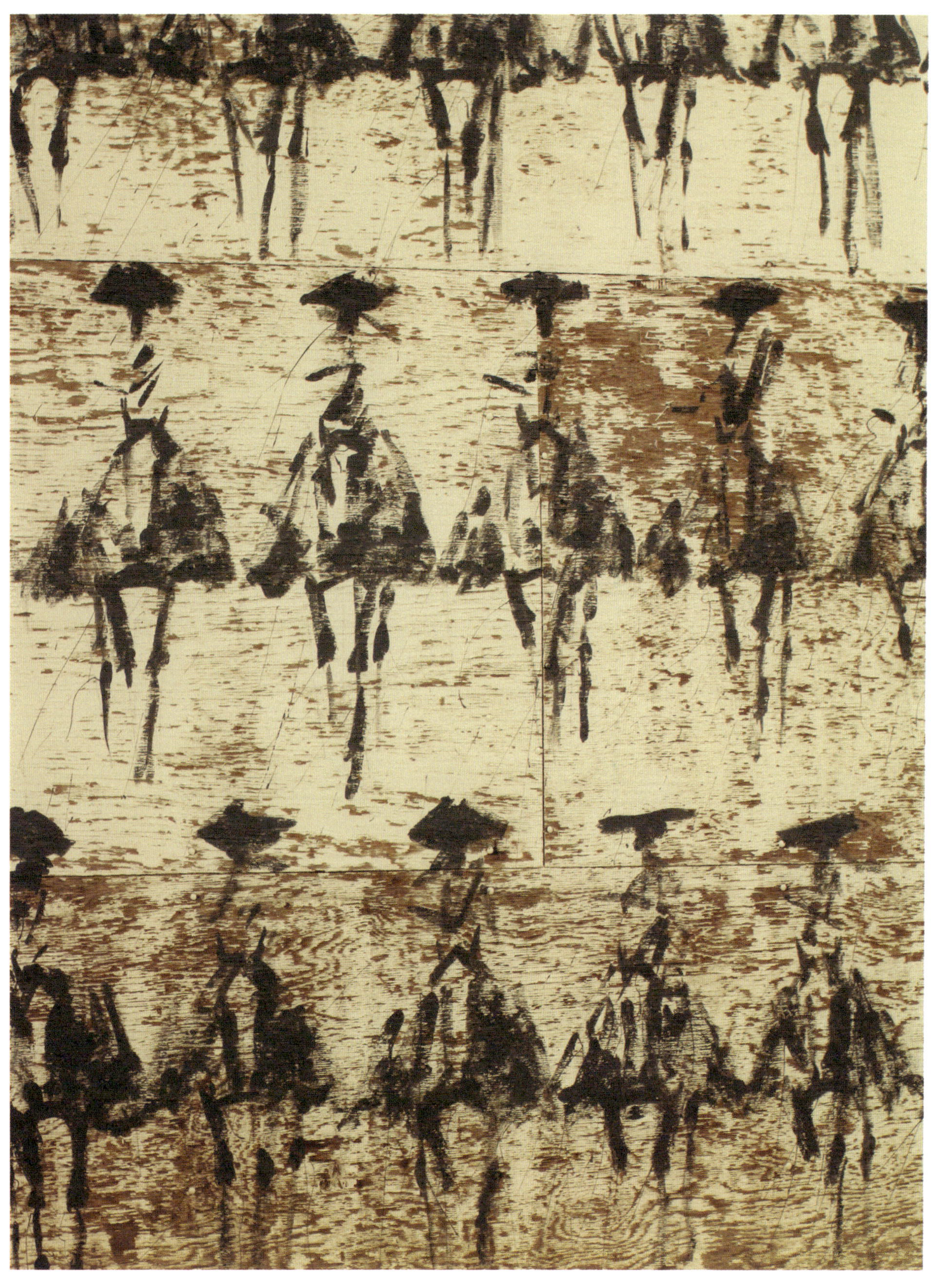

This book is dedicated to all the men and women who live and work on ranches and to all those who came before us, whose footprints we have followed. It is through your work and undaunted persistence that we can preserve our ranching legacy into the future.

A Land of Many

From the land, for the land

Our horses like Gods,
move across your skin,
hooves pressed into the soil
pull us close to you.

The wet eyes of summer
bring the promise of flowers,
a near moon and clear air,
your hands beckoning.

Who can be the ones you seek?

The beetles know us, and the
cholla cactus, sheltering the squirrel tail grass
under its skirt of thorns, from cattle
chasing the green in the spring.

The mockingbird's song, drifts
alone in the night. It echoes
your calling: what do we intend to do here?
We follow your heart beat.

Who can be the ones?

So many days have passed,
joyfull and terrifying,
living your questions, sending us
into the light that is close to everything.

For many reasons we
follow the trails that all your days
have kindled, seeking the hidden
places where you live.

Who can be the ones?
Who can braid the golden tangles of your hair?

We hear you. We are coming.
Our horses lathered, coming.

Duke Phillips
March 2024

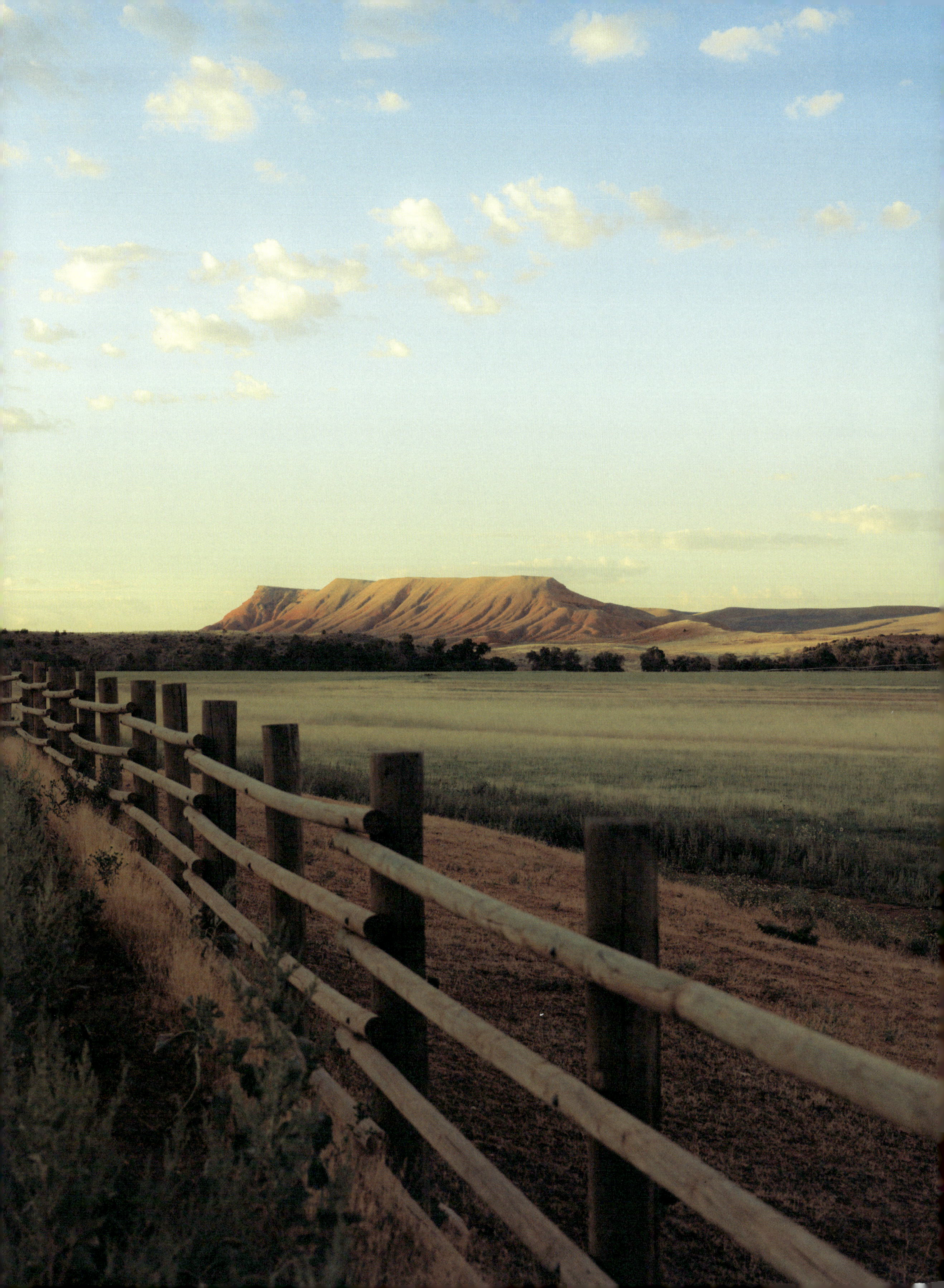

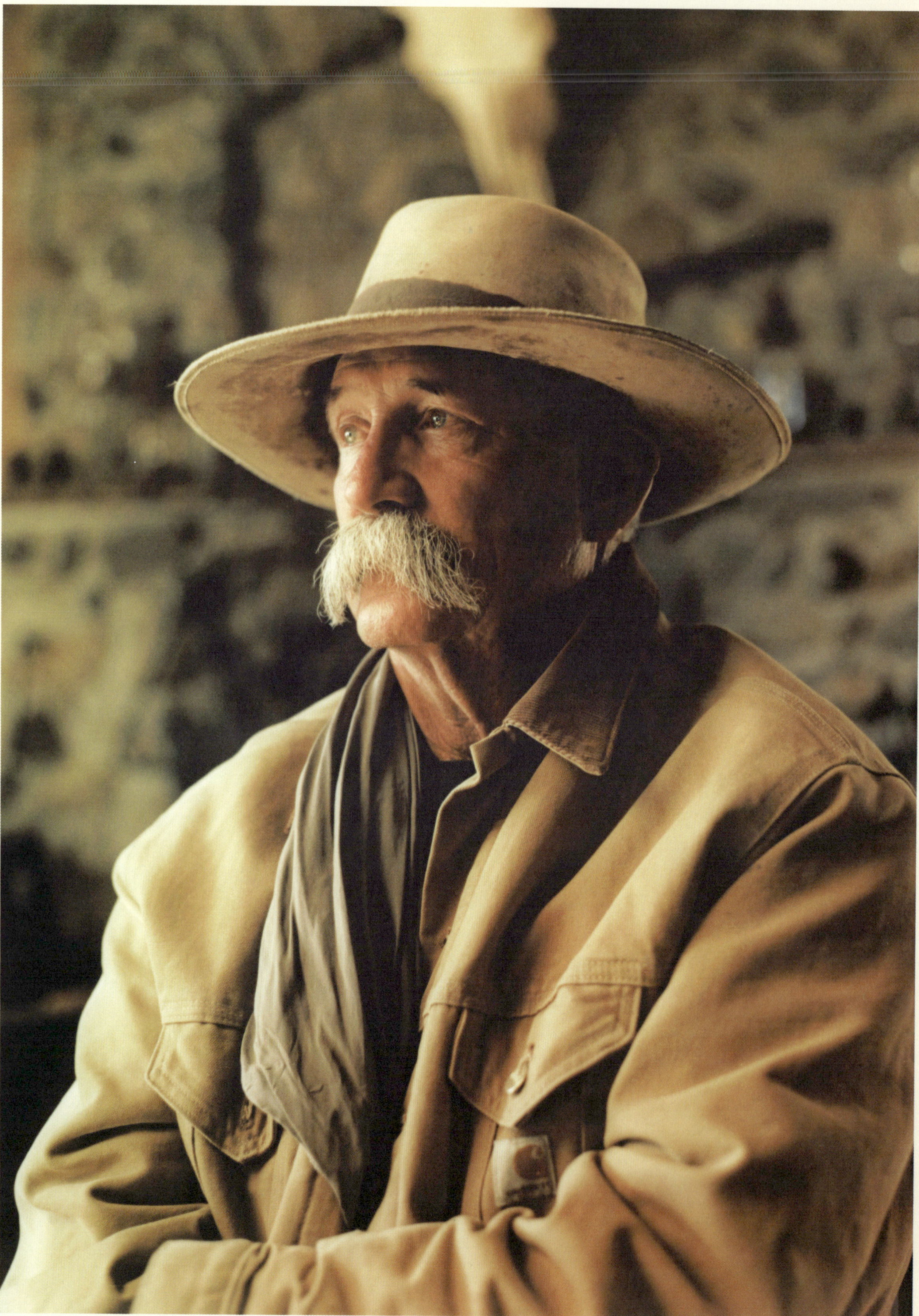

A DREAM FOR TOMORROW

DUKE PHILLIPS III

I first stepped onto the Chico Basin Ranch on November 18, 1999. I had secured a 25-year lease from the Colorado State Land Board to use its 87,000 acres to start a cattle-grazing business. I didn't know how I was going to do it because I essentially had no livestock or horses and minimal equipment. Yet, somehow, I knew I was going to make it work. I had to.

Everything happened differently than I expected. At Ranchlands, our work has transcended simply creating a successful ranching enterprise to encompass what happens when people from different walks of life work together toward a common goal. Cattle ranching is one of the most mythologized and maligned ways of life in our country today. On one hand, ranching is the birthplace of the American cowboy, an iconic figure romanticized and admired globally. Encouraged by Hollywood, the public imagines him herding cows and shooting his pistol into the sky as he gallops his horse across the desert—rough, tough, and independent. On the other hand, ranching has been decried in some environmentally conscious circles as an industry that degrades natural resources through overgrazing. Both have an element of truth, but both are ultimately oversimplified perceptions.

To me, though, ranching is about fiercely resilient men and women living lives rooted deeply in the land, in the dirt, in the wind, sun, and cold. Ranching is uninterrupted by today's fast-changing modern life, while simultaneously upholding sacred values for the integrity of the land, for family, and for community. The consciousness that comes from this deep connection—created by living and depending on the land through generations of hard work and trial and error—is an important, if not essential, element in our cultural identity today, a founding wellspring of pride for Americans.

The global fascination with ranching gives it an intrinsic power, and when coupled with the connection between nature and people who are deeply grounded in it, our work has the potential to impact the ecological woes that we face today in a tremendous way that is bigger than we can imagine. It is this convergence that gives Ranchlands' story its incredible reach. When I won the lease on the Chico Basin Ranch, I thought I was getting into the business of grazing animals. Instead, Ranchlands has become more about cultivating and growing a group of people working together on a mission to create a better world through the legacy of ranching.

The life and blood of this medium is not only the land but also the herd of cattle that has grown to populate the five ranches we currently manage. This beautiful, home-raised herd is the embodiment of a management philosophy that selects and culls animals to be adapted to their natural environment an ideology that represents our personal values and perspective on managing the land and wildlife that support us. We see ourselves not as masters of nature but part of it. For example, if one of our cows loses a calf to a predator or becomes anemic due to pest infestation, we sell the cow; we do not shoot the predator or treat our herd with pesticides. This process of simulating natural selection has created a herd that is well adapted to life on our pastures.

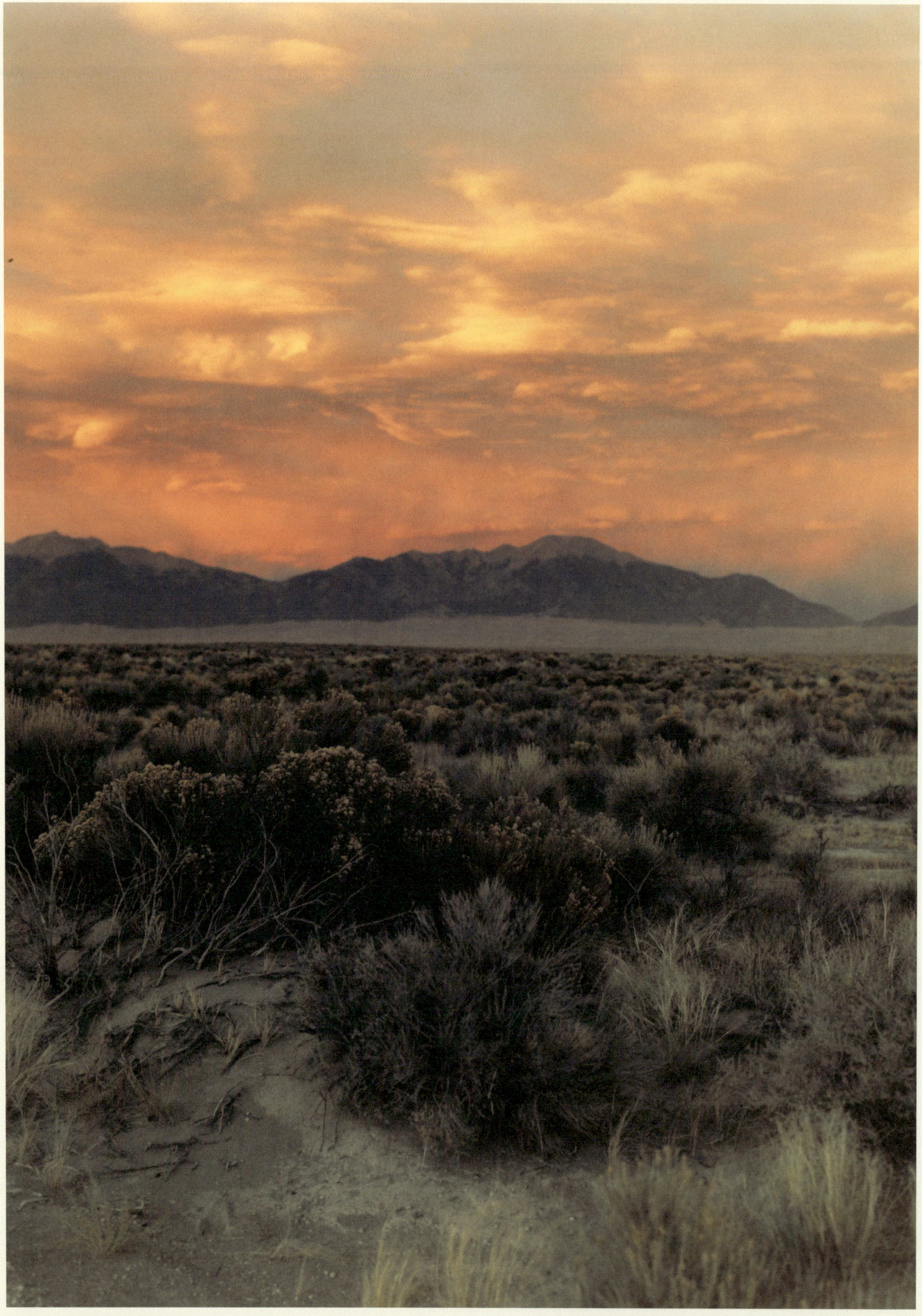

We use our cattle as a tool to disturb the ground surface to achieve conservation strategies, emulating the giant bison herds' symbiotic relationship with the great expanse of North American grasslands. The herd represents our connection to the past but also to the future of ranching, which is an evolving practice of living on and working the land. Our traditions hold land stewardship at their core, generating conservation outcomes that will outlive any single animal in our herd. And if we do it right, this will be the world-changing legacy of our work, outliving each of us as well.

Conservation is one of the most important and defining elements of ranching in the United States. But our story is distinguished by the overwhelming number of people who have become Ranchlands: the close-knit circle of people working on our ranches and the bigger, growing network of environmental organizations, students, guests from around the world, neighbors, and like-minded ranchers. How could I have known that there would be such a magnitude of enthusiasm from so many people who support us through social media, educational events, field days, guest stays, art exhibits, and concerts? How could I have known that these events on our ranch would become rich venues for people from various backgrounds to gather, talk, dance, exchange perspectives about ranching and land stewardship, and build bridges between our community and theirs—that these important interactions would become guiding lights in our mission?

I have chosen to live my life on a ranch for so many reasons. Horses. Cows. Cattle culture. The freedom that comes with being endlessly surrounded by open land. I grew up within a ranching community in northern Mexico that felt set back in time, a hundred years behind modern-day life. We lived five hours from the closest village and used horse-drawn wagons every day to go out and harvest dead oak tree branches for heating and cooking. We used horses for working the cattle herd and to travel between the outposts spread across the ranch that housed working men and their families. Their only contact with the rest of humanity was the monthly trip to the headquarters commissary to restock dry goods—coffee, sugar, flour, and beans—and, from time to time, to the ranch dance that was held on a swept-off patch of dirt behind the cook house. It was a holy place for me, where I lived isolated from the world with my people, my horses, and my land, a refuge for my deep love of the wildest places. Yet it was also a curse because I knew I would never be able to duplicate it.

Today, I find myself in a place that is the opposite of my home in the mountains of old Mexico in so many ways, yet also strikingly the same. Ranchlands has built on my experience as a boy by inspiring people from all walks of life to work together and show the world our way of life and how it nurtures and connects us to the land—a holy place for each of us.

This story shows the beauty, resilience, and juxtaposition of the two ranching worlds that my family and I have traveled. It's a daily celebration of life, working the land, and doing what we love, while bringing real possibilities that will ultimately lead to living in grace and harmony with the natural world.

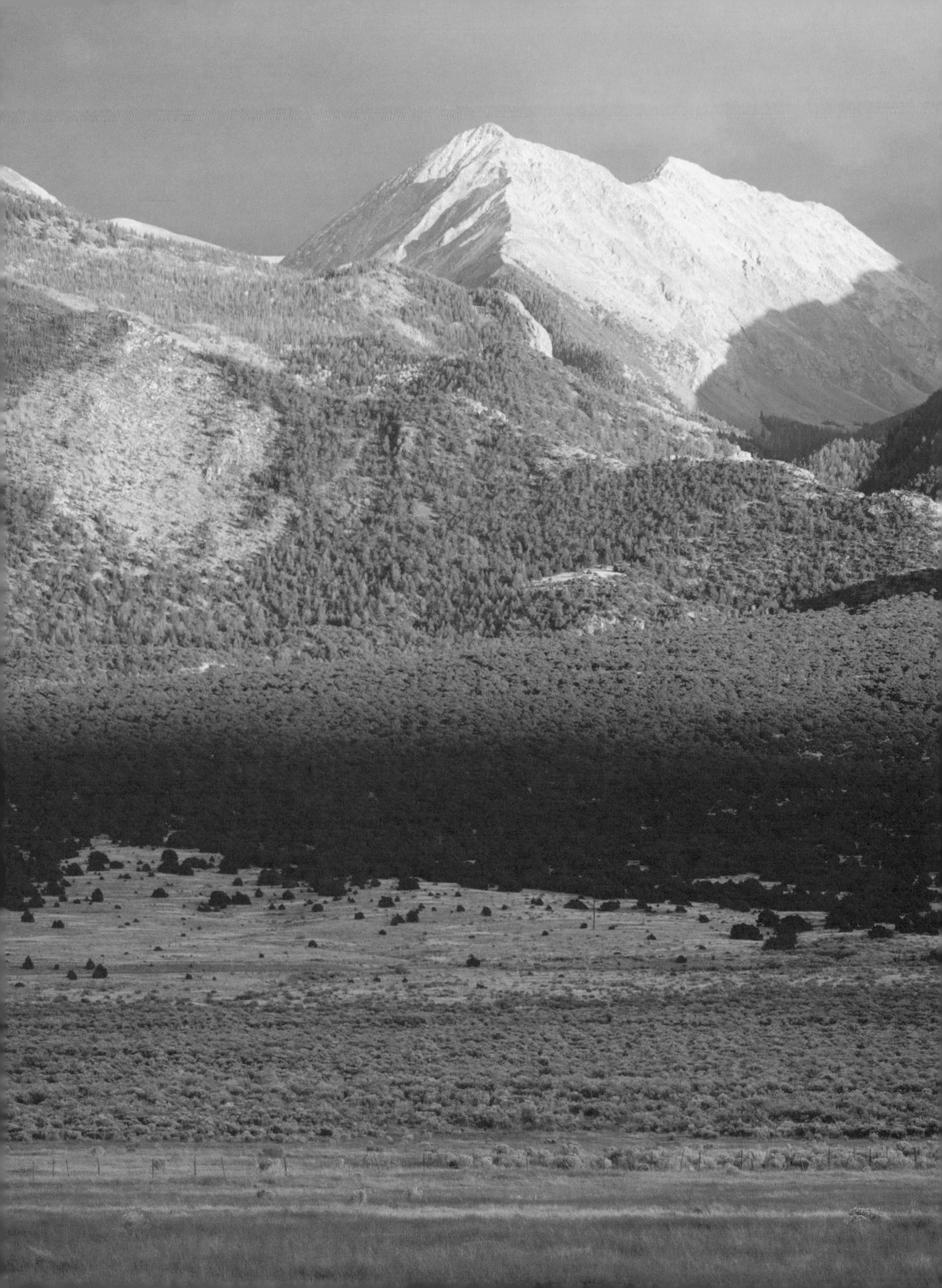

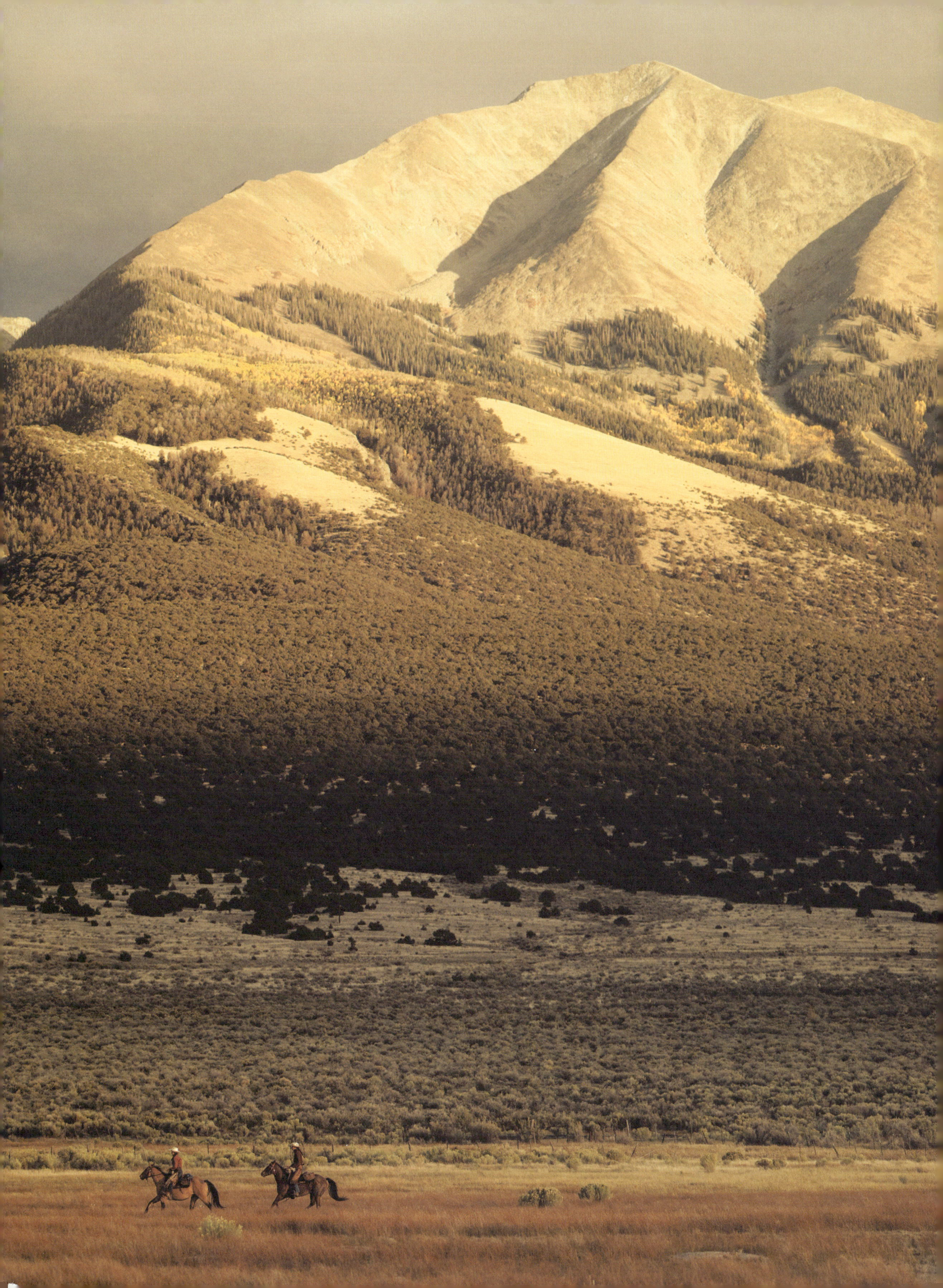

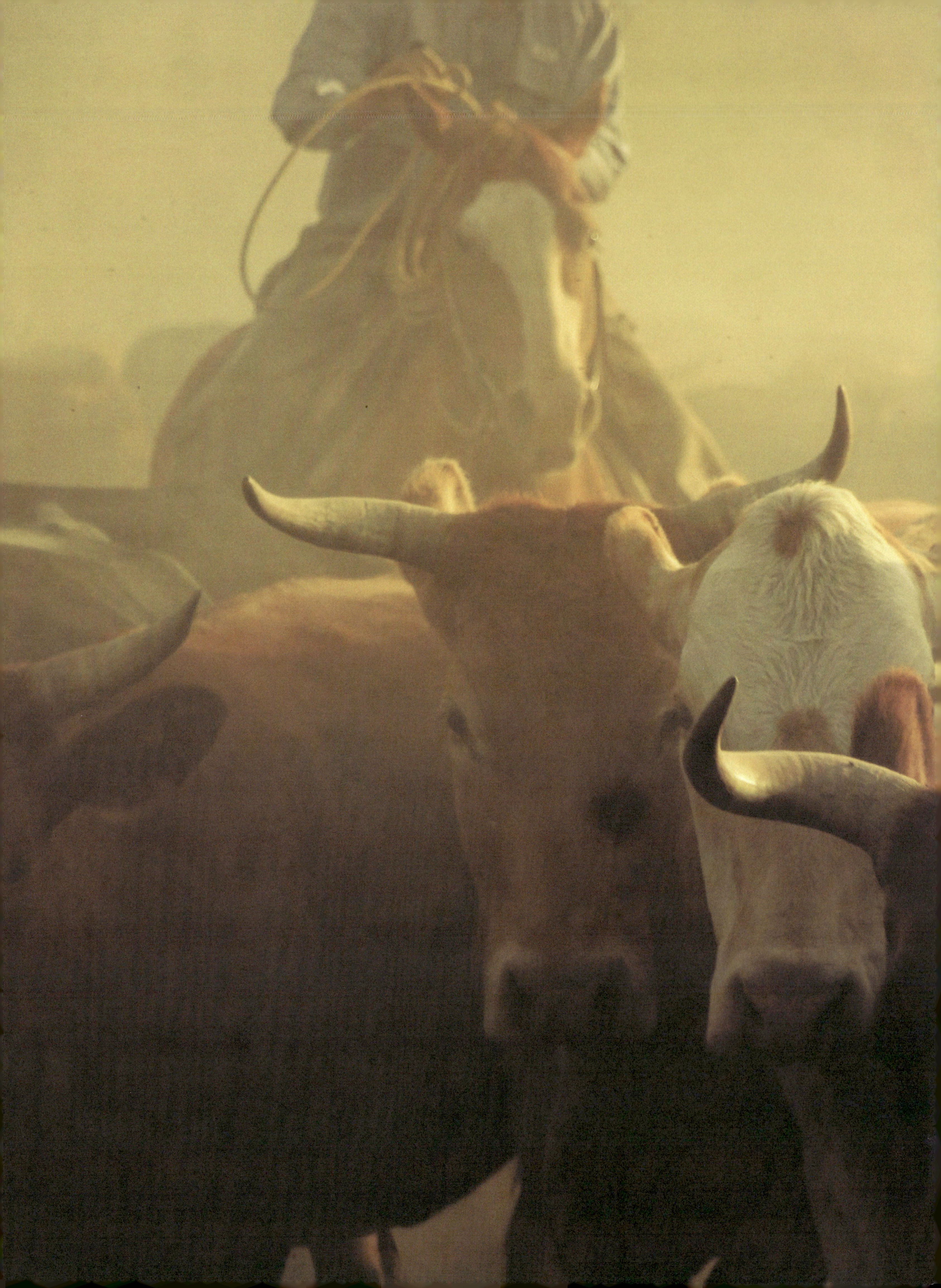

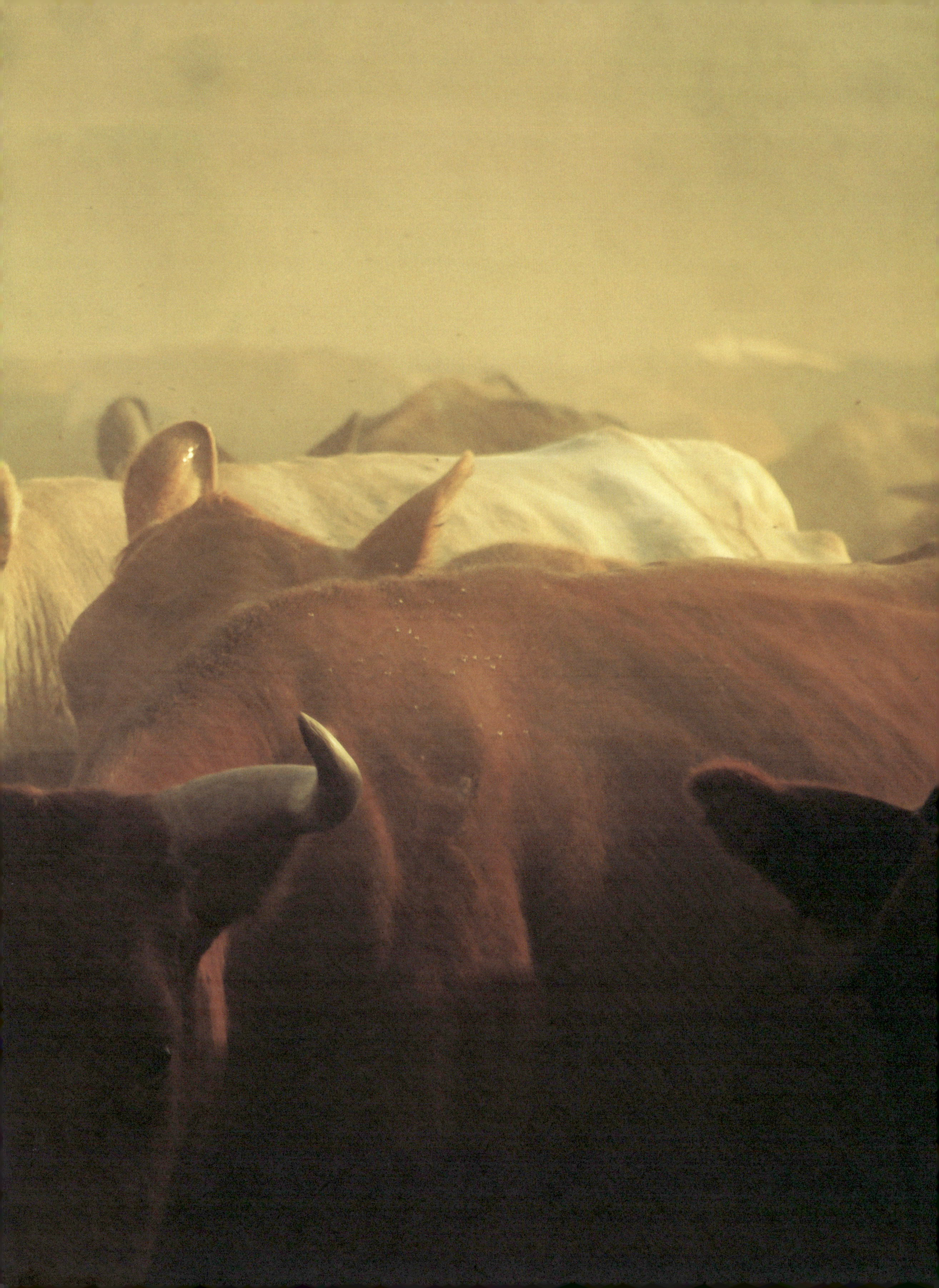

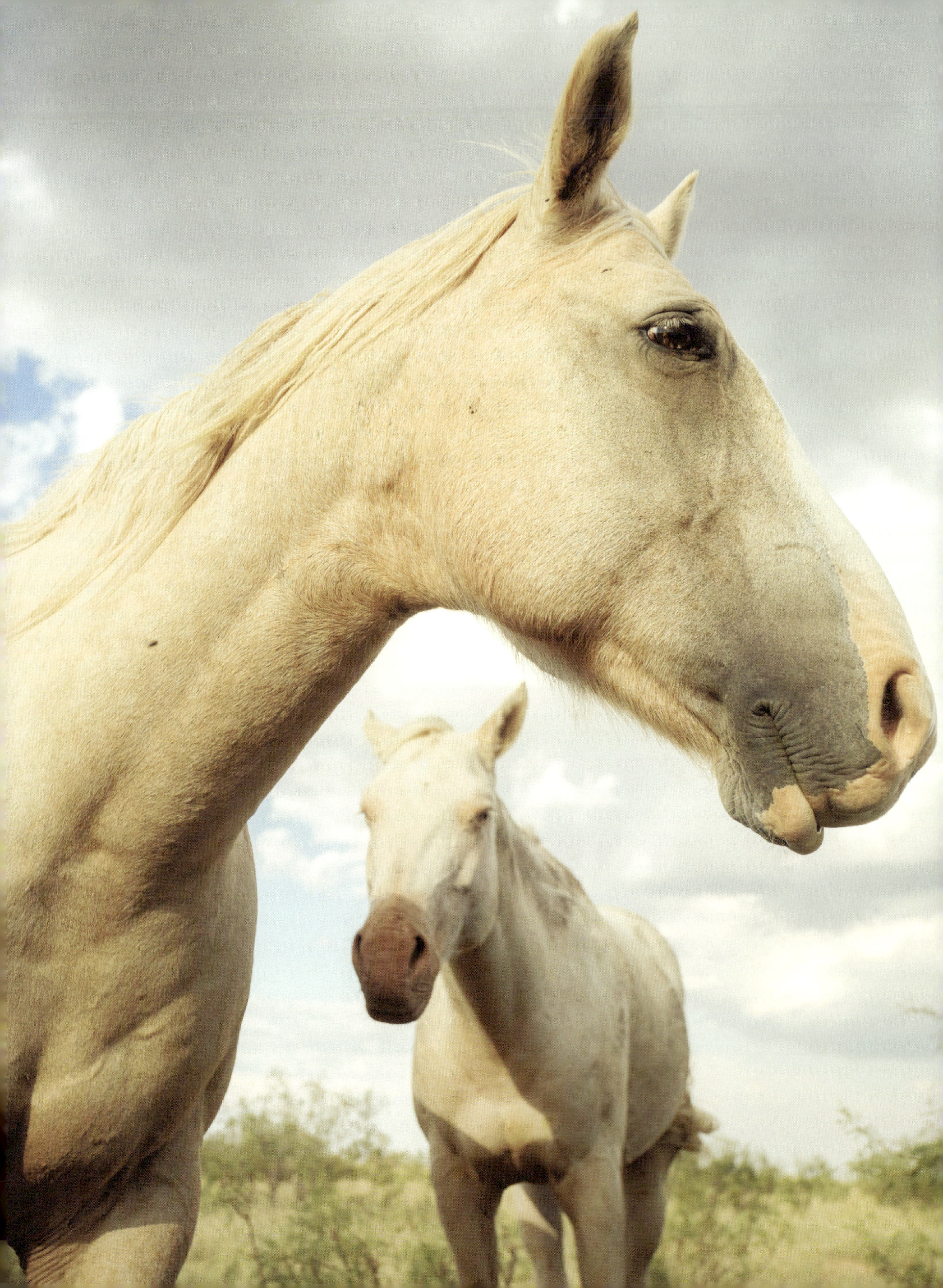

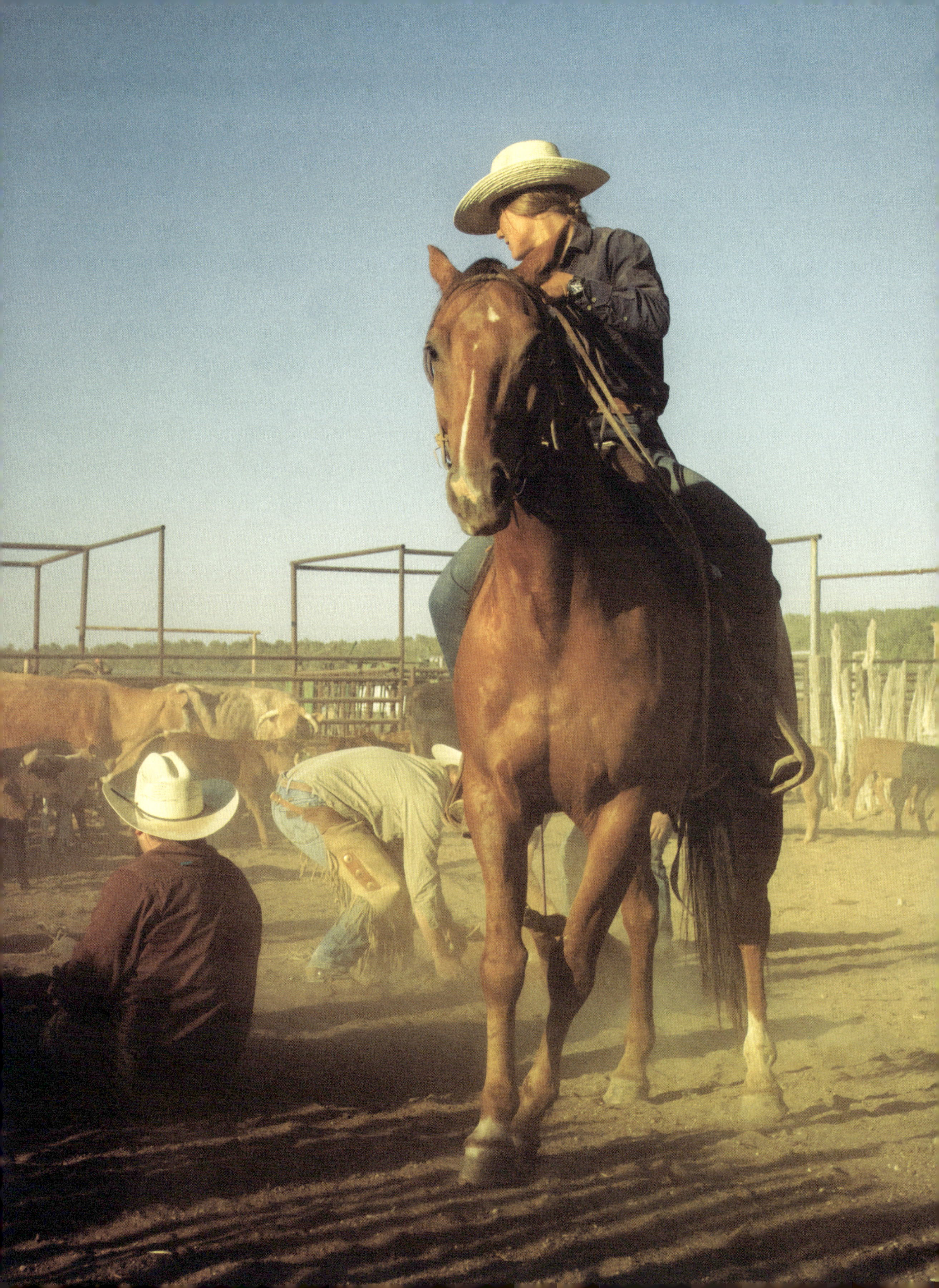

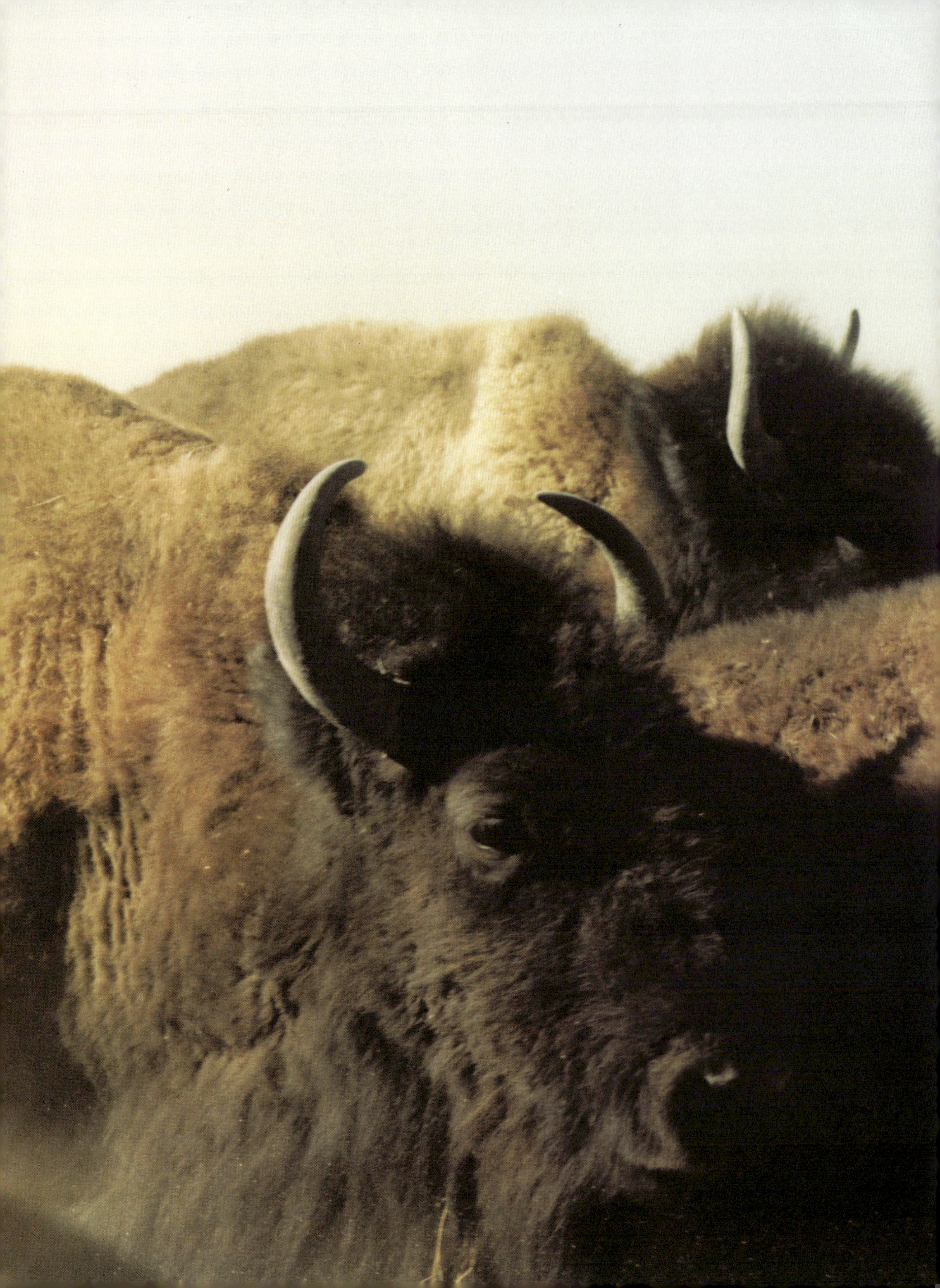

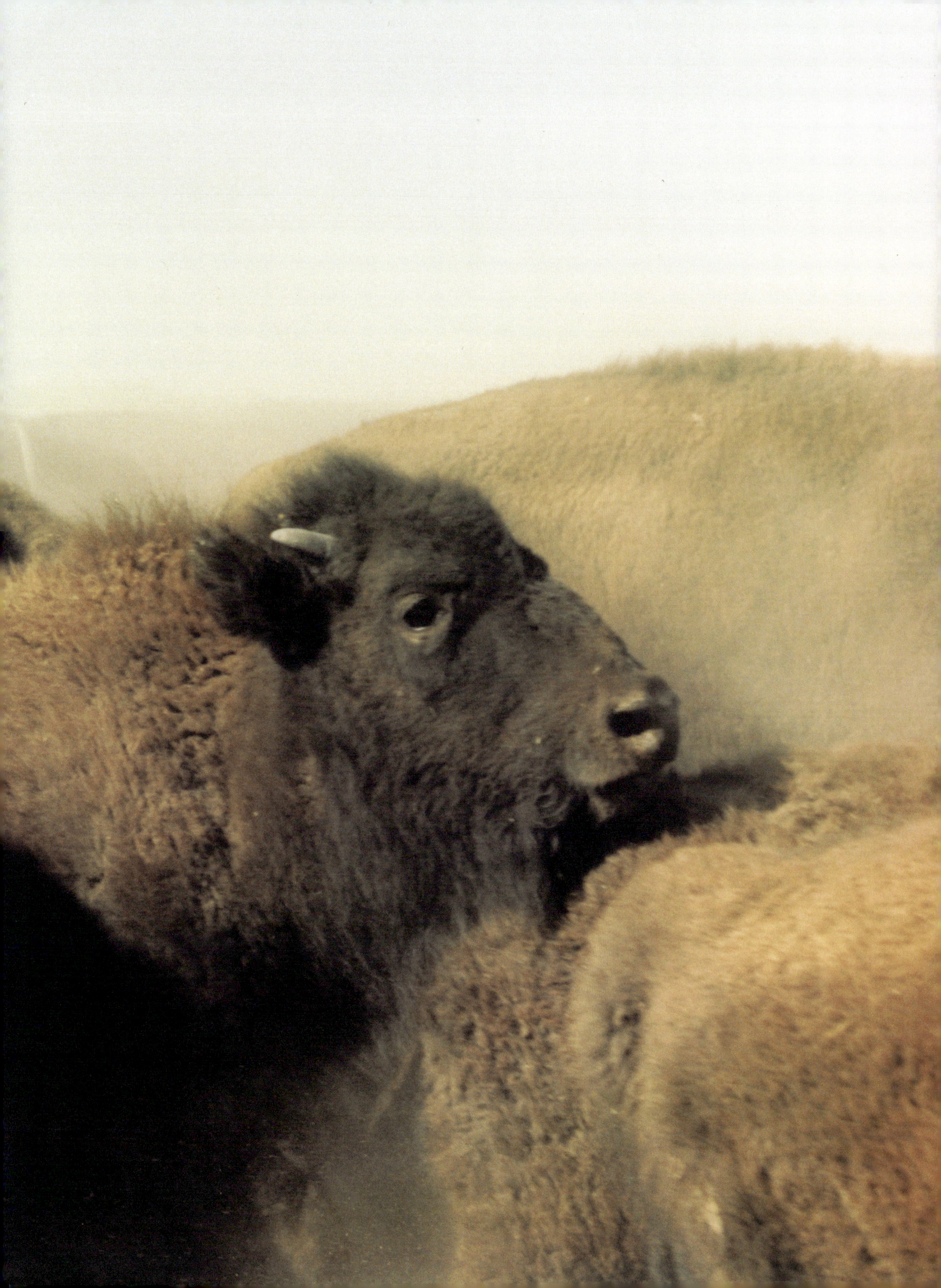

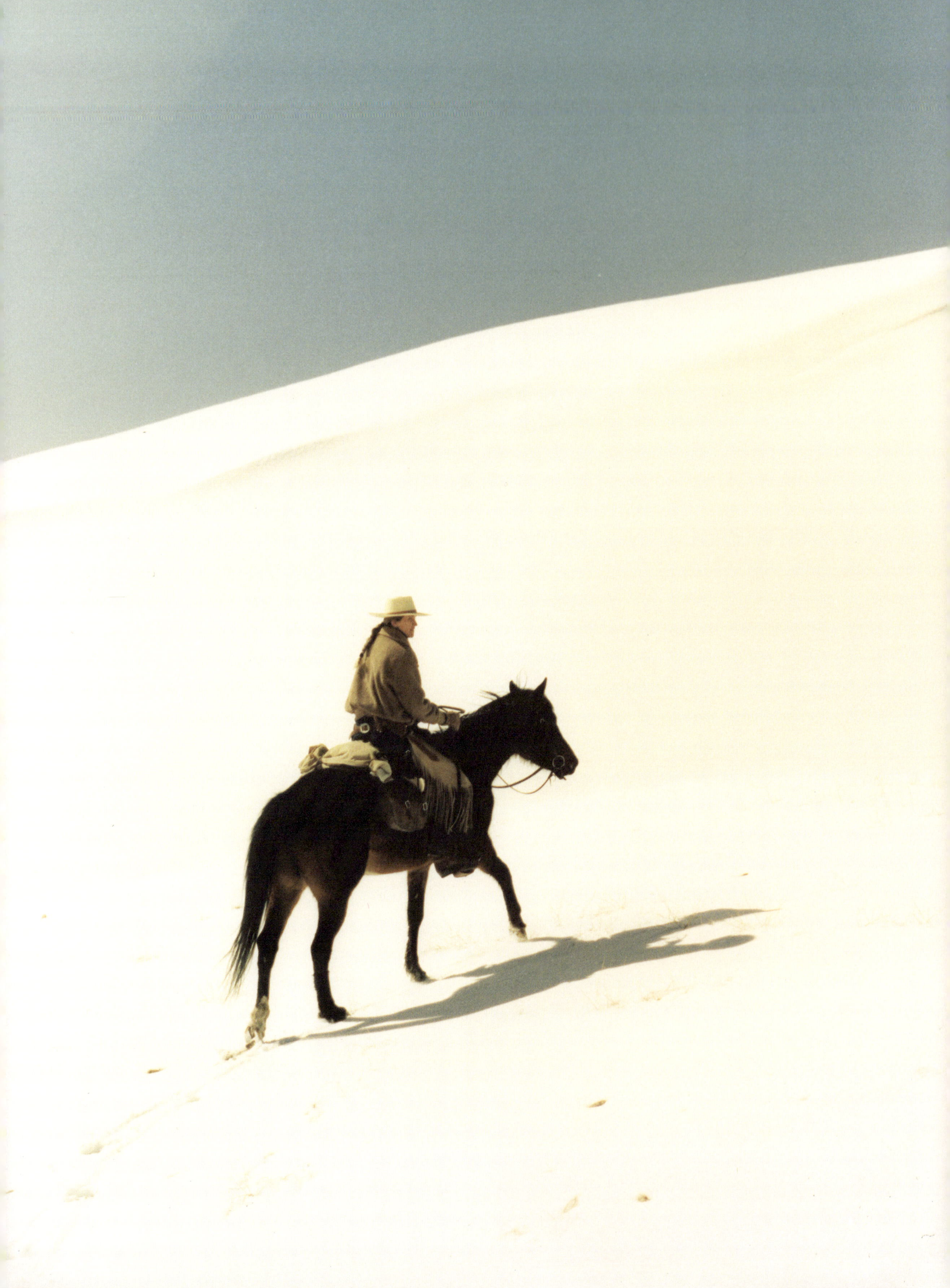

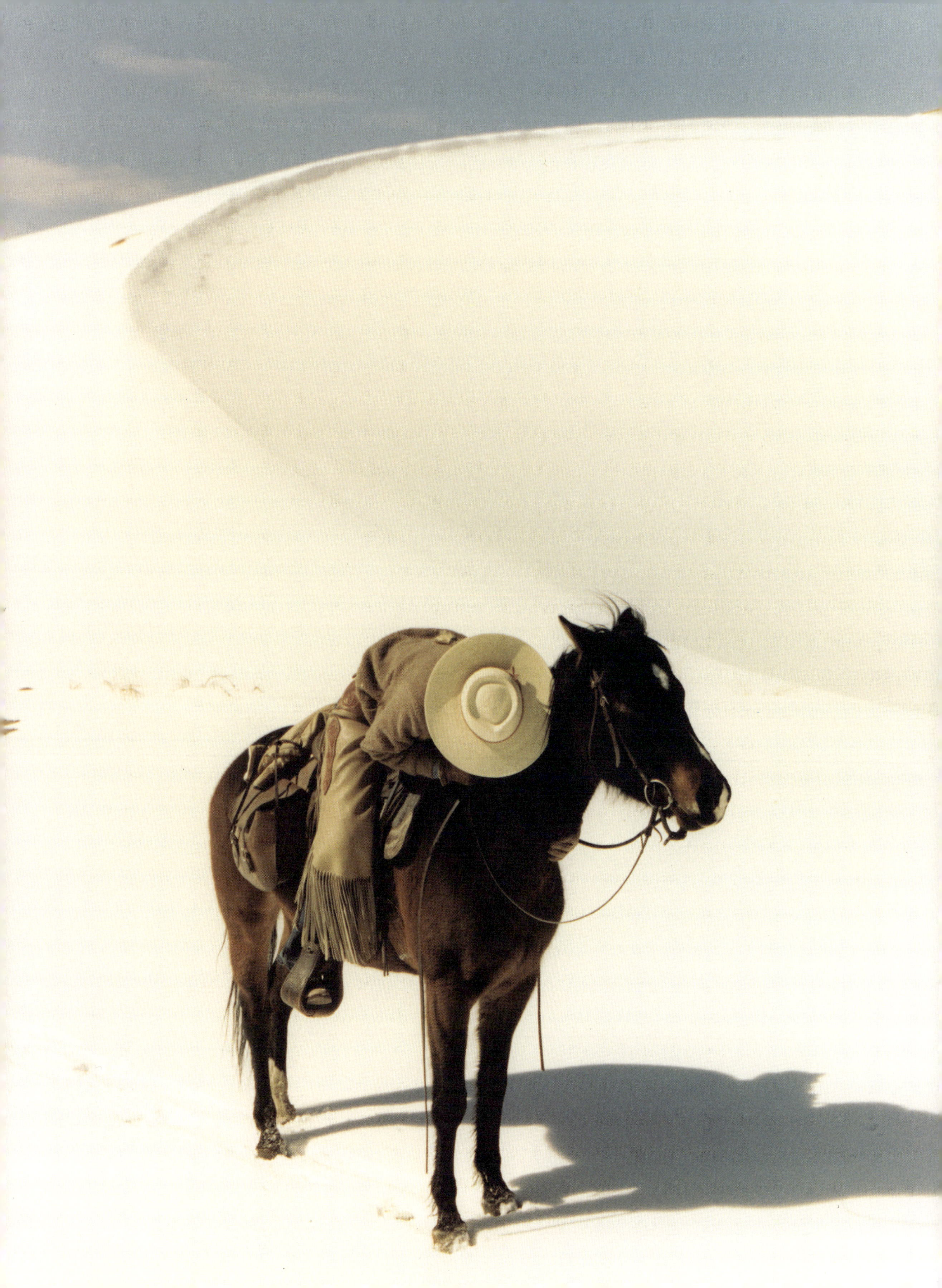

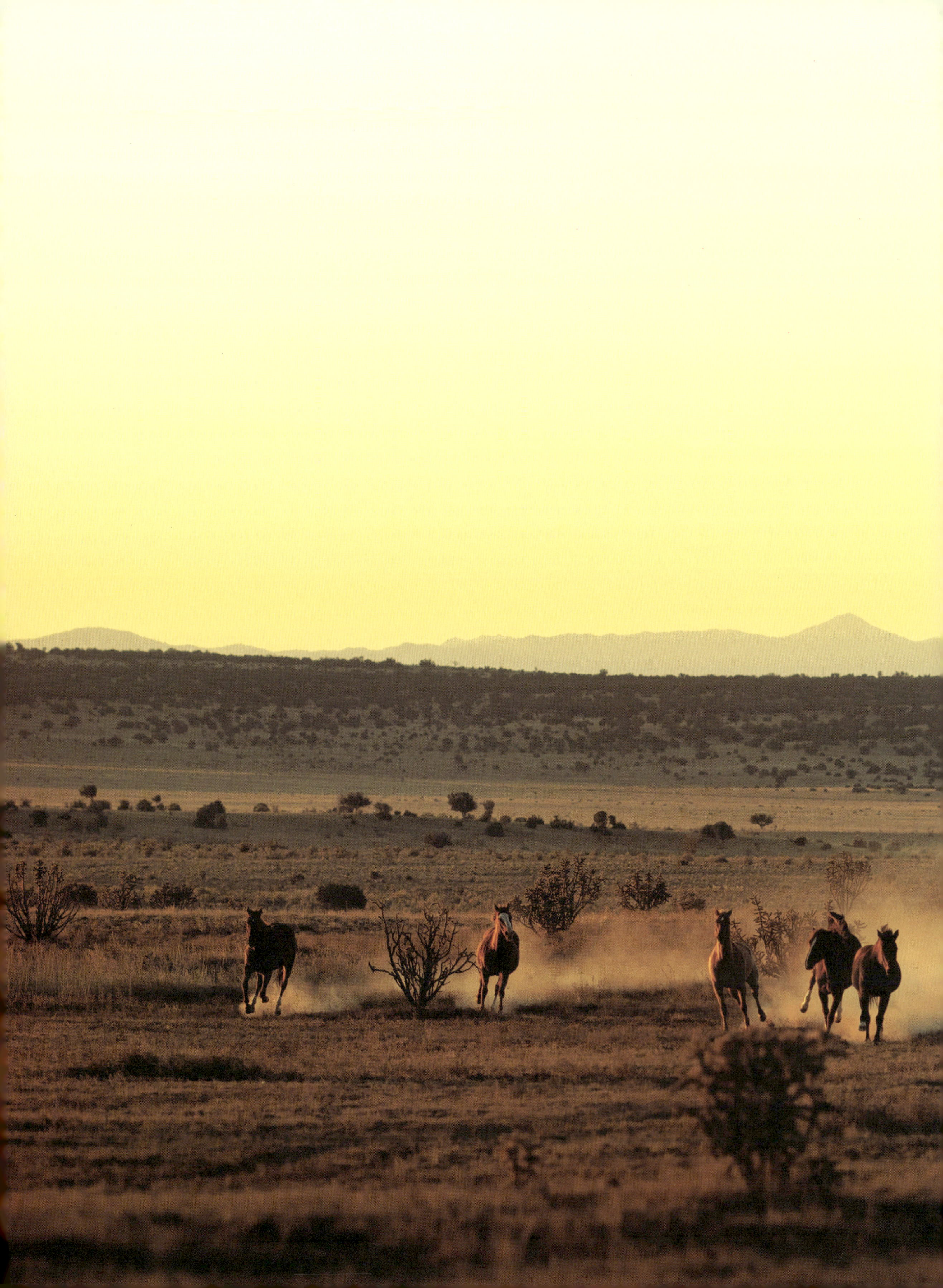

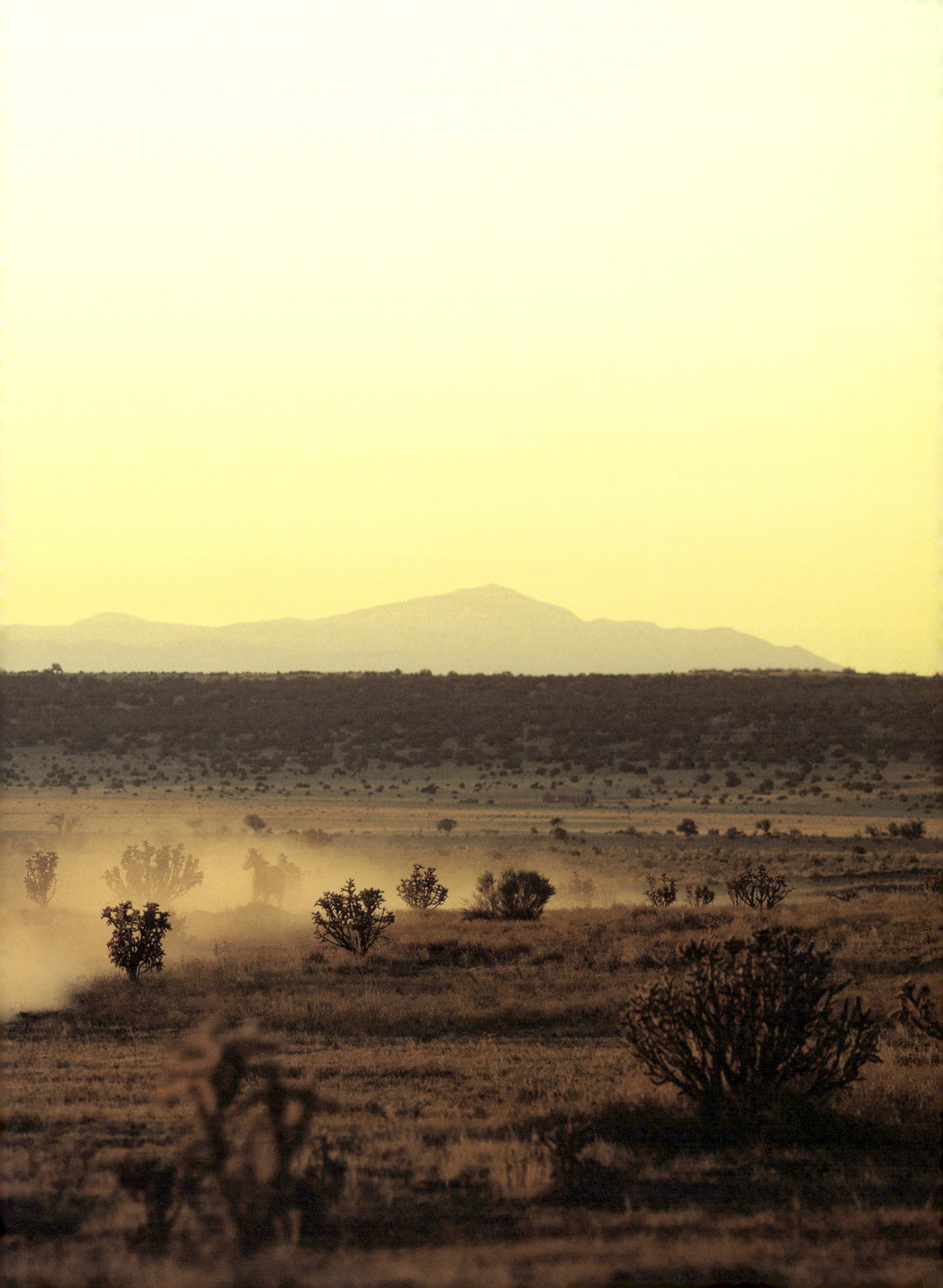

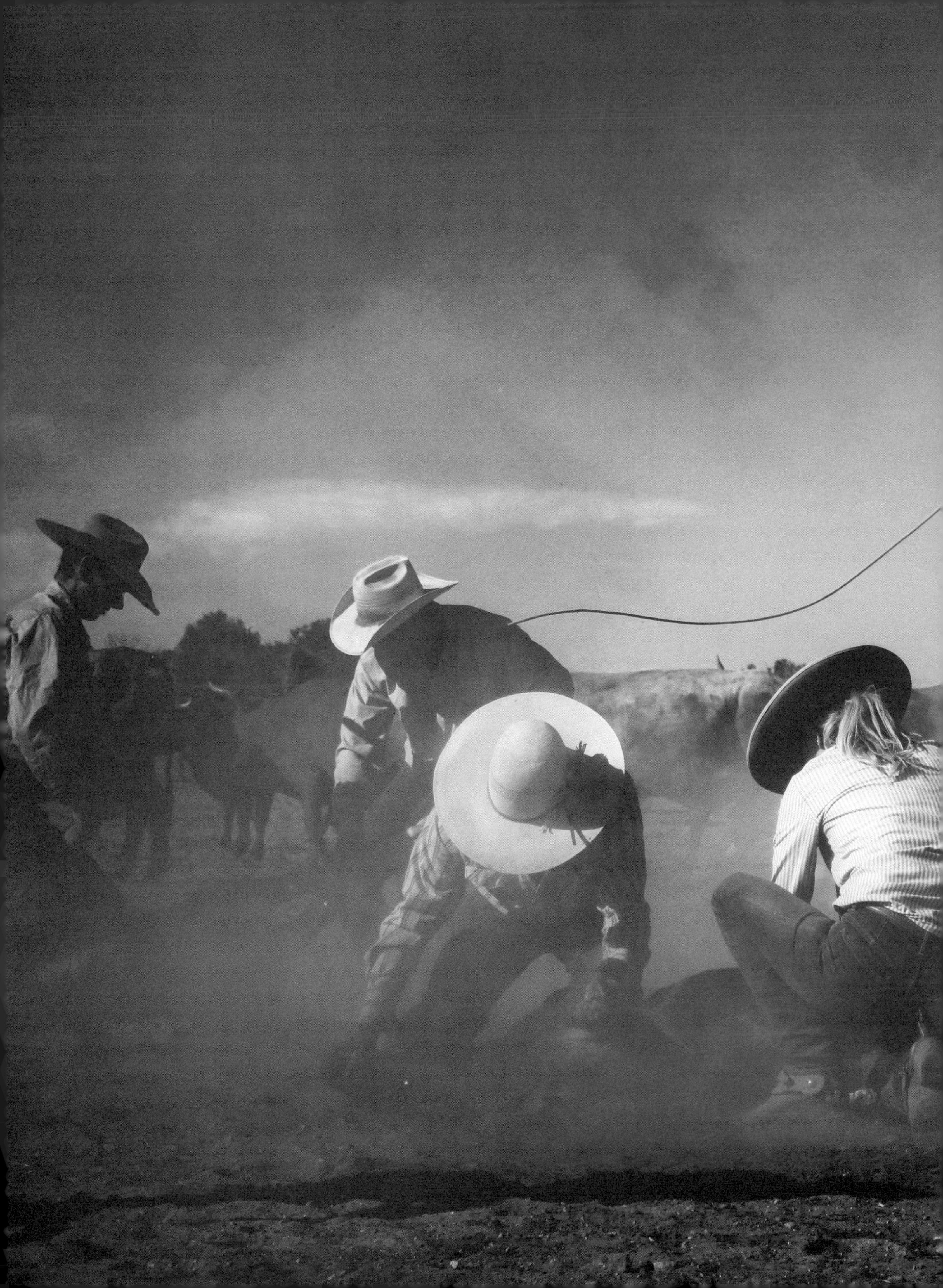

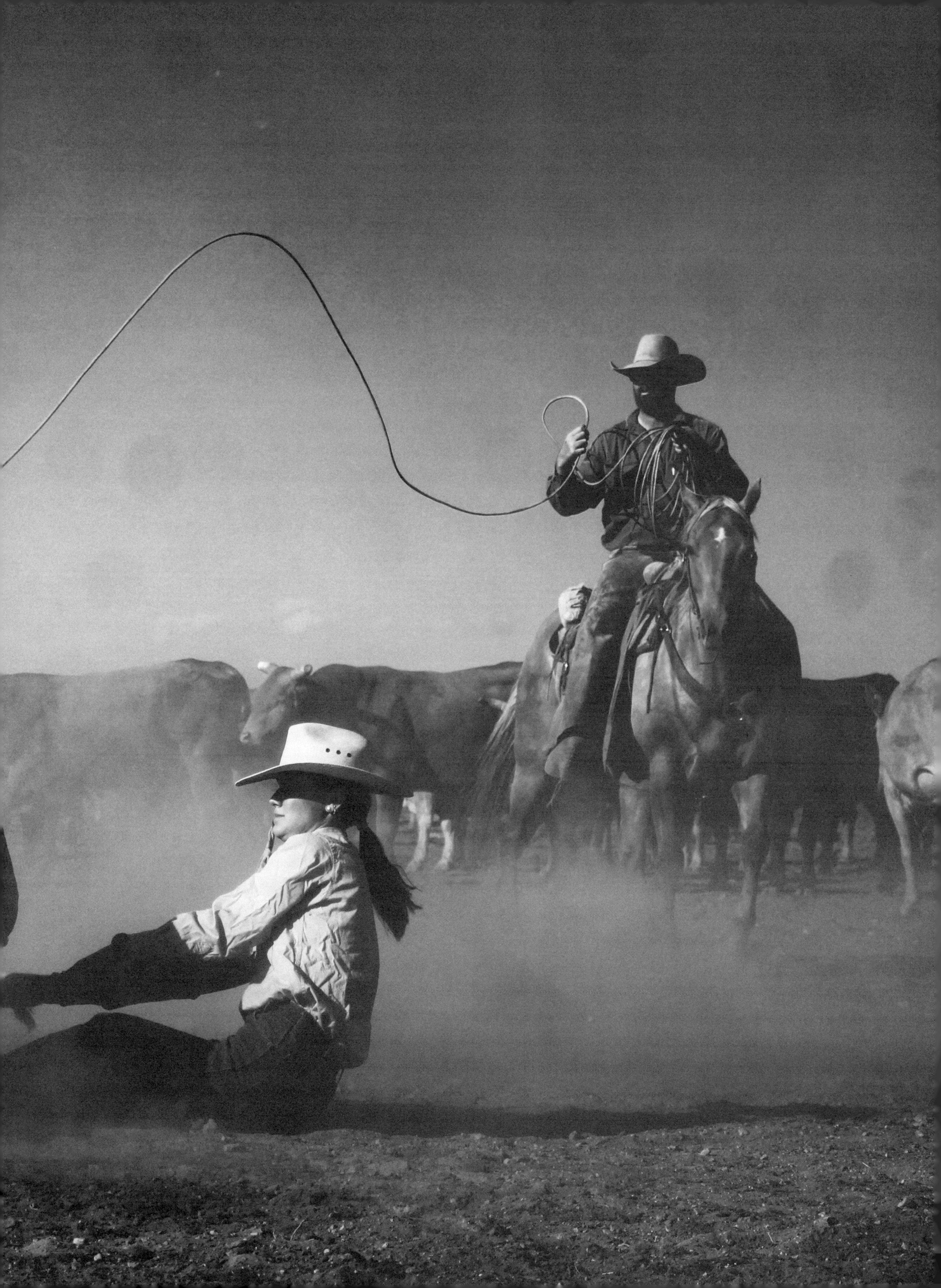

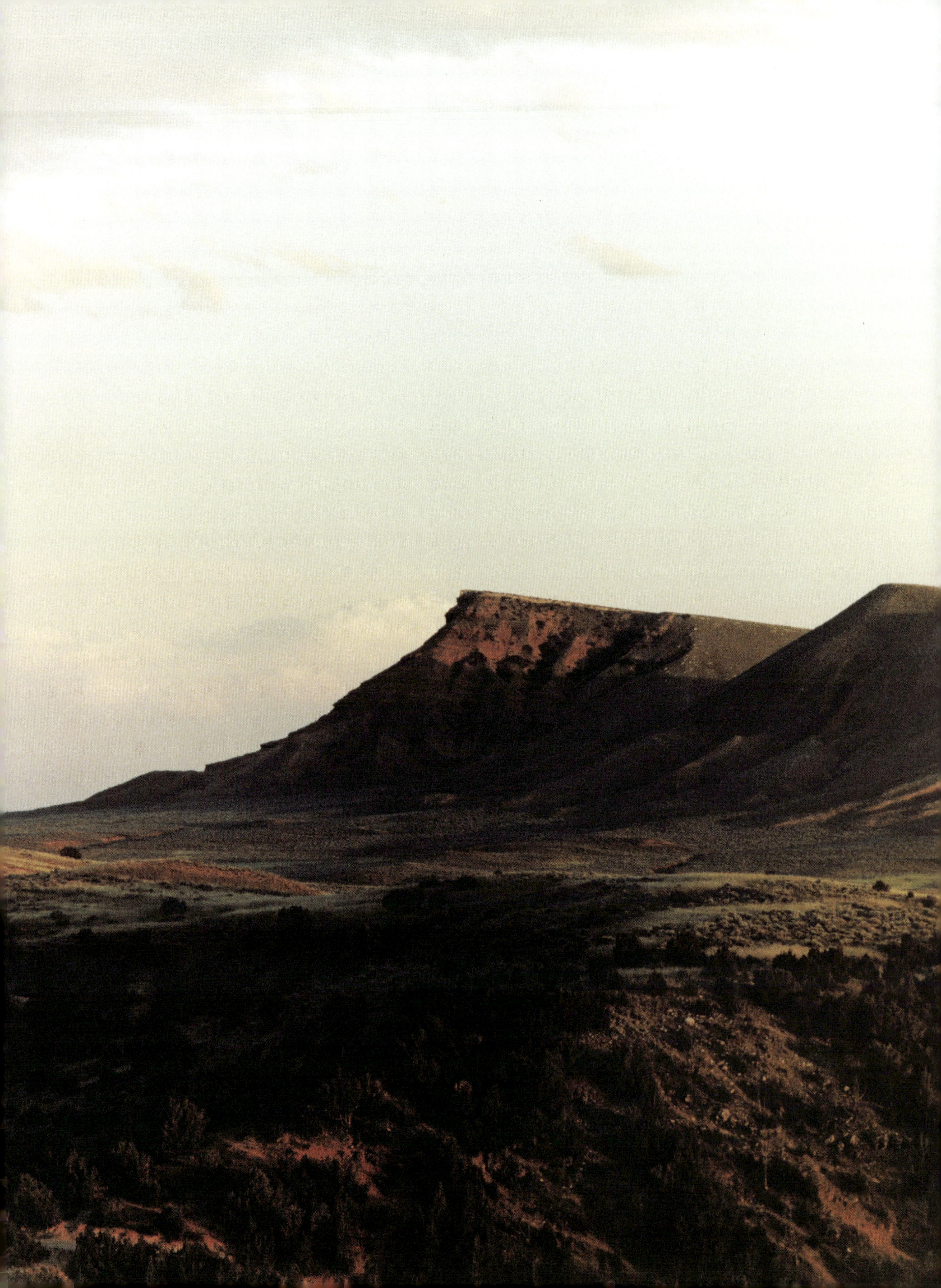

THE

LAND

Land is precious. It is where life renews itself. Essentially, it is everything.

The open prairie, rolling hills, draws, springs, hidden meadows, and mountain valleys across the West are our homes. Across the hundreds of thousands of acres where we operate, from Wyoming to Texas, the land is our refuge. When the sun sets every evening, it feels like the front doors of our homes are closing behind us.

This land is the place we have chosen to share with our loved ones, where we cook our food, sleep, think, make babies, raise children, and invite people close to us to share our lives. It is the most intimate of all places, both private and personal. We share this home with cattle, horses, deer, antelope, ants, and birds. We are a community living together and depending on one another.

The scale of our ranches is big—each property we manage is between 30,000 and 100,000 acres. We are used to this vastness. What outsiders see as huge and seemingly impossible to keep track of is ordinary to us. The main ingredient of this feeling is the peace that comes from knowing we are a part of the land and that the land defines who we are, what we stand for, and what we want for the future of the world.

This land has trapped us but not in a bad way. It is difficult to think of land as something we can own, like a knife in our pocket or a car. It is beyond ownership. Instead, the land owns us. If it fails, our life here is over. In many ways, taking care of the land is a natural instinct that drives us, as it does all ranchers, as natural as putting our pants on in the morning. Why us? Because we spend all day immersed in the landscape, whether it's boiling hot, freezing cold, windy, or the most beautiful day of the entire year. Land provides us with the most fundamental, important, and life-sustaining functions that we need as humans. Managing the wellbeing of this life source is a responsibility that we do not take for granted. We are entrusted with keeping the land and the wildlife whole and healthy in order to grow our cattle and preserve the future for our kids.

We feel the best way to do our work is to look toward nature and emulate its processes. We try to graze our cattle as

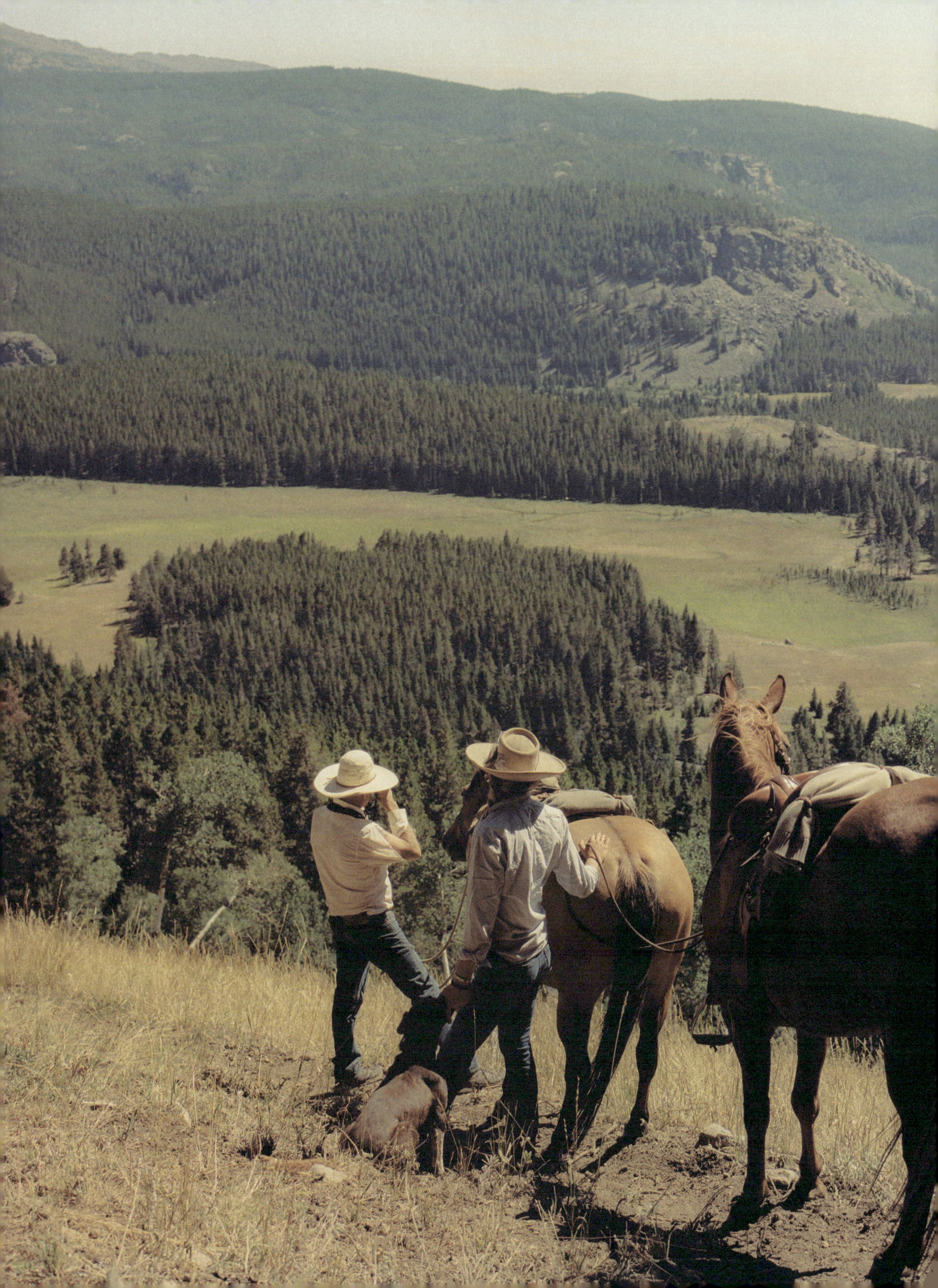

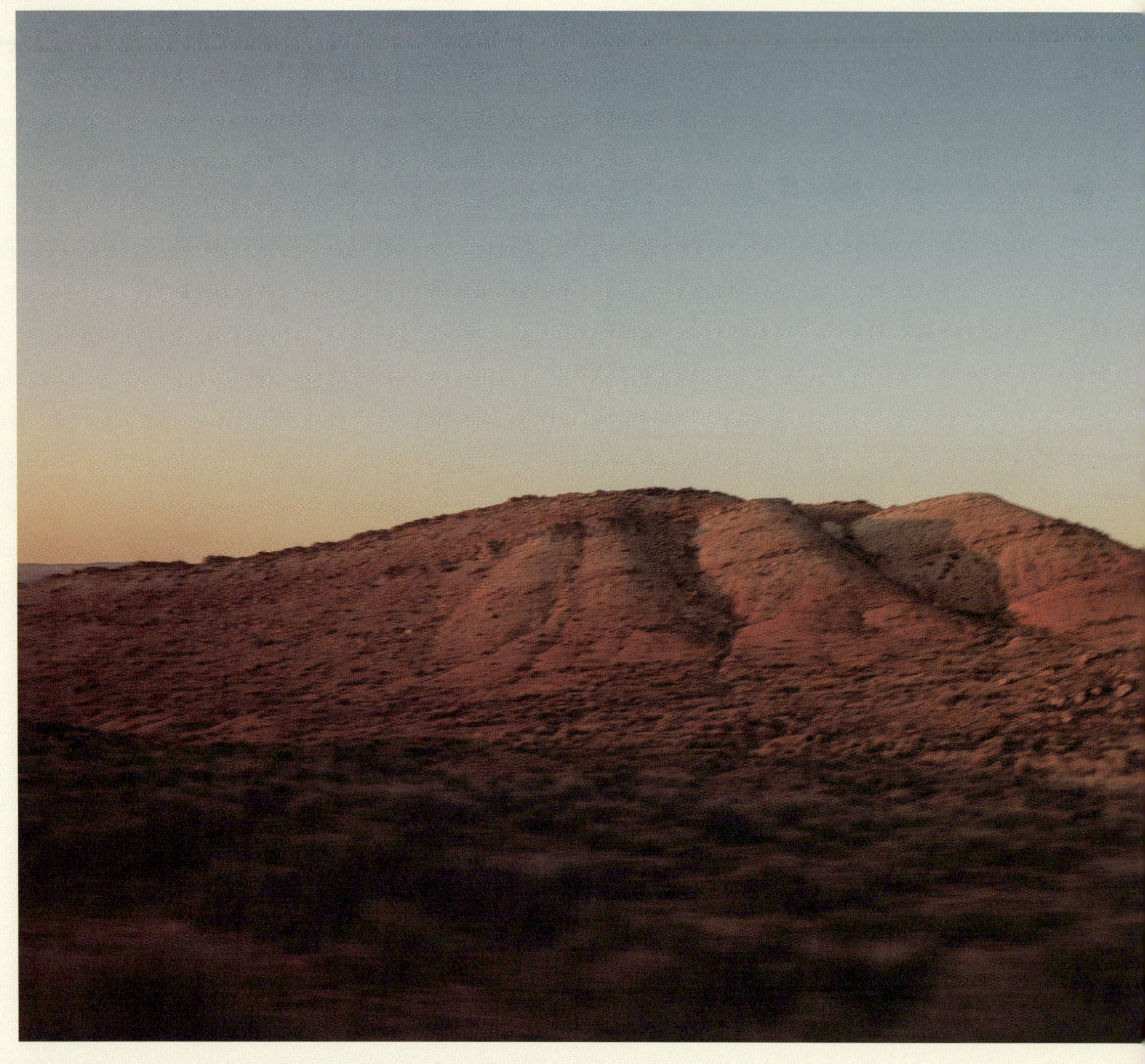

PAINTROCK CANYON RANCH
Wyoming

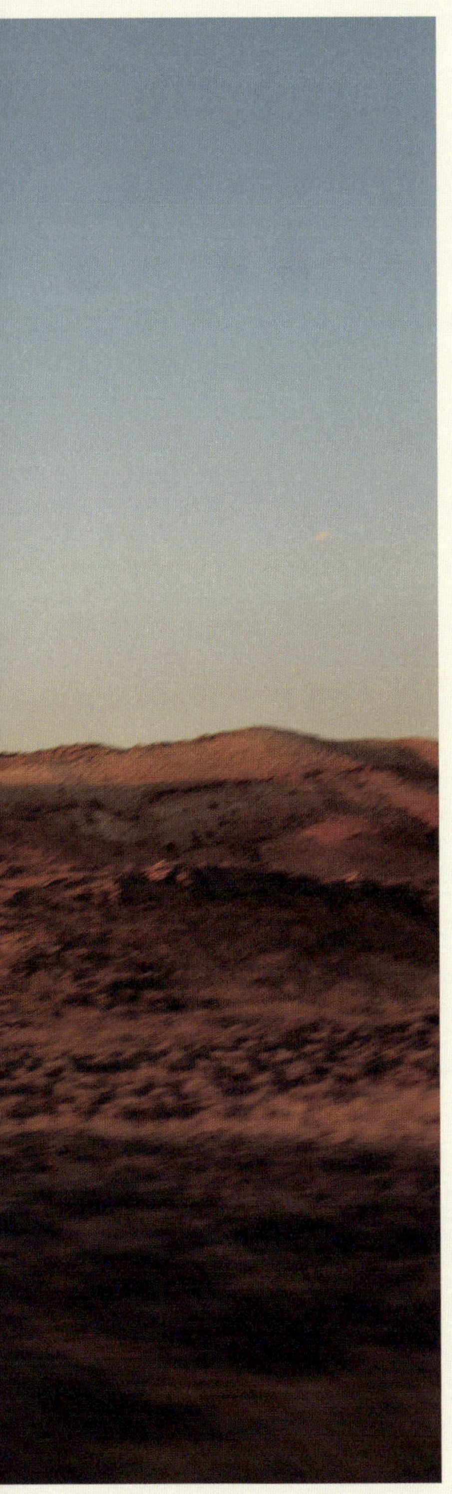

the enormous herds of bison once did when they migrated across the plains between Canada and Mexico, spreading their manure and urine over the ground. What they didn't eat, they trampled, so the grass would decay and return to the soil. Covering the ground, the trampled forage protected the soil from the heat of the sun and the direct impact of rain, helping preserve moisture for seedlings and the trillions of living organisms within the soil.

After the bison moved through, the surface of the ground must have looked disheveled, like soil freshly tilled by a tractor plow. In actuality, the millions of hooves were doing their job as nature planned it, giving the soil a good fertilization and stir. And then the miracle would occur. The entire landscape would return to a fabulous, rich, green garden, fulfilling the needs of every single being above and below the ground. The whole community of plants could then recover from what seemed like devastation and instead grow bigger, stronger, and richer. It is hard to imagine anything more miraculous. It is so simple and natural. And so good.

By amalgamating our cattle herds, we create the same energy and powerful impact as the bison herds. Every year, we sit around a table, mapping out a migratory path that the cattle will graze, carefully planned to ensure the plants have sufficient time to grow and replenish before the next grazing period. We adapt this plan based on the amount of rain and sunlight we receive. If the growing conditions are better than expected, we move the herd faster so we can leave the pastures to be watered and grow. If it does not rain, we slow the herd down to create more rest. When there is a need for laying more organic material down on the ground, we increase densities of cattle by fencing the pastures into smaller parcels. Every move and every decision relative to the grazing plan is made from a deliberate process that's continuously built on by the records we keep, the things we observe, and the data we collect from monitoring. It's a true working landscape.

Most importantly, we are in direct contact with the land every day—we observe it and scratch our heads. We ride horses as much as possible instead of trailering them out behind trucks. This takes us off the roads and into the draws and hills where we need to be in order to truly get a sense of what is happening out there. It has been said that the land's best fertilizer is the caretaker's footsteps. As we ride along, our horses' hooves slide where it has rained. We see baby deer and antelope, we notice where water is running across the ground instead of soaking into it, and we take time to observe seedlings growing. Having this direct contact with the interlinked chains of life provides a deeper understanding of the land than what is possible by looking from a pickup truck window or at a map.

The romance of living and working on our ranches is undeniable. When we find joy and fulfillment in what we are doing, and when our work is helping make the natural world a healthier and happier place for the future, everyone wins. In today's world, 98 percent of the population either lives in urban areas or places that have been increasingly removed from nature. As ranchers, our connection and dependence on the land bridges this growing chasm, hopefully for generations to come.

COLORADO

CHICO BASIN RANCH

The Chico Basin Ranch has been our home base for many years. Located just outside of Colorado Springs on the high prairie, it sprawls across three different ecosystems, with two small creeks and large lakes filled with warm-water fish, ducks, and rich aquatic wildlife. It has a front-row seat to the Rocky Mountain Front Range, showcasing Pikes Peak, which towers 14,000 feet high. It's the home of Ranchlands' seedstock Beefmaster cow herd, which produces the bulls used on all of our ranches. It's the home of our largest commercial cow herd. The Chico Basin Ranch houses our main leather shop, and it's where we hold art shows and concerts. Tens of thousands of school kids have participated in our educational programs over the last two decades, and the ranch is home to the second-oldest bird-banding station in Colorado, initiated and funded by Ranchlands. We're proud to have developed this place into one of the foremost ranches in Colorado by building a state-of-the-art livestock-watering system and sectioning off the rich riparian areas in order to graze cattle effectively, protecting the corridor that runs through the middle of the ranch.

TESS LEACH

"The way the sunlight creeps across the prairie in a straight line, lighting up Pikes Peak first, then the grasses, the cholla, like an army marching across from west to east—it's something I look forward to each morning."

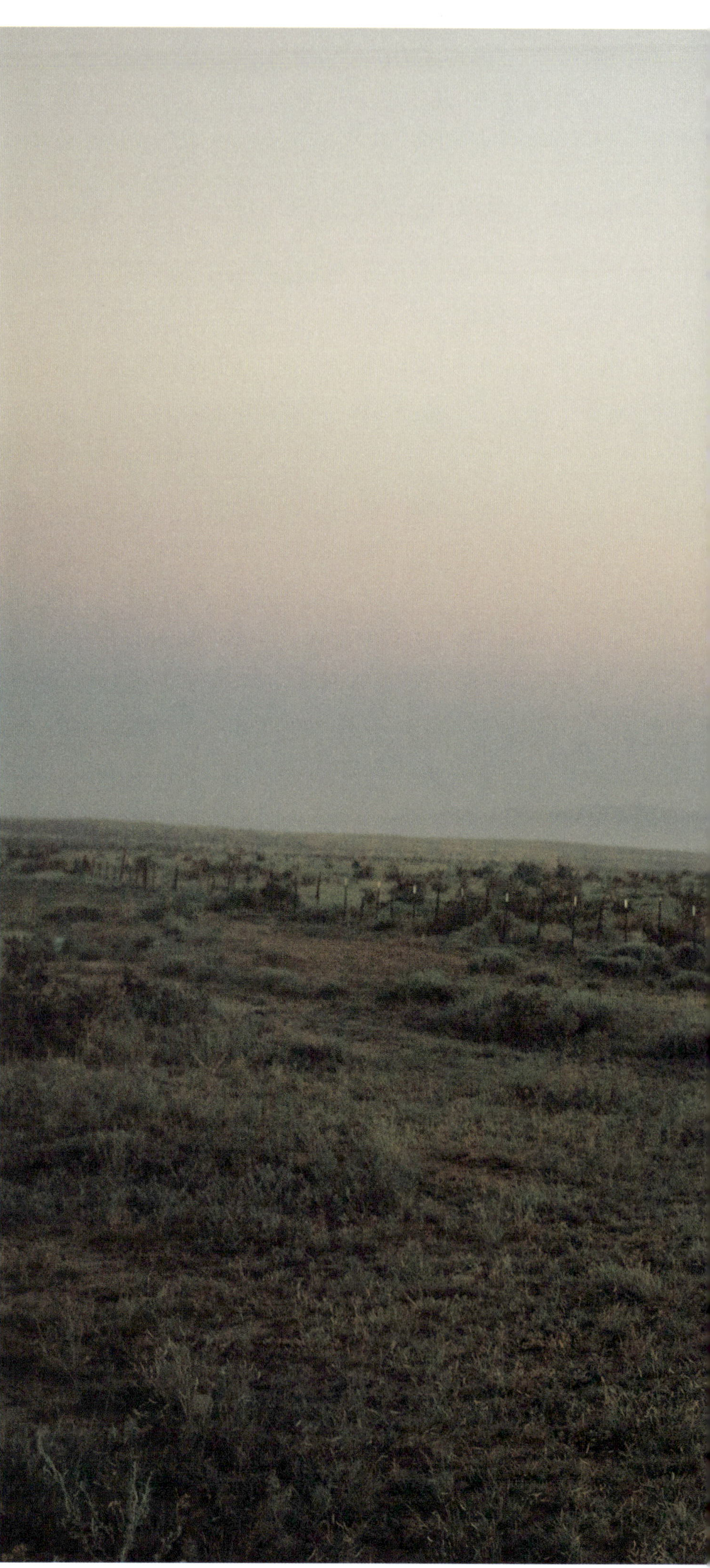

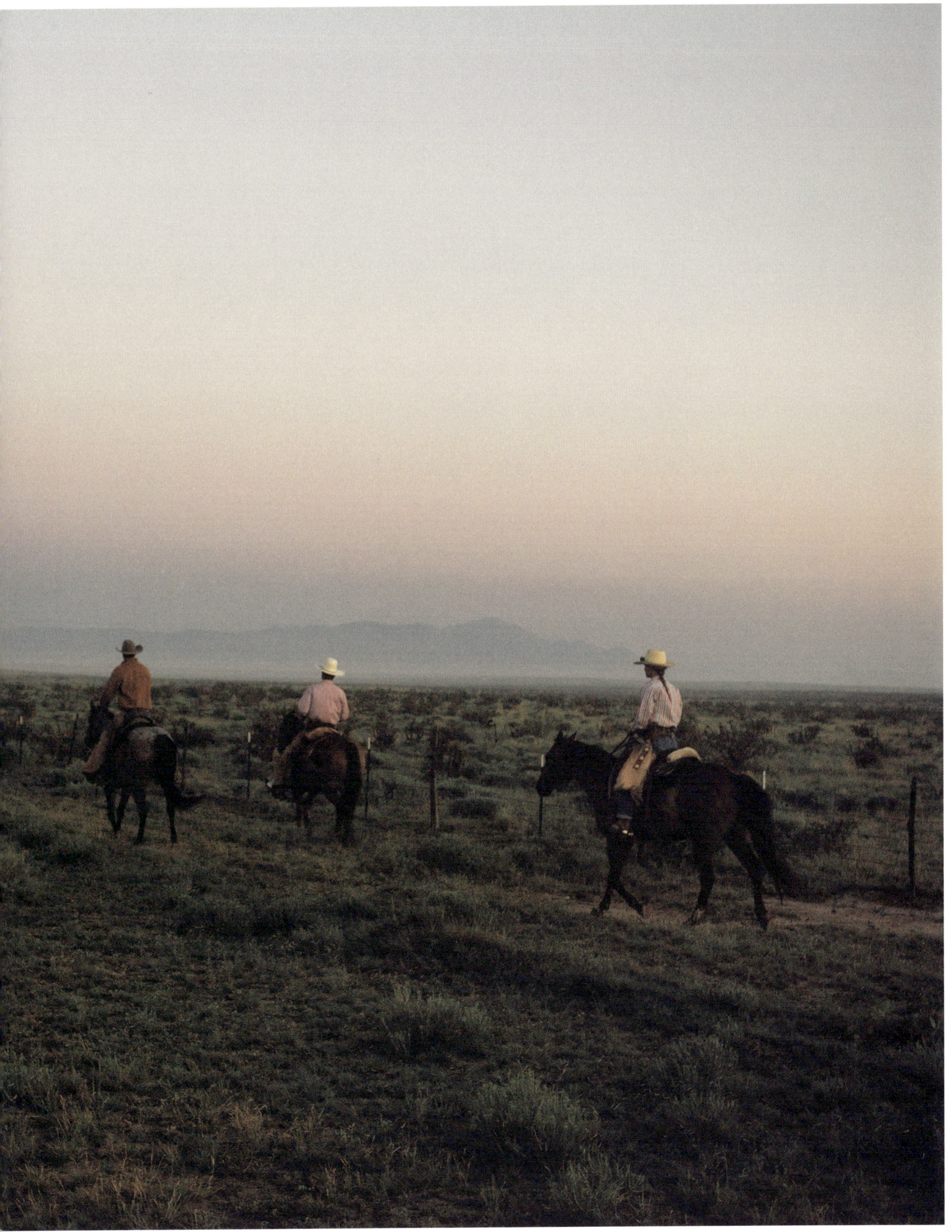

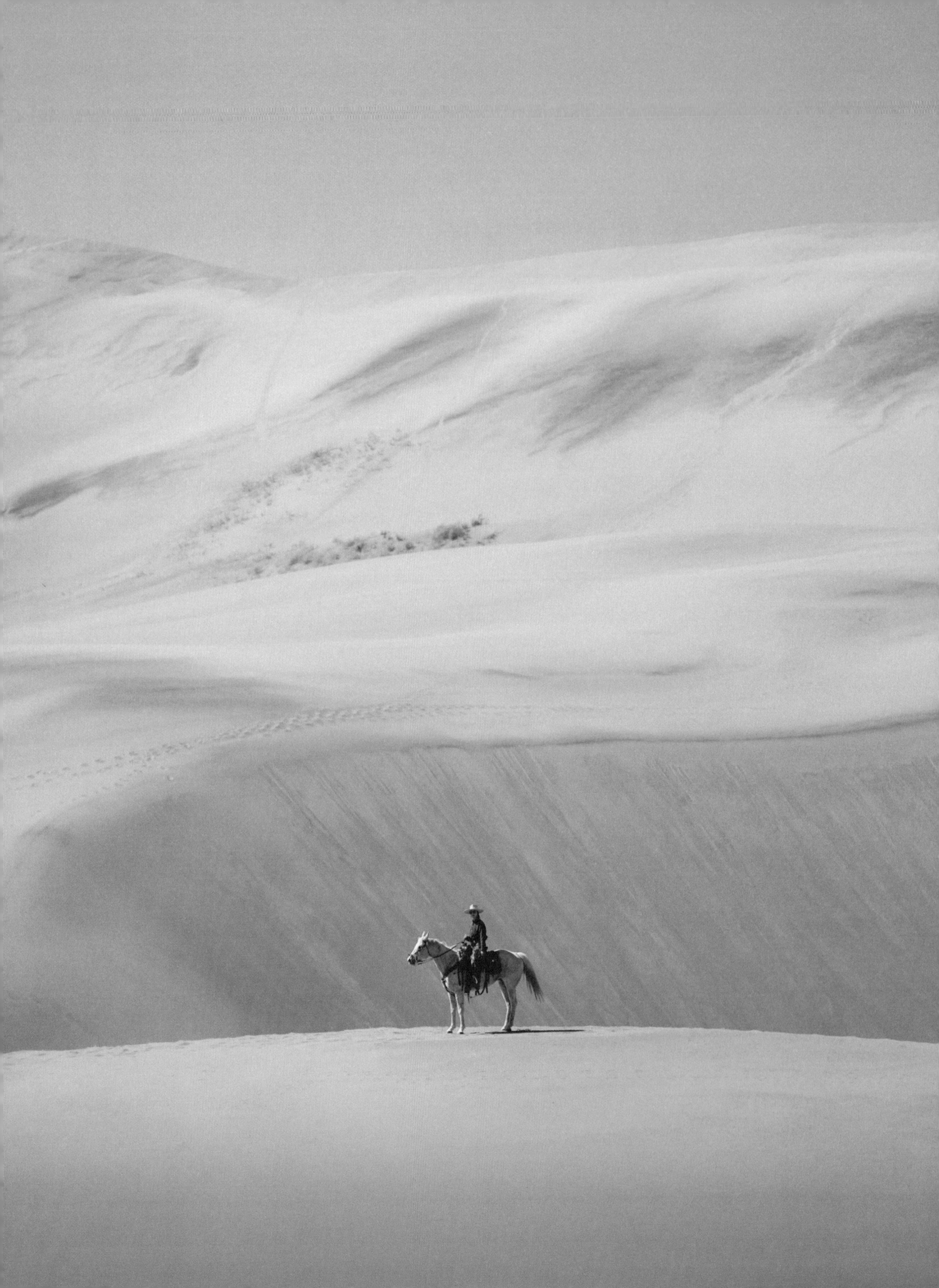

COLORADO

ZAPATA RANCH

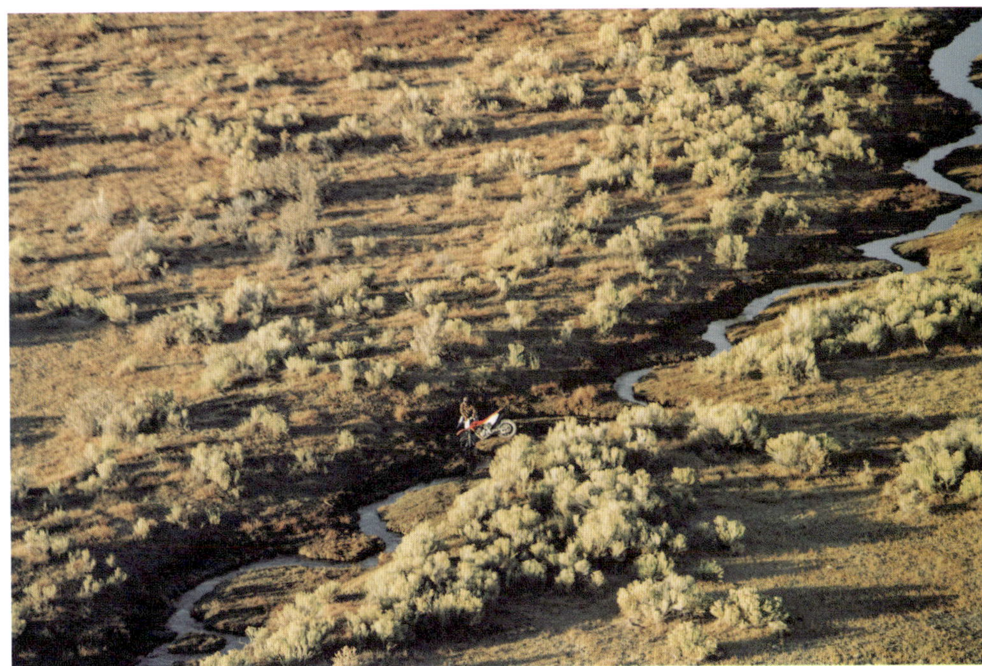

In spring, the big meadows on the Medano burst open with a profusion of grass and flowering Rocky Mountain bee plants, all fed by underground water coming from deep under the sand dunes.

The Zapata Ranch borders Great Sand Dunes National Park, home of the highest inland sand dunes in North America. It is made up of mostly sand and rabbitbrush, with cottonwood trees lining the Sand Creek streambed through the entire ranch. The wildlife is rich, with vast herds of elk, gigantic flocks of sandhill cranes in the spring, and ducks flying from the hidden lakes and small streams. One of the largest conservation herds of bison in the world roams freely in a 50,000-acre pasture. Guests from all over the world visit the Zapata Ranch to learn about ranching and its role in preserving natural resources and to ride horses through the sand dunes and bison herd.

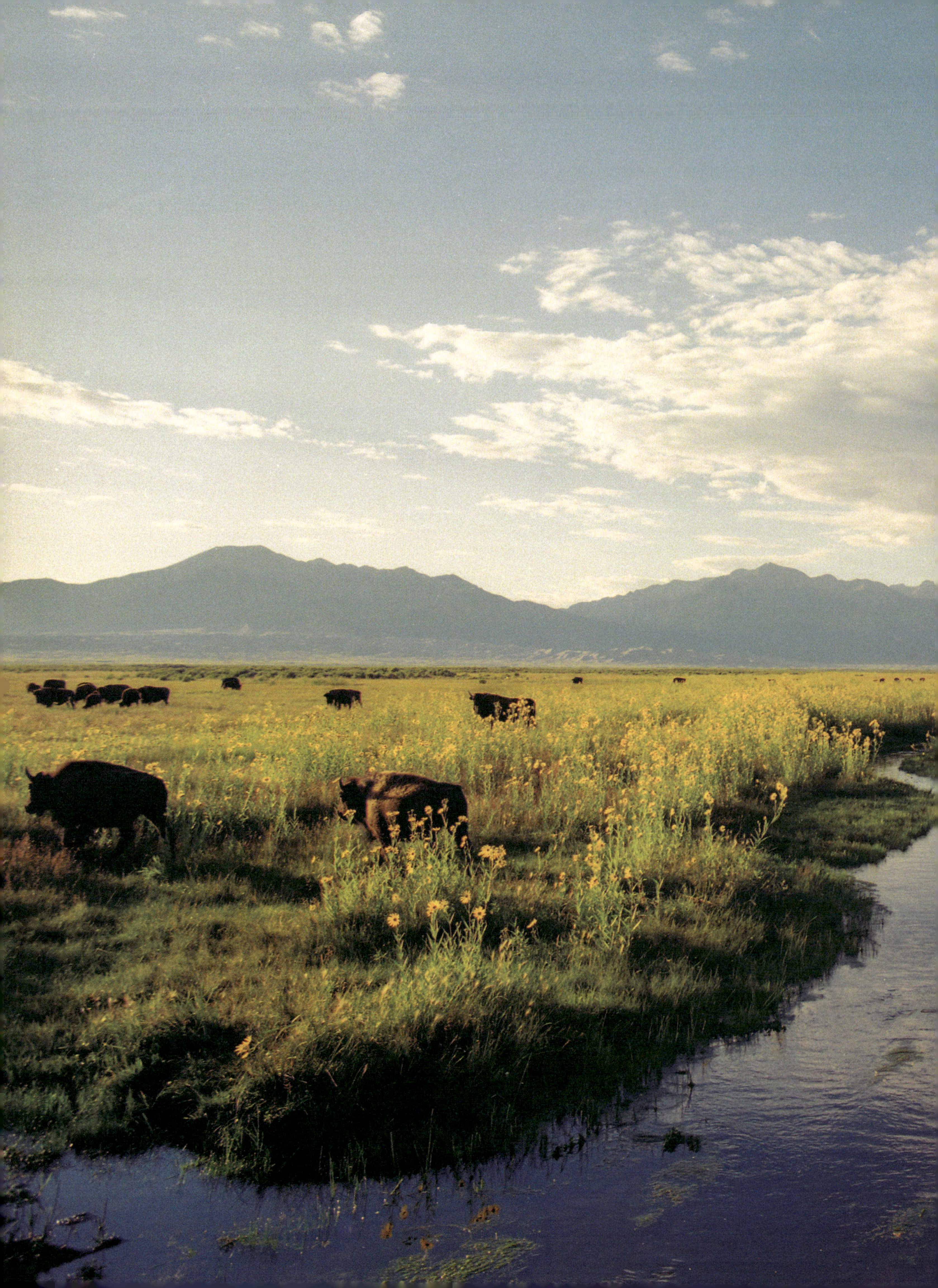

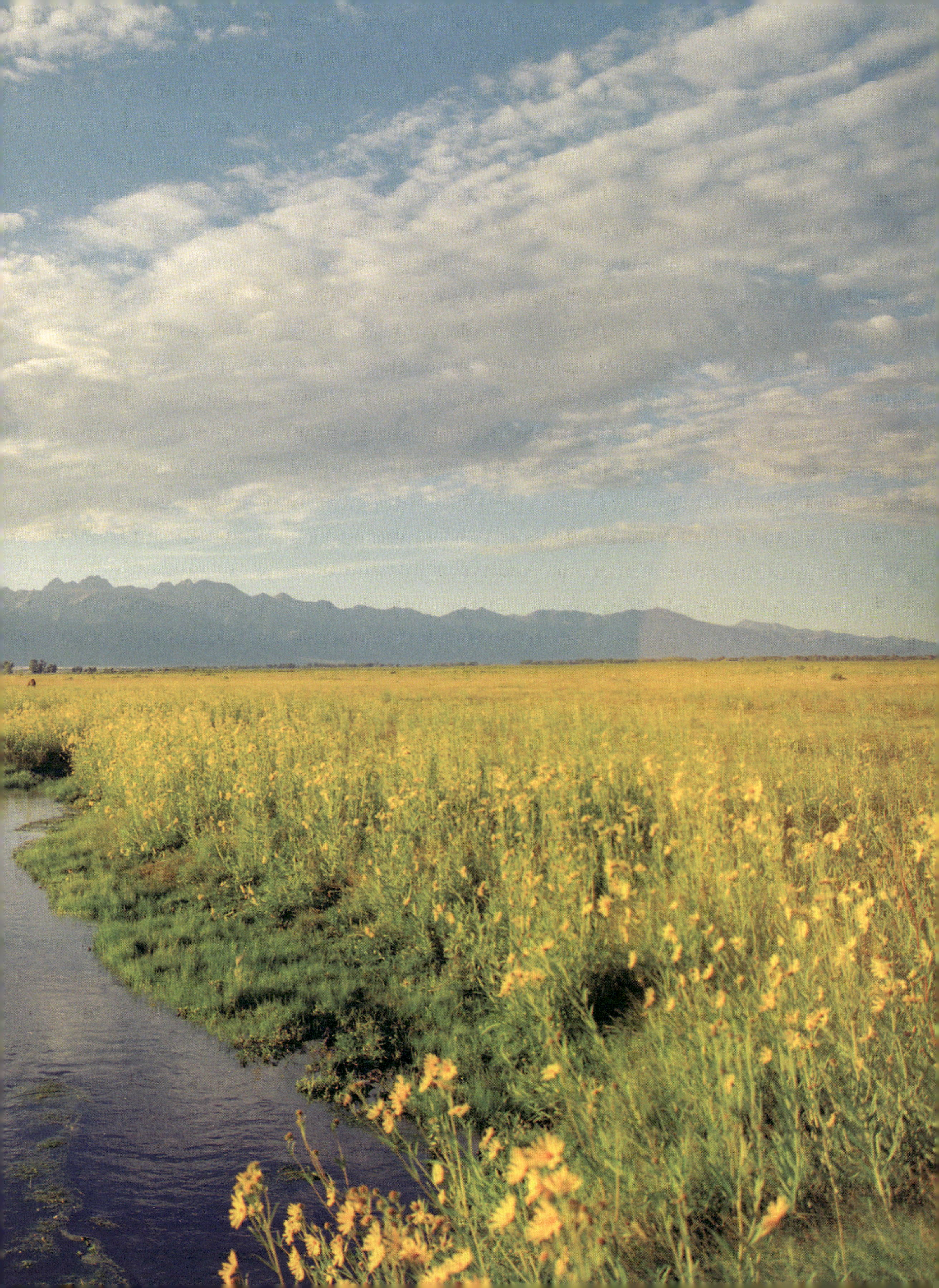

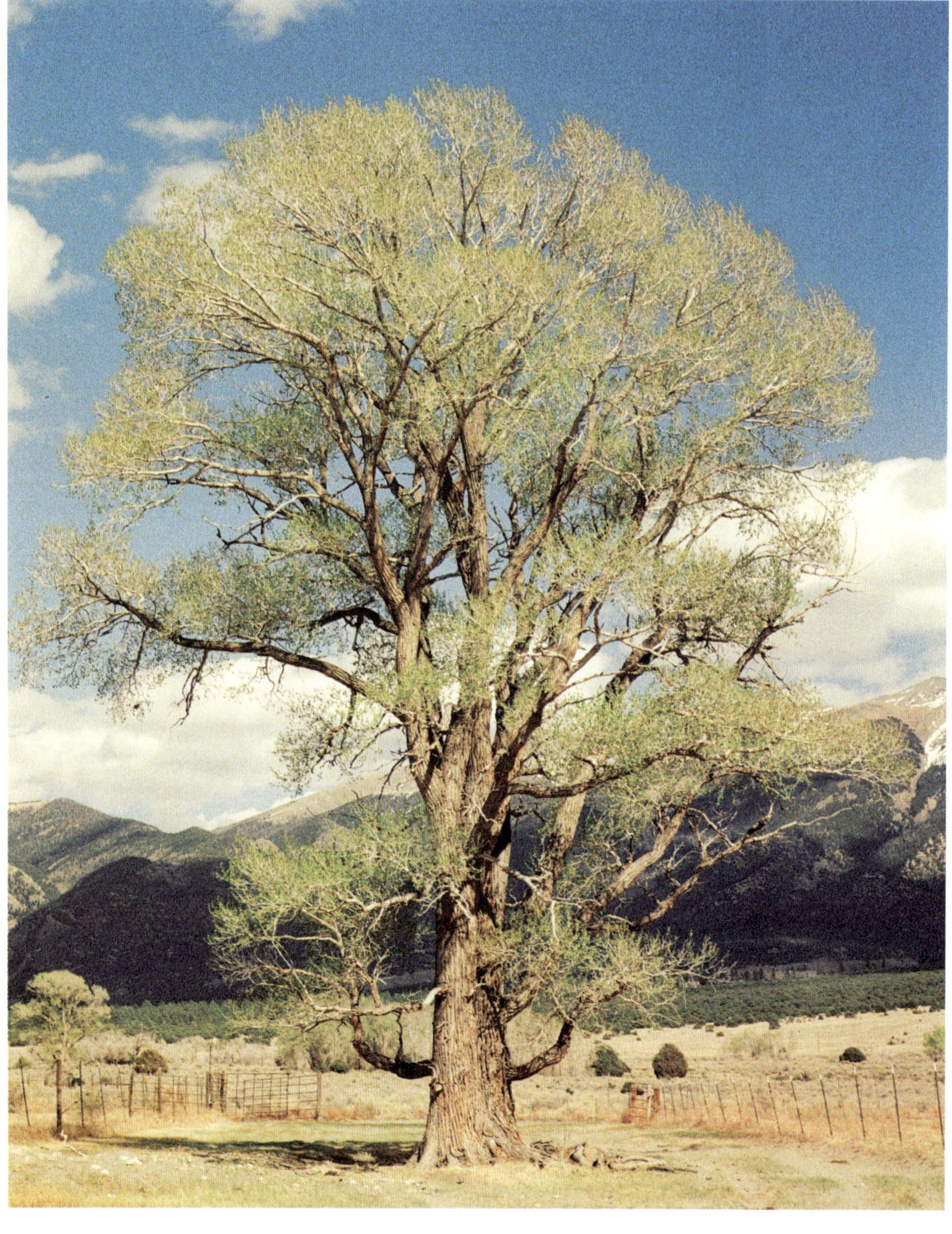

During the Westward Expansion, as European settlers made their way across the prairie, cottonwoods were one of the first trees they encountered.

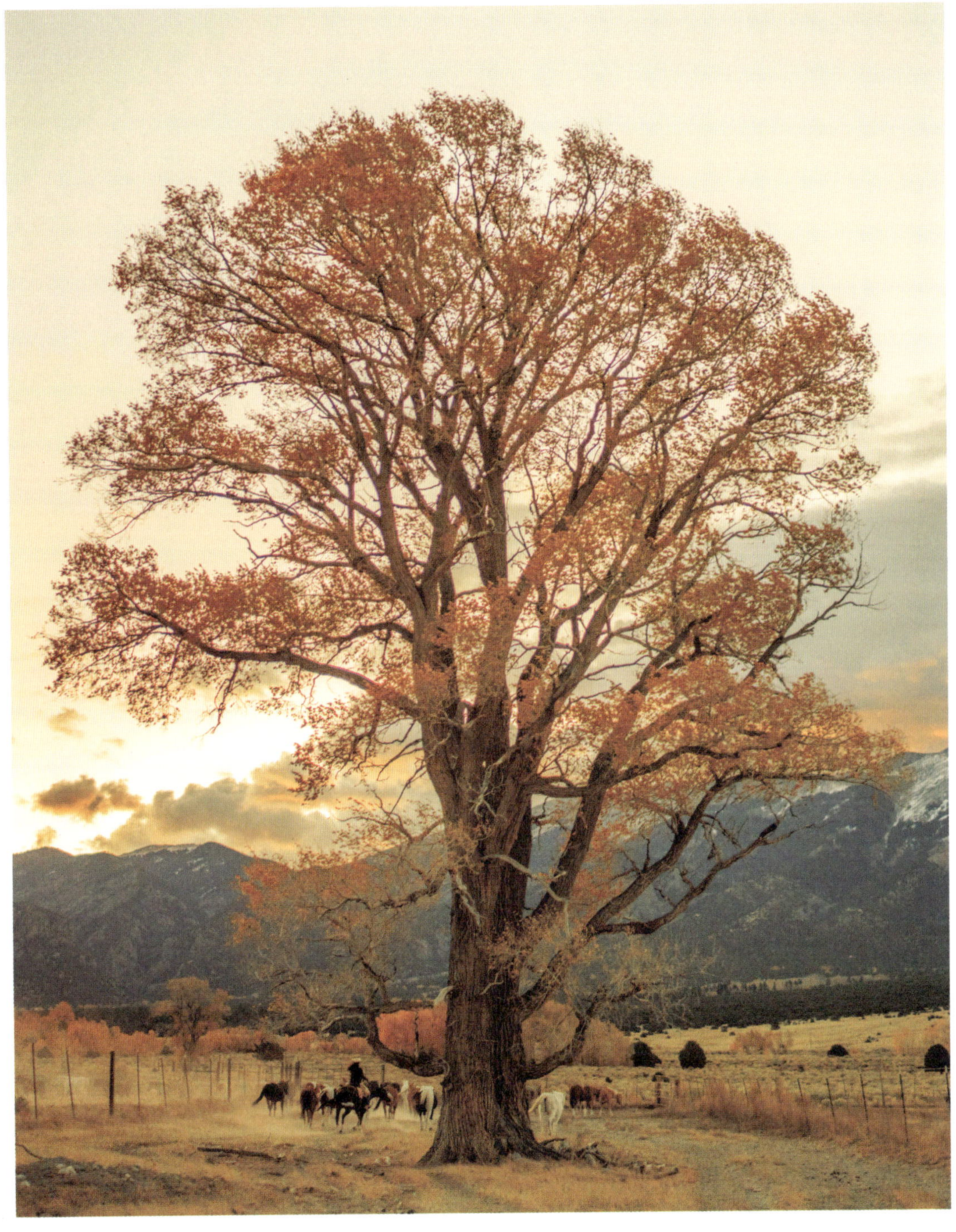

Cottonwoods are an important resource in an otherwise mostly treeless ecosystem.

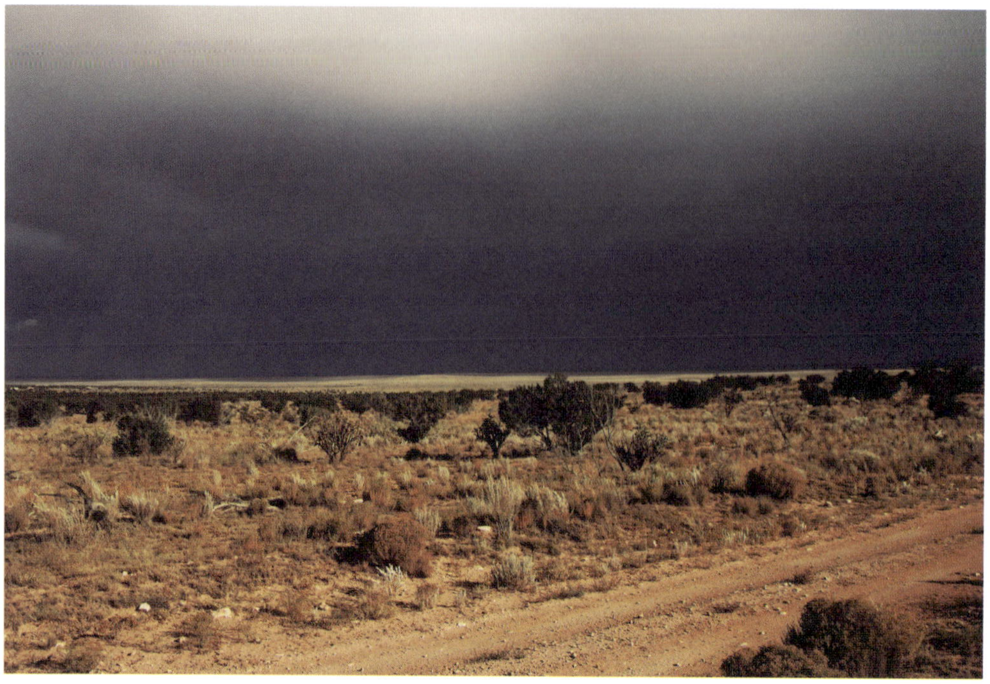

NEW MEXICO

MP RANCH

The MP Ranch is a wild, remote landscape in central New Mexico. It remains one of the increasingly rare places that still feels remote from modern life. The MP Ranch is characterized by the high desert, a stark and beautiful terrain of rough, forested hills and hidden draws and meadows. It is home to a rich population of elk, African gemsbok, antelope, mule deer, and much more. Because of its isolation, the night sky is untainted and bright with crystal-clear stars. The ranch ranges from shortgrass prairie to pinyon-juniper woodlands. The highest point of the ranch is Brushy Mountain, where layers of purple mountains can be seen disappearing into the west. It sits almost 6,625 feet high, with the ranch headquarters at 6,200 feet. Climbing to the top and looking out over the Chupadera Mesa is one of the most inspiring views imaginable. On the ranch, we led the design and construction of a livestock-watering system. Today, gravity feeds and distributes a large volume of water across the entire ranch, elevating the land's ability to efficiently use its diverse landscape as a means to create a healthy ecosystem.

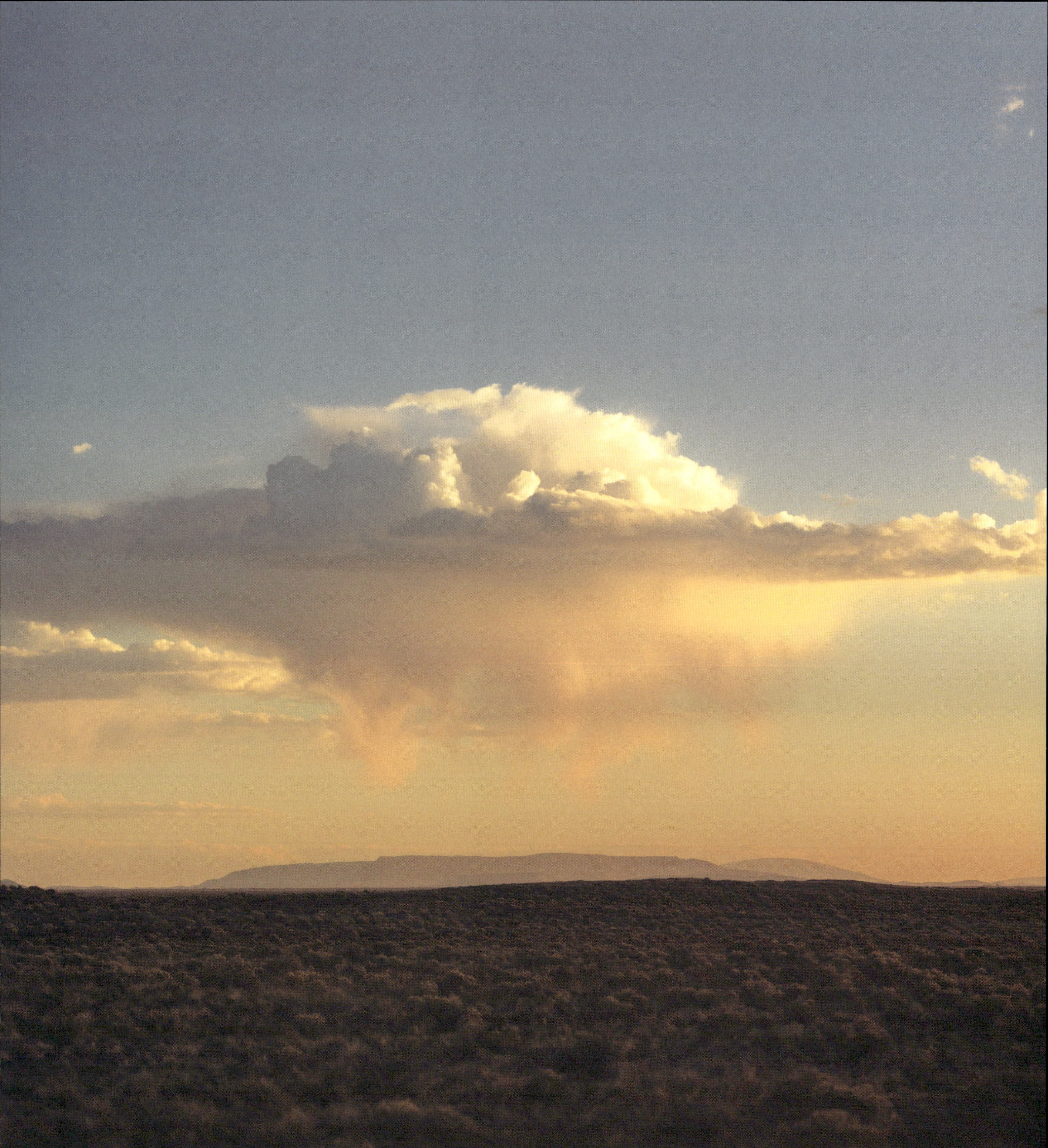

I remember once again what I have always known: this land is where the wind is born.

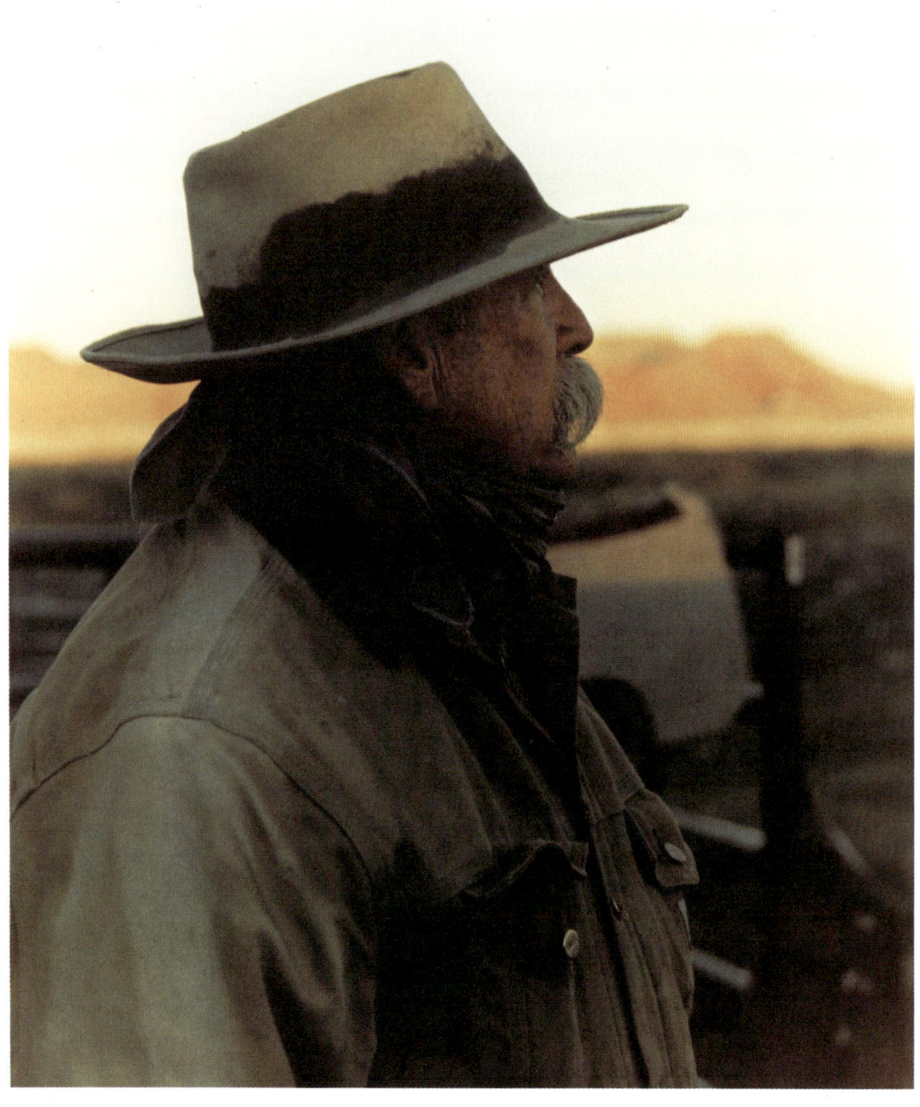

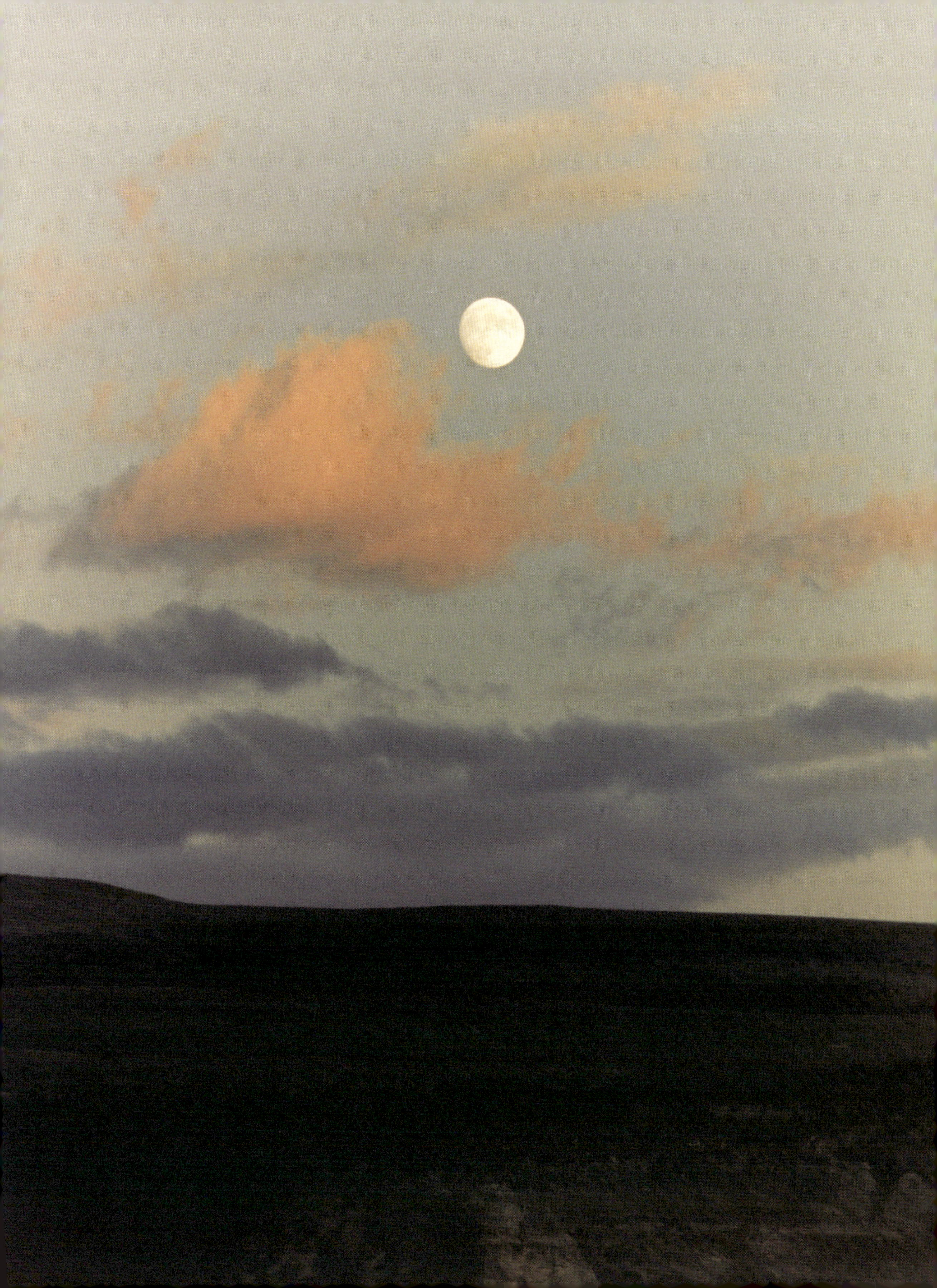

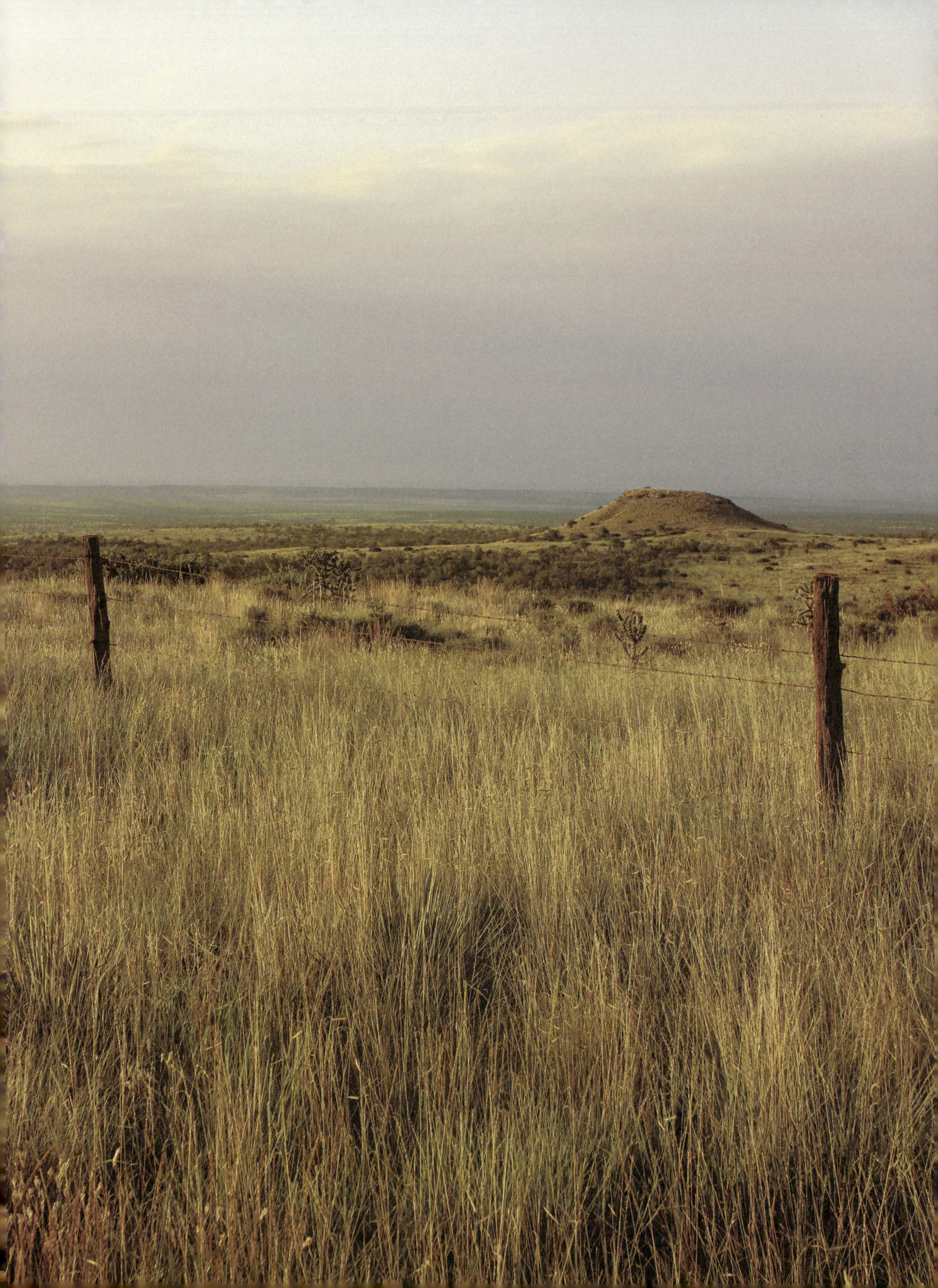

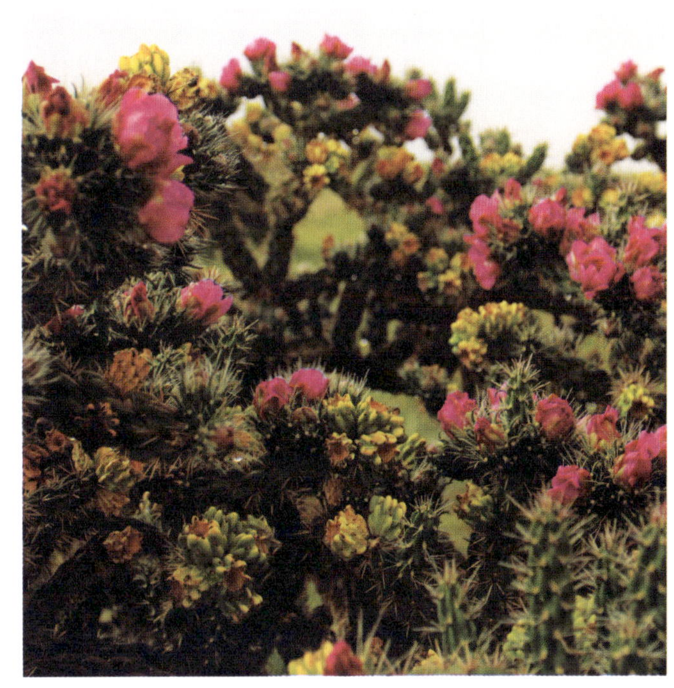

TEXAS

FRYING PAN RANCH

The Frying Pan Ranch is located in the Texas Panhandle. Here, we run a large commercial cow herd that produces calves that are shipped to Wyoming the following summer season for fattening. The landscape consists of rolling hills and dense grassland plains that drop off sharply into canyons and basins of mesquite brush. In the middle, a small spring feeds a pond surrounded by large trees that are home to wild turkeys and great blue herons. As you travel farther north through the ranch, the landscape shifts again to red-dirt mesas and sprawling hills peppered with juniper and sage. Creeks created by seeps trickle with water coming up from the ground. The ranch has many faces, all of them ancient, sculpted largely by the cattle that have historically been managed across this rich landscape. A large pipeline runs down the center of the ranch, feeding livestock with watering troughs and a high volume of water, which enables us to run large herds.

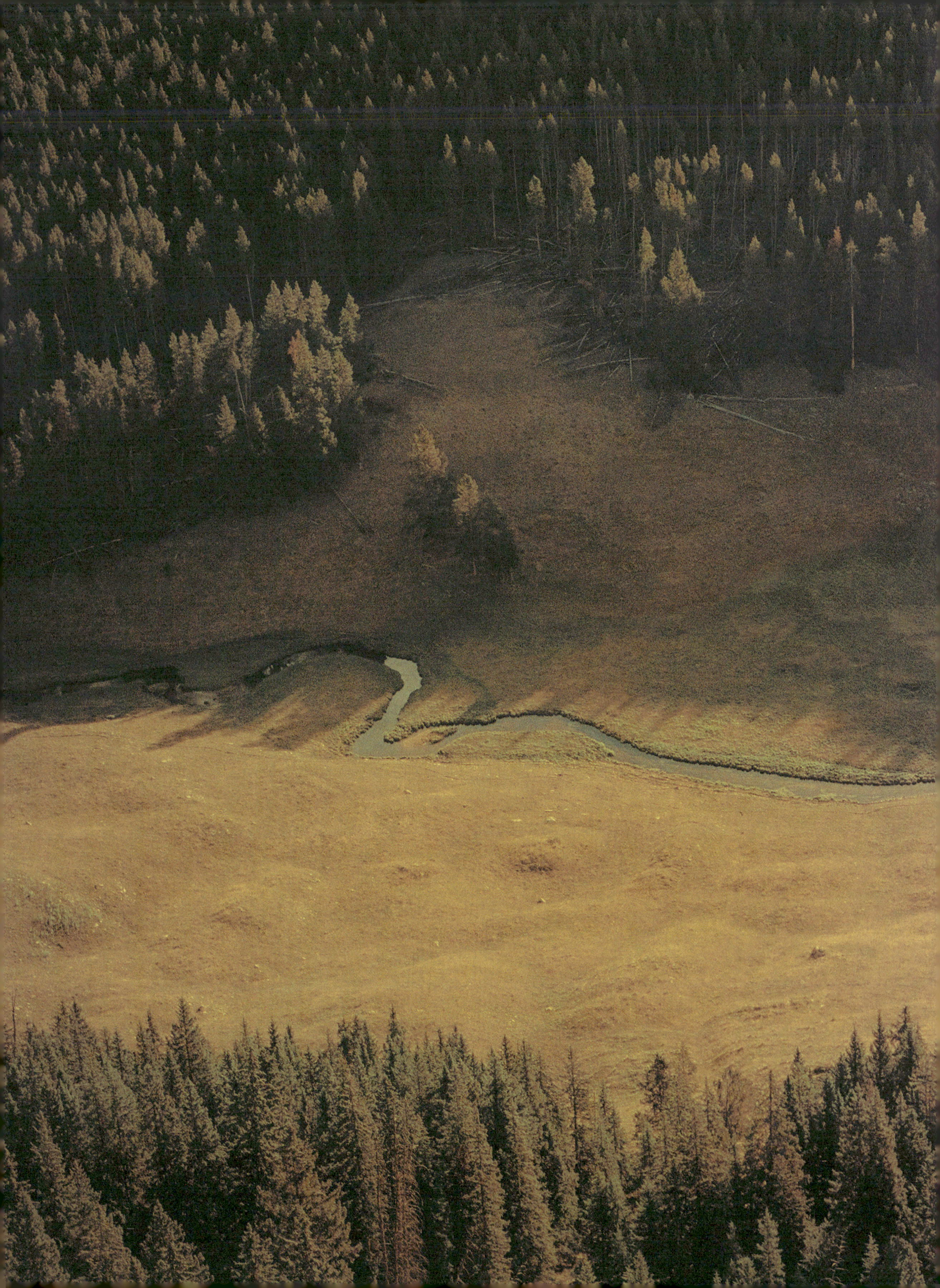

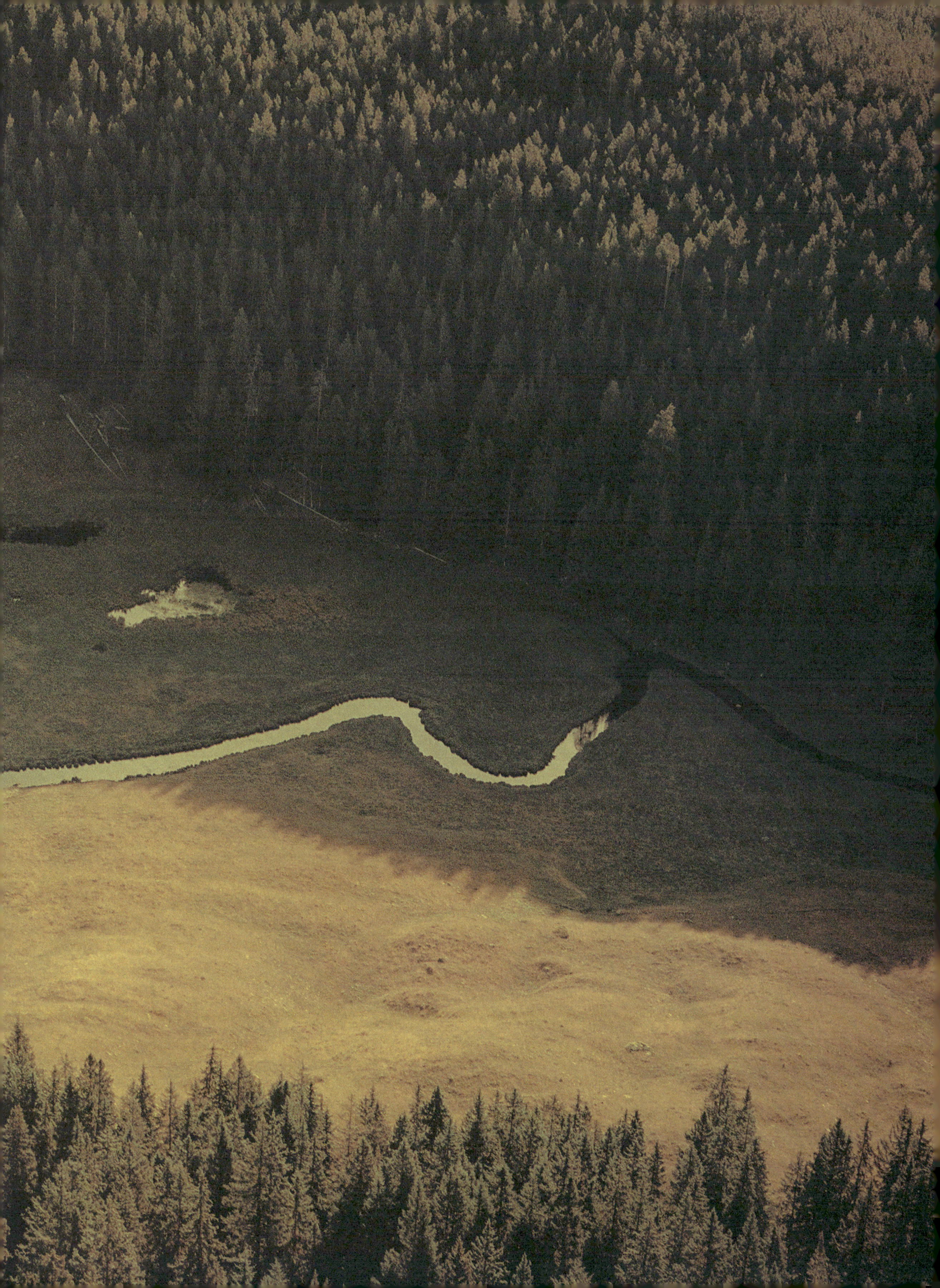

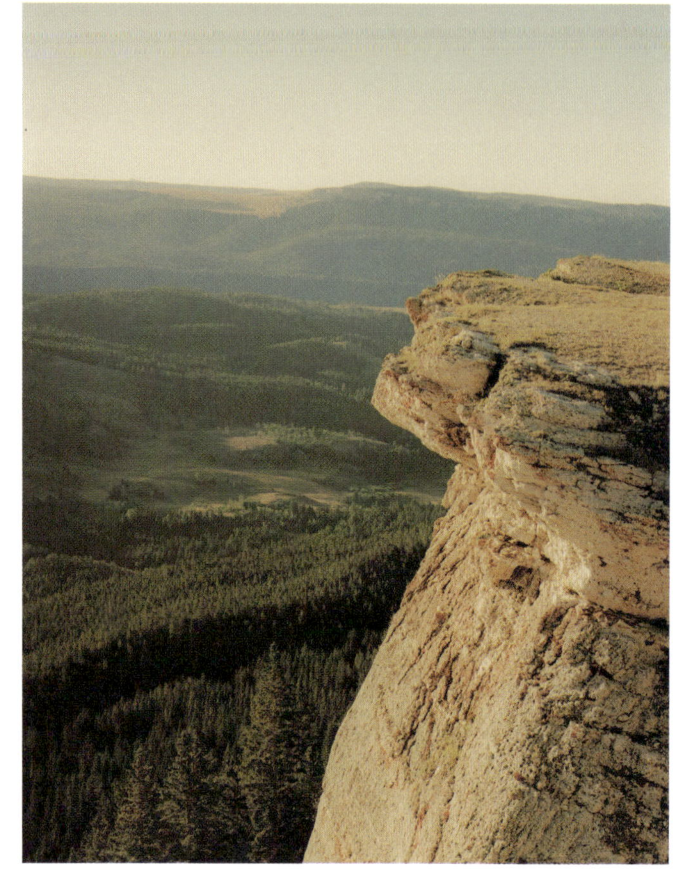

WYOMING

PAINTROCK CANYON RANCH

Located at the western base of the Bighorn Mountains, the Paintrock Canyon Ranch is Ranchlands' headquarters. Composed of an 80,000-acre range that reaches up and around Cloud Peak—the highest peak in the Bighorns—these lands boast creeks, springs, meadows, and canyon walls thousands of feet high. The terrain changes from deep orange and burgundy sandstone hills to monolithic rock formations, high-mountain meadows, pine forests, and lakes. The ranch's namesake, the Paintrock Canyon, begins at the ranch headquarters and meanders for eight miles. At the bottom of the canyon is Paintrock Creek, widely known to be one of the best trout fisheries in the region. Wildlife is aplenty; bobcats, elk, mules, whitetail deer, pronghorns, and bears roam freely.

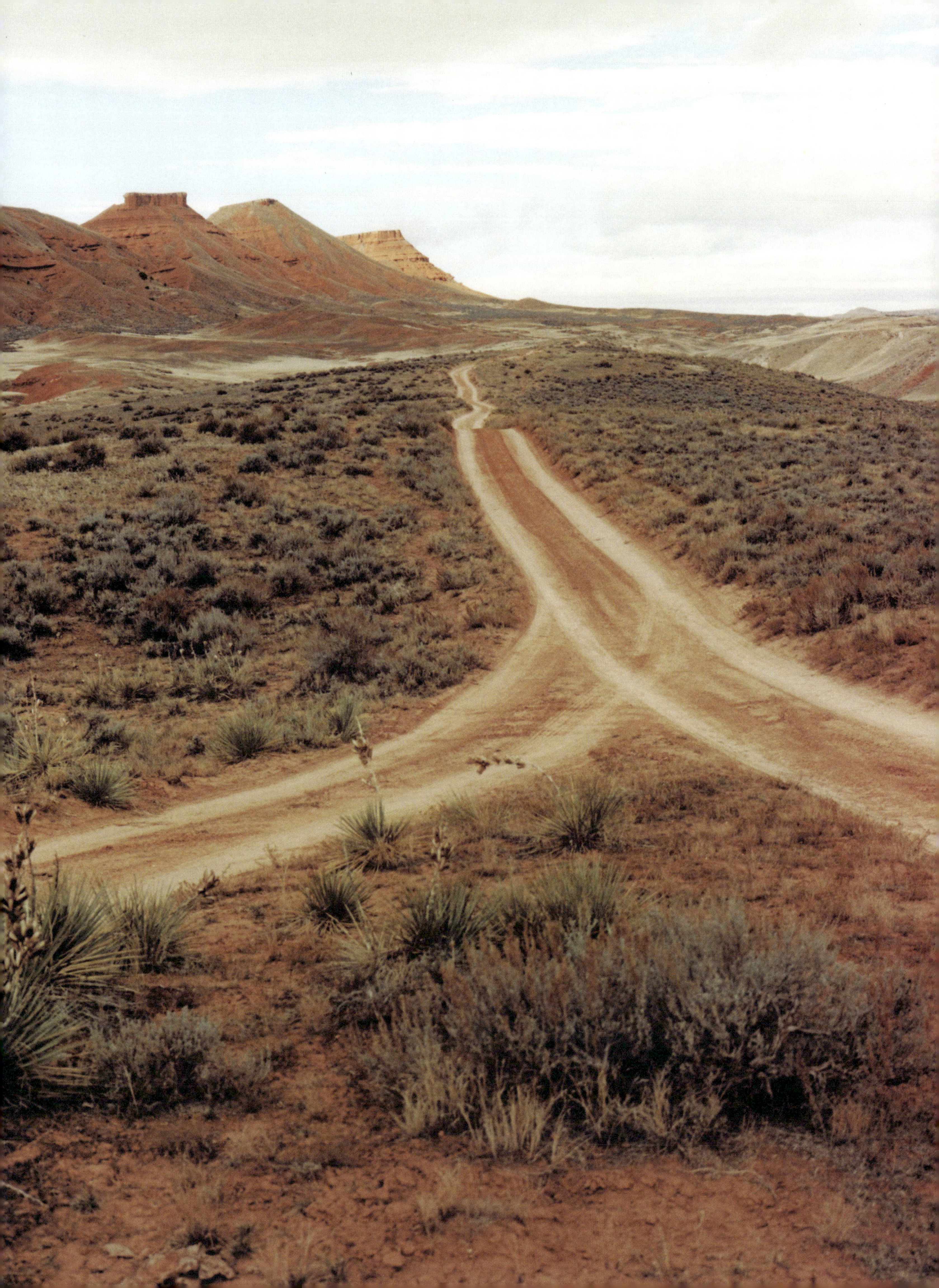

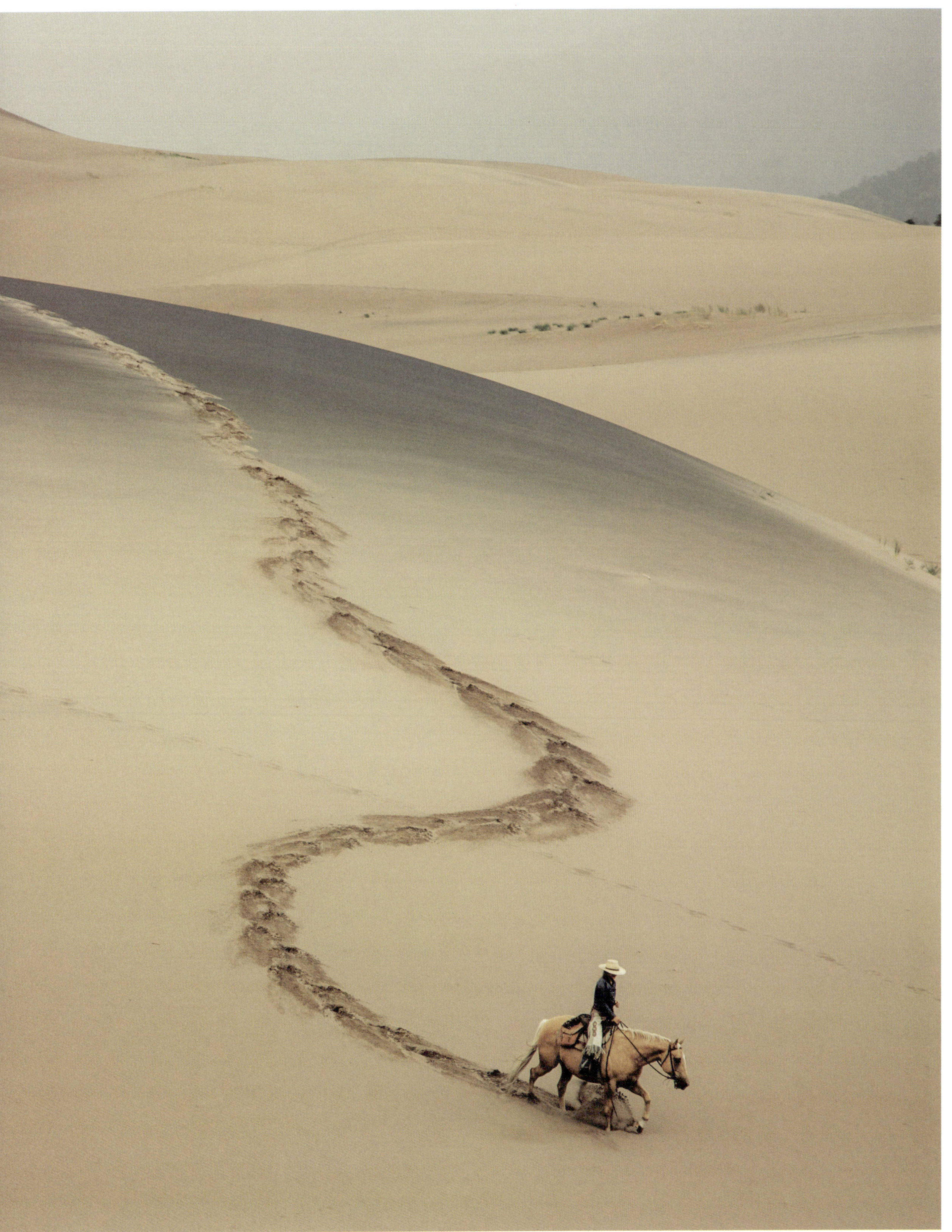

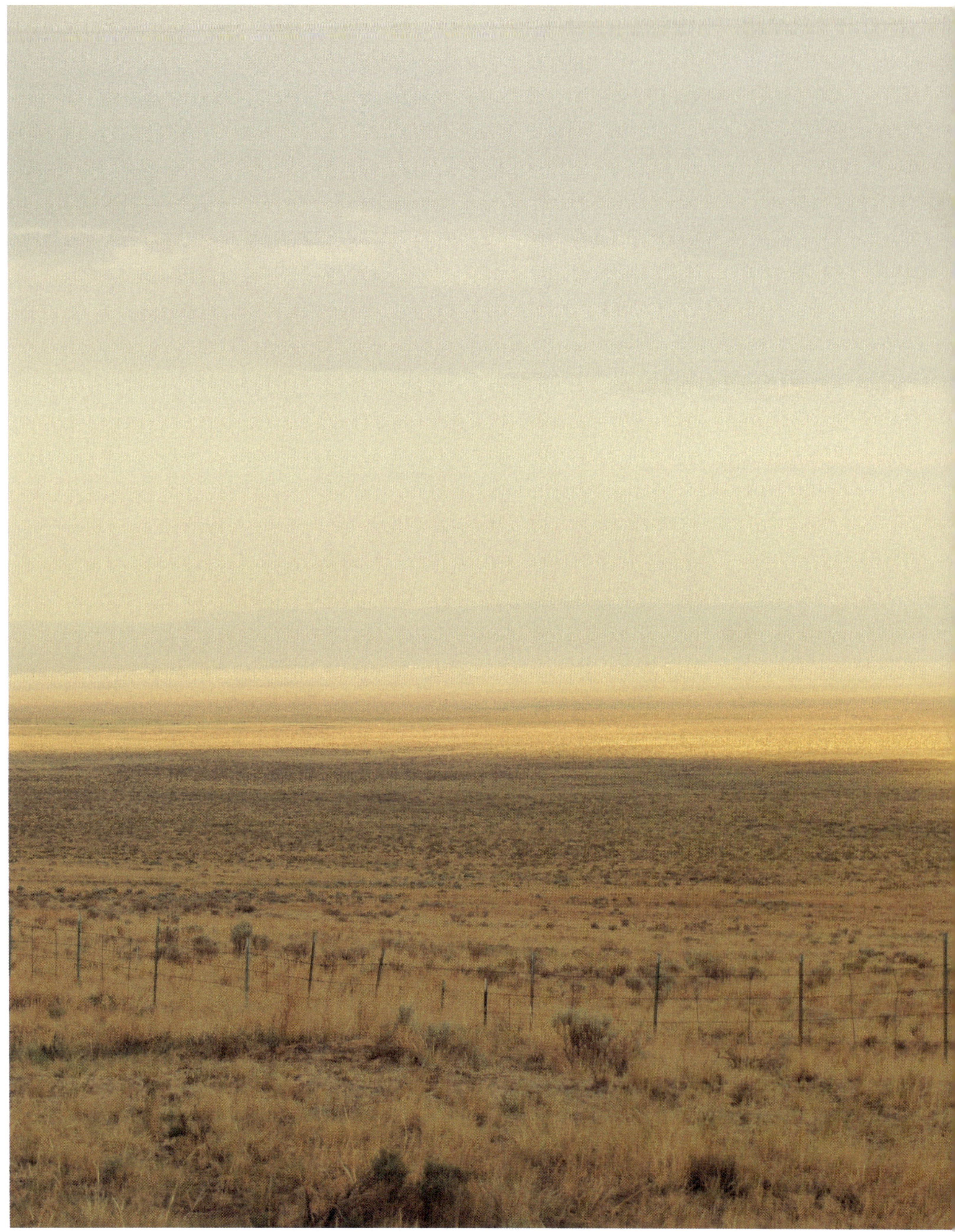

THE PRAIRIE

Grasslands make up nearly 40 percent of our country, and yet they are the least populated biome and, arguably, the most mistreated. This negative stigma gained traction in the early 20th century, when immigrants from Europe tried to farm sections throughout the West and Midwest and failed. The aridity, variety of grasses, and lack of elevation were a shock to the European standards of beauty and fertility. Many could not adapt to the landscape and instead tried to conquer it. Native grasses were dug up, exotic plants replaced them, and overgrazing became the norm.

Large swaths of native prairie have now been reduced to monocrops as a result of the Great Plow-Up, and Americans associate grasslands not only with barrenness but also with dullness. It's a dullness that goes beyond environmental degradation and bleeds into the subconscious. People think flat, open landscapes are wastelands, but that could not be further from the truth. If you look closely, you begin to see movement (life!) on multiple levels—a grasshopper on a single blade of sand bluestem, sandpipers scuttling about on silty ground, or shallow patches where mule deer had bedded down the night before. These rich tapestries of forage provide diversified communities of plant species, habitat for wildlife and for grazing cattle. Grasslands are a resource that cannot be harvested in any way other than grazing, which is a disturbance that, if managed correctly, can increase the health of our natural places.

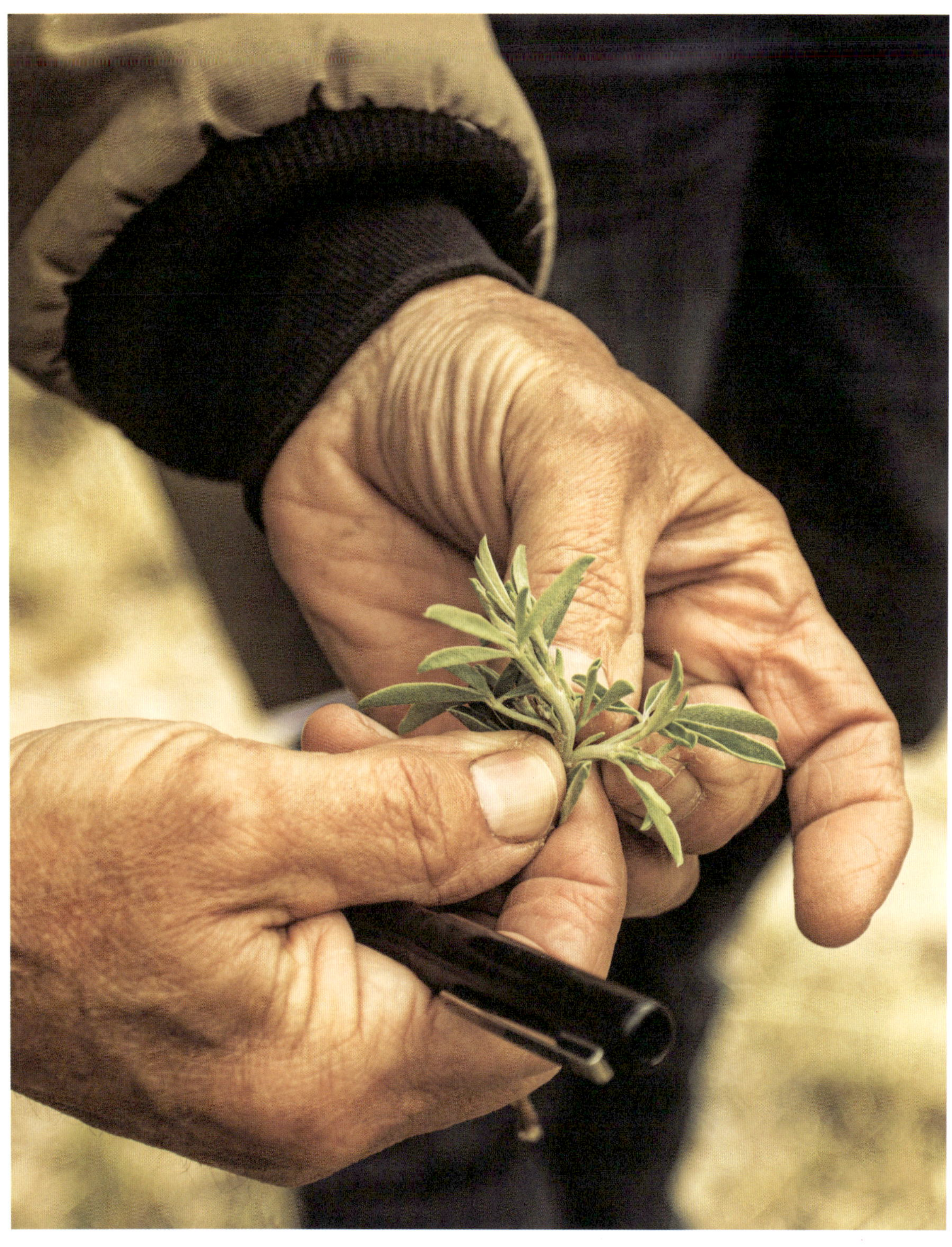

Range ecologists bring a scientific perspective to managing grasslands that complements the stewardship methods of ranchers.

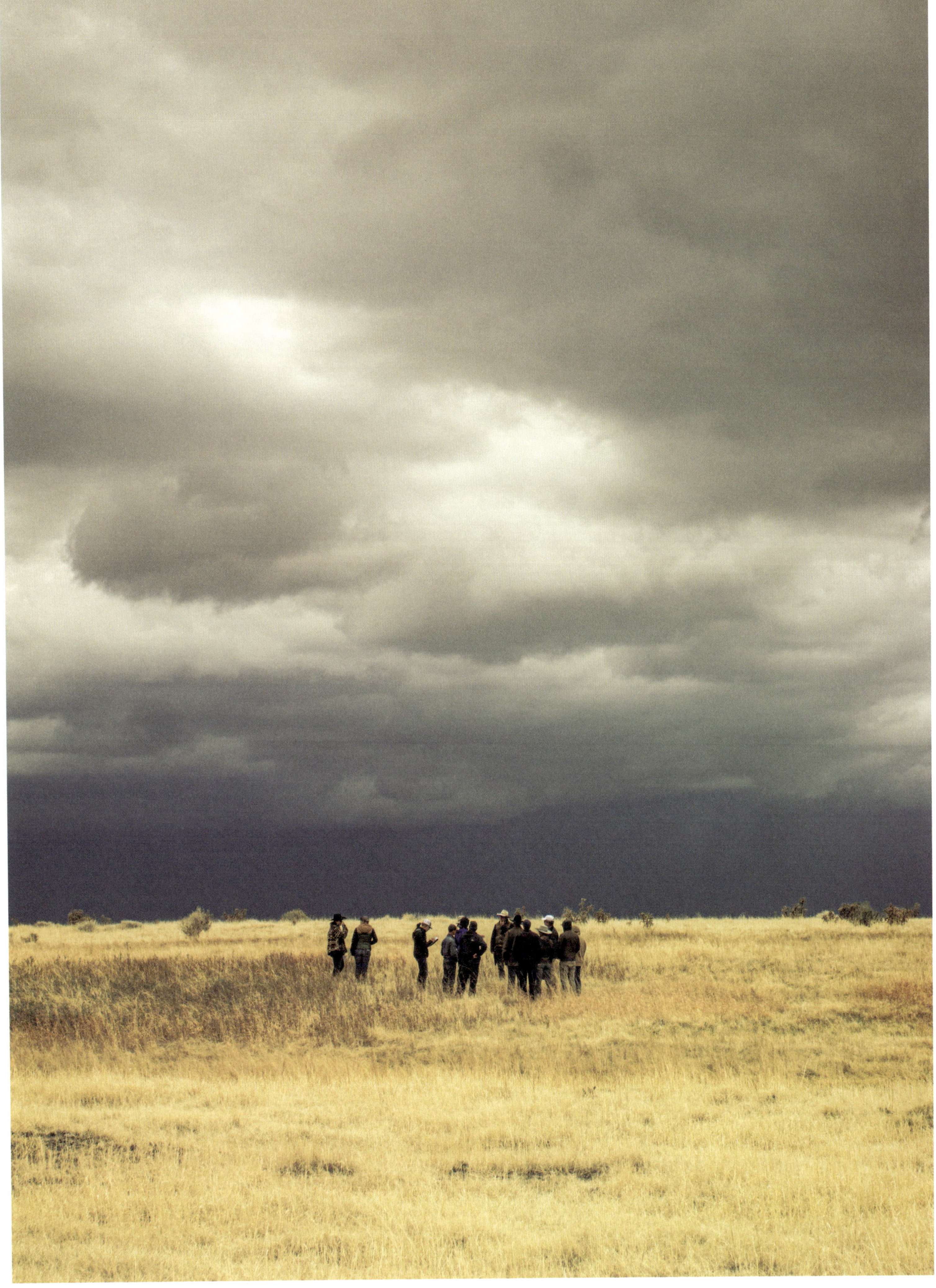

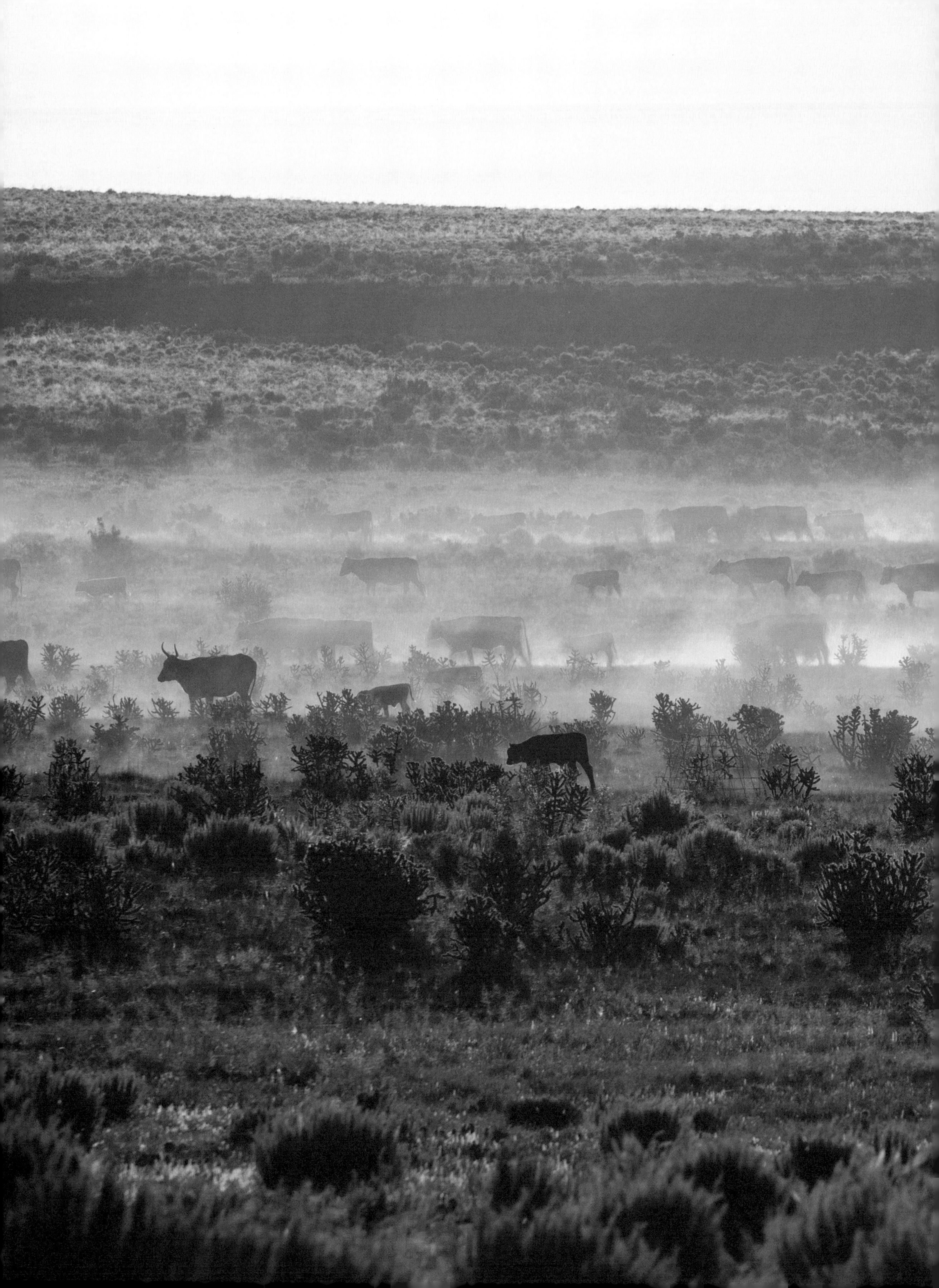

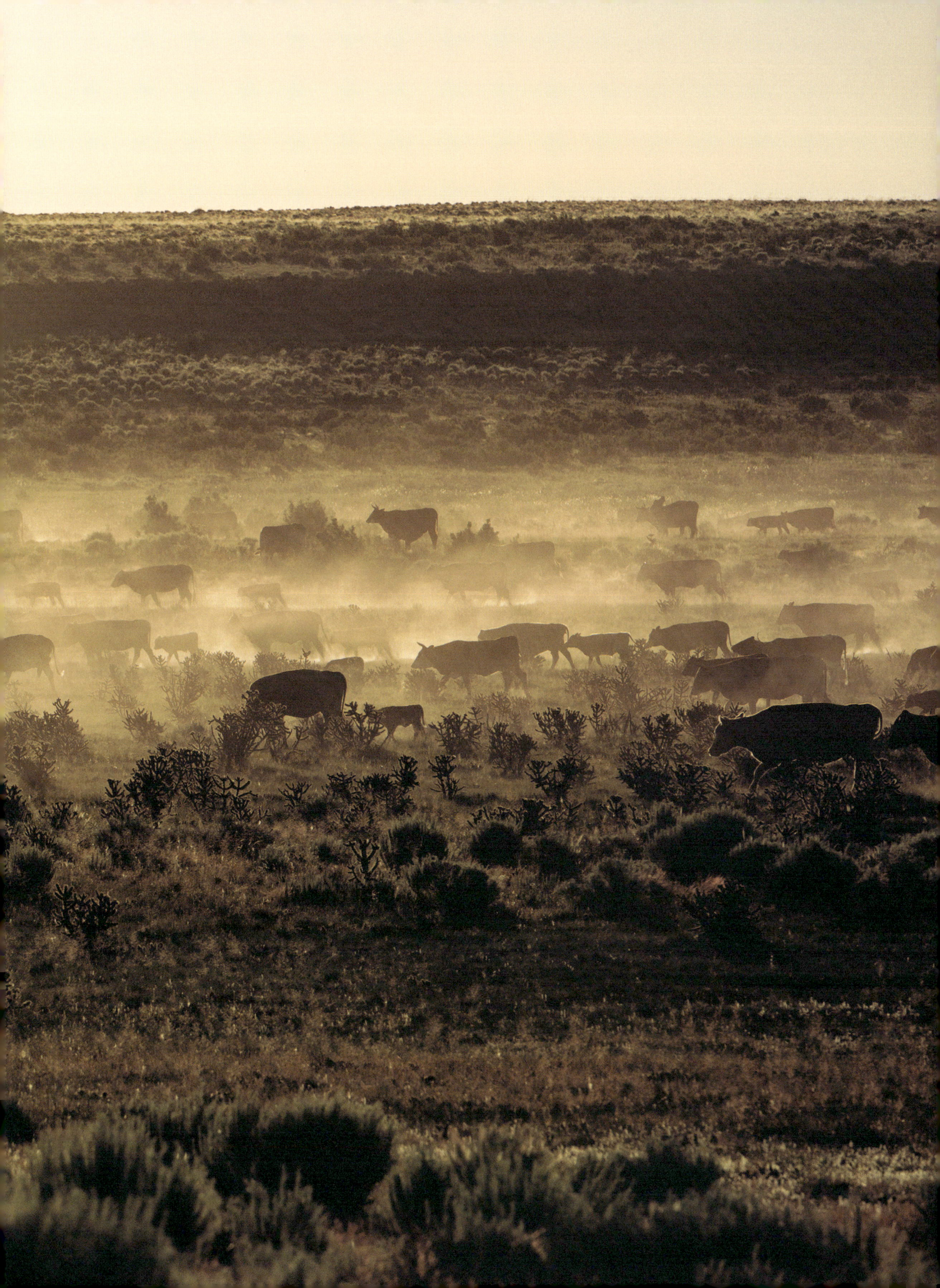

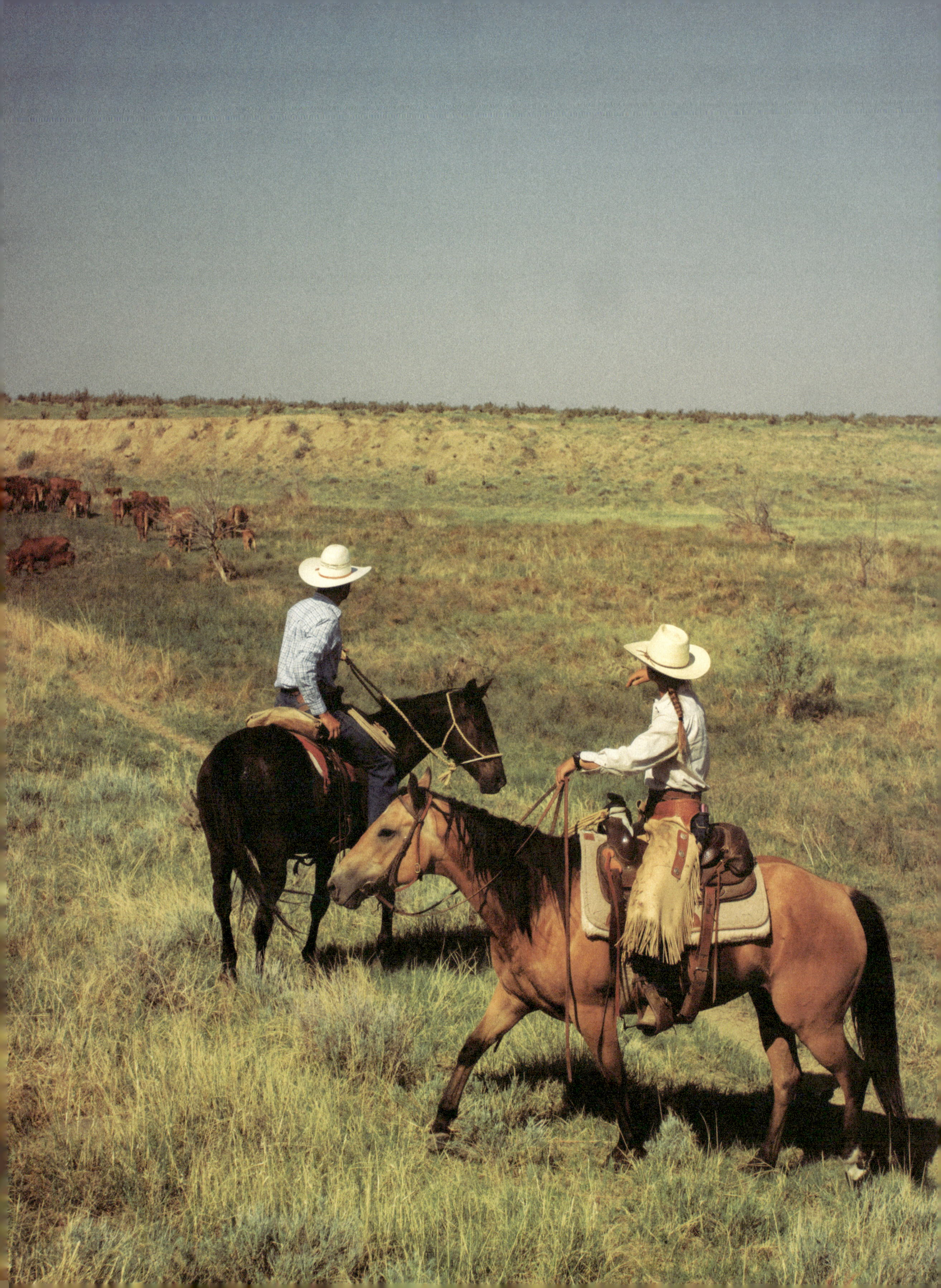

The rains return in the spring, and the grasslands come alive once again. Pastures become verdant, providing habitat for wildlife and cattle herds. This regeneration brings tremendous peace to us all, reminding us of the sun's life-giving qualities and an understanding that the rainfall will always come.

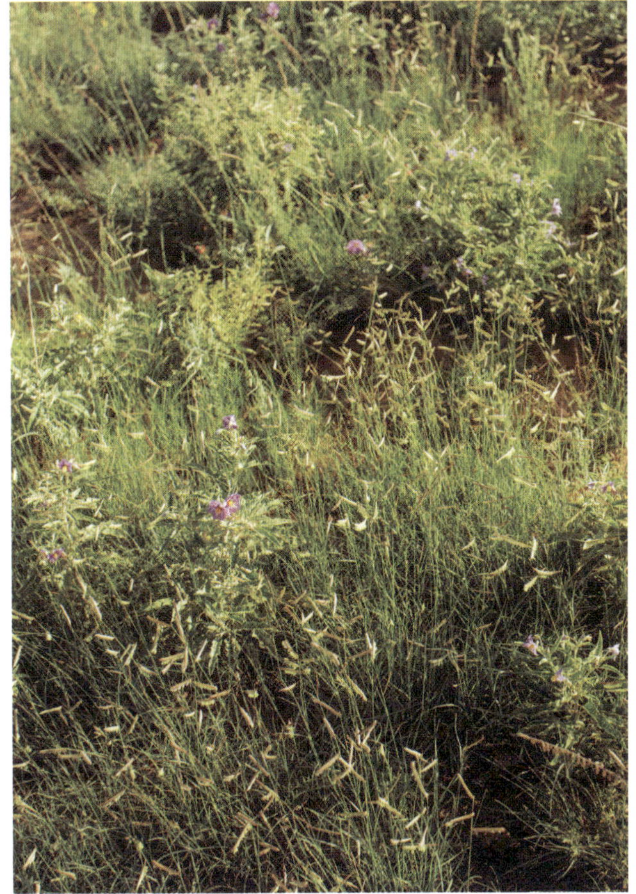

LAND-MANAGEMENT PHILOSOPHY

Many ranchers have long practiced responsible land stewardship—what we might call "regenerative." Today, the term *regenerative agriculture* has gained a cachet, as if it's something new or exclusive to the few who are using the latest ideas and technology. Of course, the recent developments in land-management philosophy and methodology are vital to the process of healing the earth and addressing the ever-increasing ecological challenges we face, but the soul or active consciousness that comes from a deep connection with the land is also important, if not essential. This new vision of regeneration, alongside the old methods of the ranchers who have come before us, offers hope that we can share this message: ranching is one of the most compelling forms of large-scale conservation today.

The wet eyes of summer bring the promise of flowers, a new moon, clean air, and eternal openness.

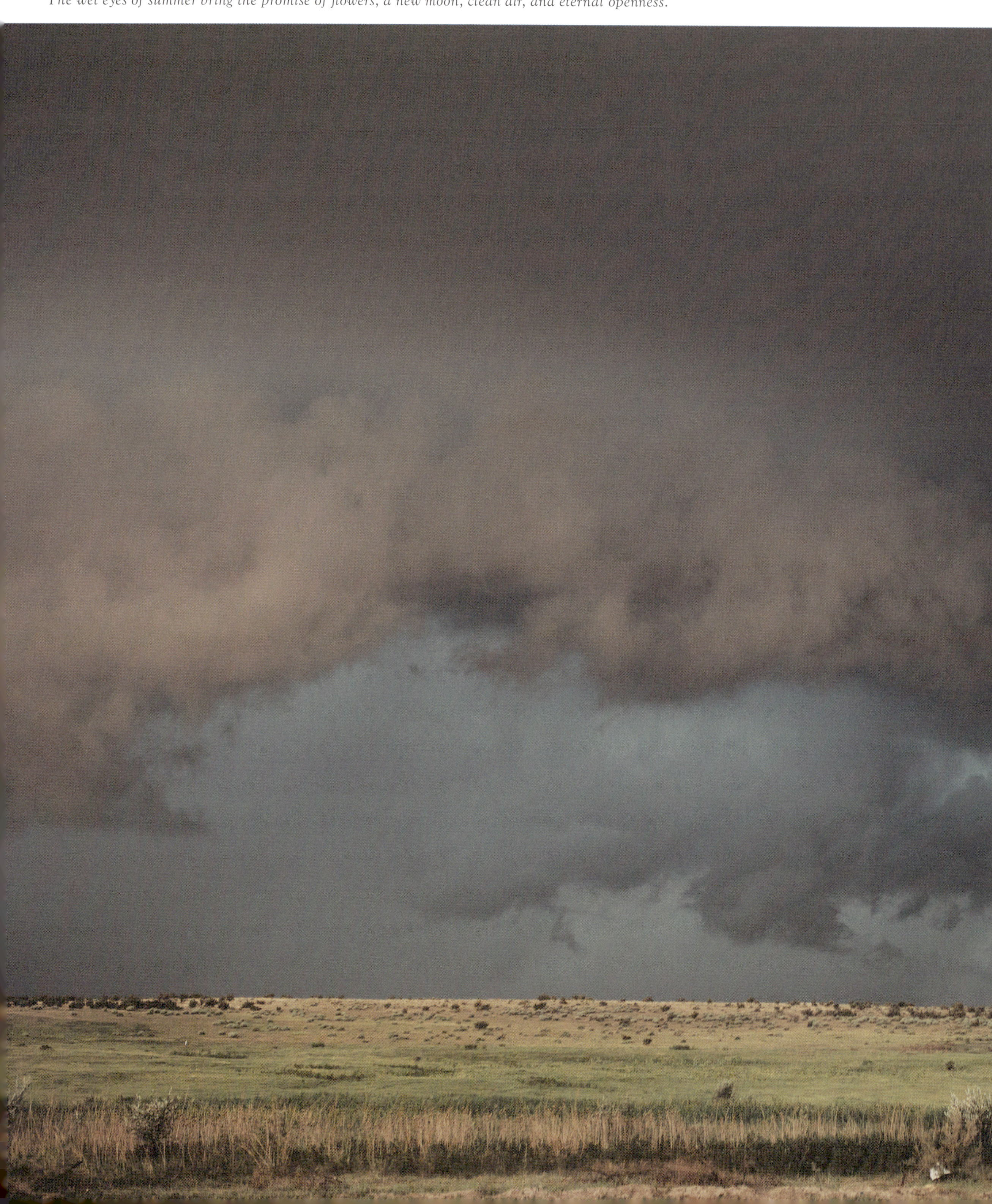

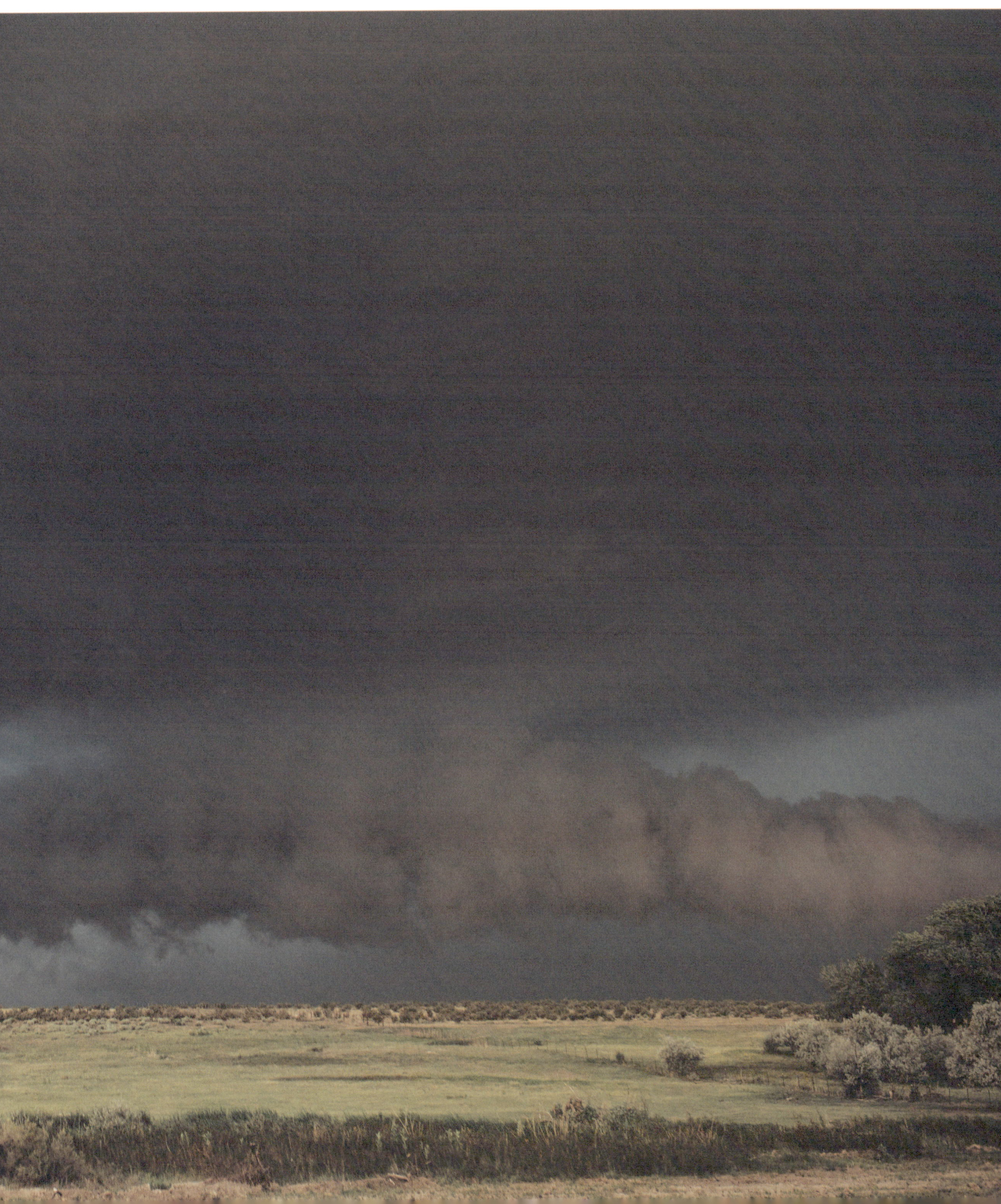

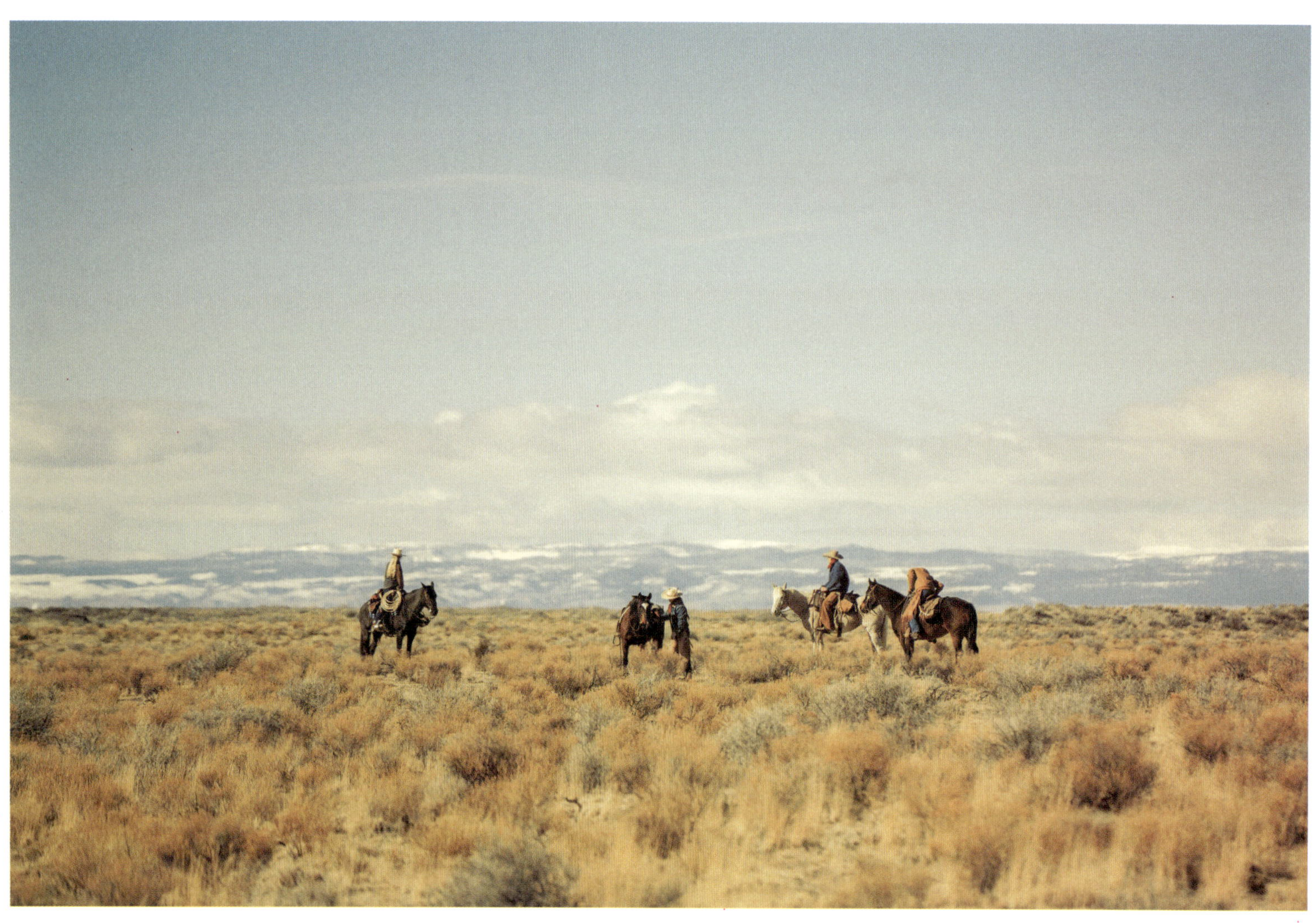

The changing seasons always give us a sense of time passing, and with it, the accomplishment of projects,

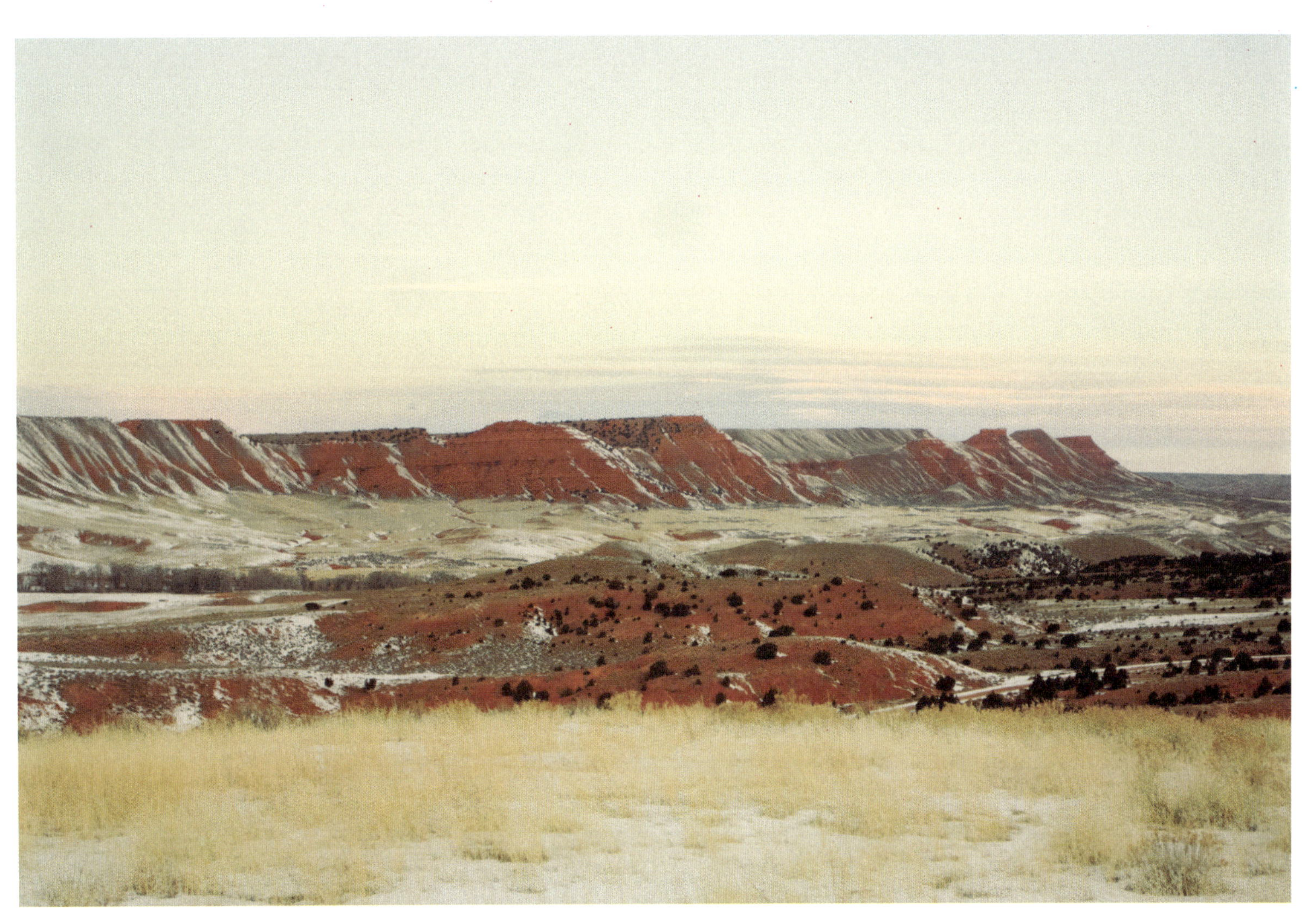

enjoying days with our family and friends, and a reminder of the wonder of nature's cycle of growth and decay.

A Chico Sunday In The Drought

Its Sunday morning
on the porch.
The sun pulls
into the pale sky
a golden open sore.

I see smoke
from the fires
curling over the ground
thick,
like a gorging larva.

I search for movement
in the dry pasture, and
see thin antelope families threading
outward from the water tank
zigzagging through the yucca stands,
while mares stood inward
with foals stuck to their flanks
like butterfly adalvas.

The leaves from the apricot trees
fence them as motionless
as dry bones,
waiting as if
for the skies to call.

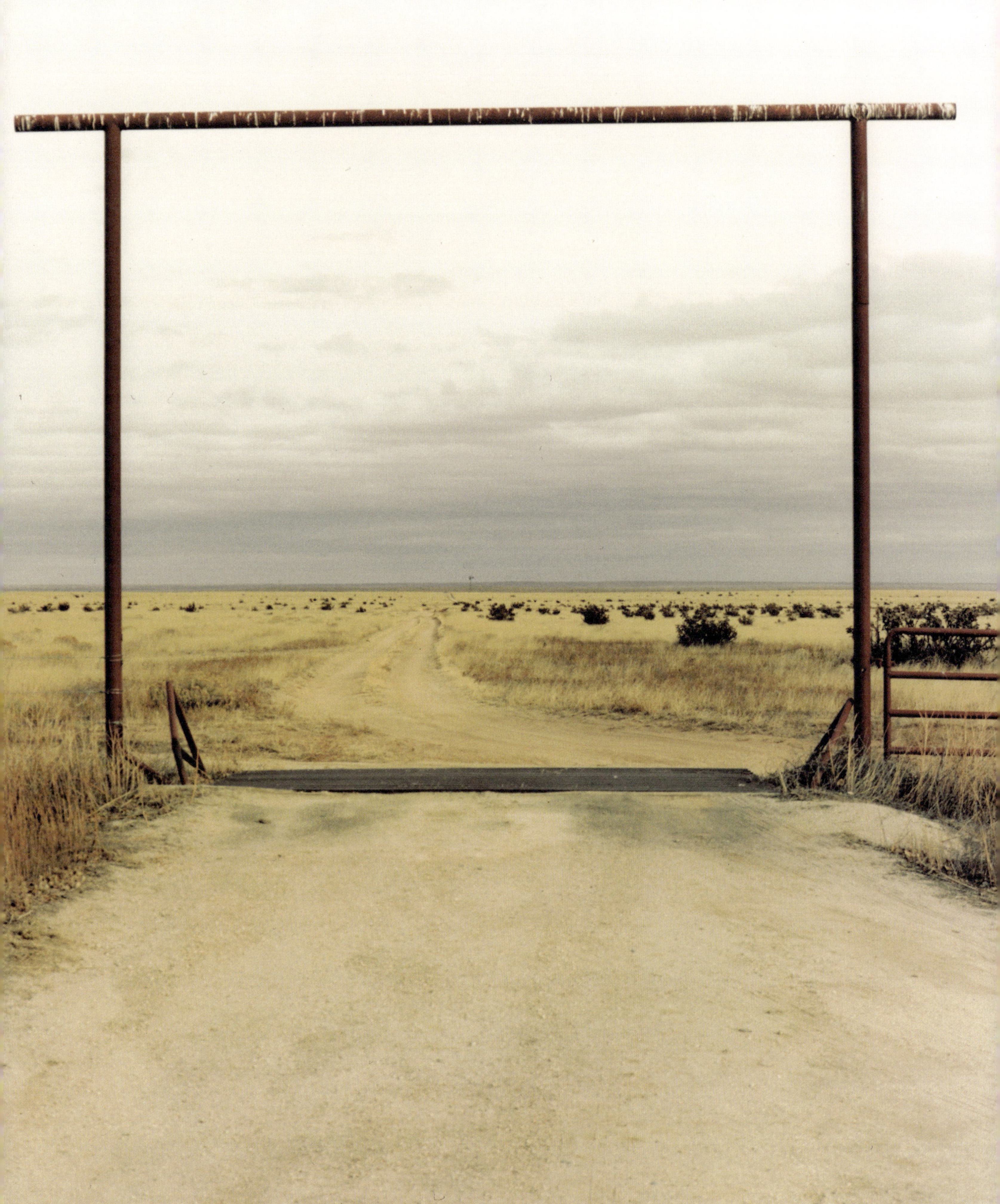

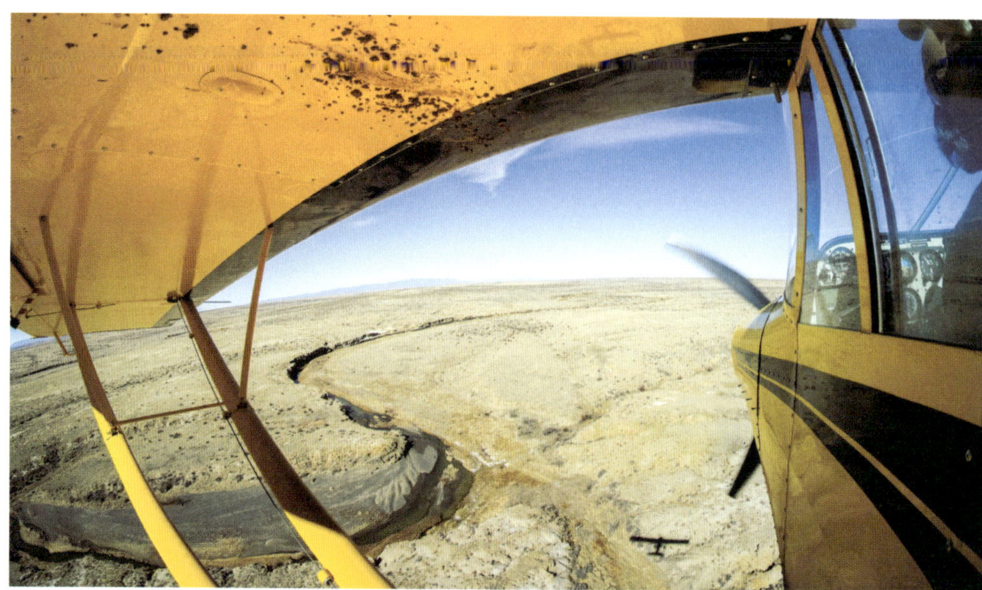

Aviation is an important tool that we use for many things on a ranch, such as checking pastures for rainfall or observing changes in plant communities. The view from high up gives us a different perspective that allows us to see the entire dynamic that we are managing.

DROUGHT MANAGEMENT

Not enough rainfall is essentially one of the greatest threats to ranchers—no rain means no grass and no way to properly feed the cattle. For an industry that operates on thin margins to begin with, this can largely affect the bottom line for the year and put a rancher out of business. Part of our model at Ranchlands works to combat these effects. For starters, we have multiple properties so that a ranch in one part of the country can potentially absorb cattle from another ranch if conditions are challenging. We work to diversify the business so that taking a hit on the cattle side is not the be-all and end-all of circumstances. In this way, diversification is also conservation friendly because it gives us flexibility in destocking if we aren't getting enough rain rather than keeping those cattle and doing damage to the land. Less rain and less grass means there's greater potential for overgrazing, even with the same amount of cattle. Instead of looking at land as strictly a home for our cattle, we look at it as a multidimensional resource to make our business stronger and more flexible in times of significant drought.

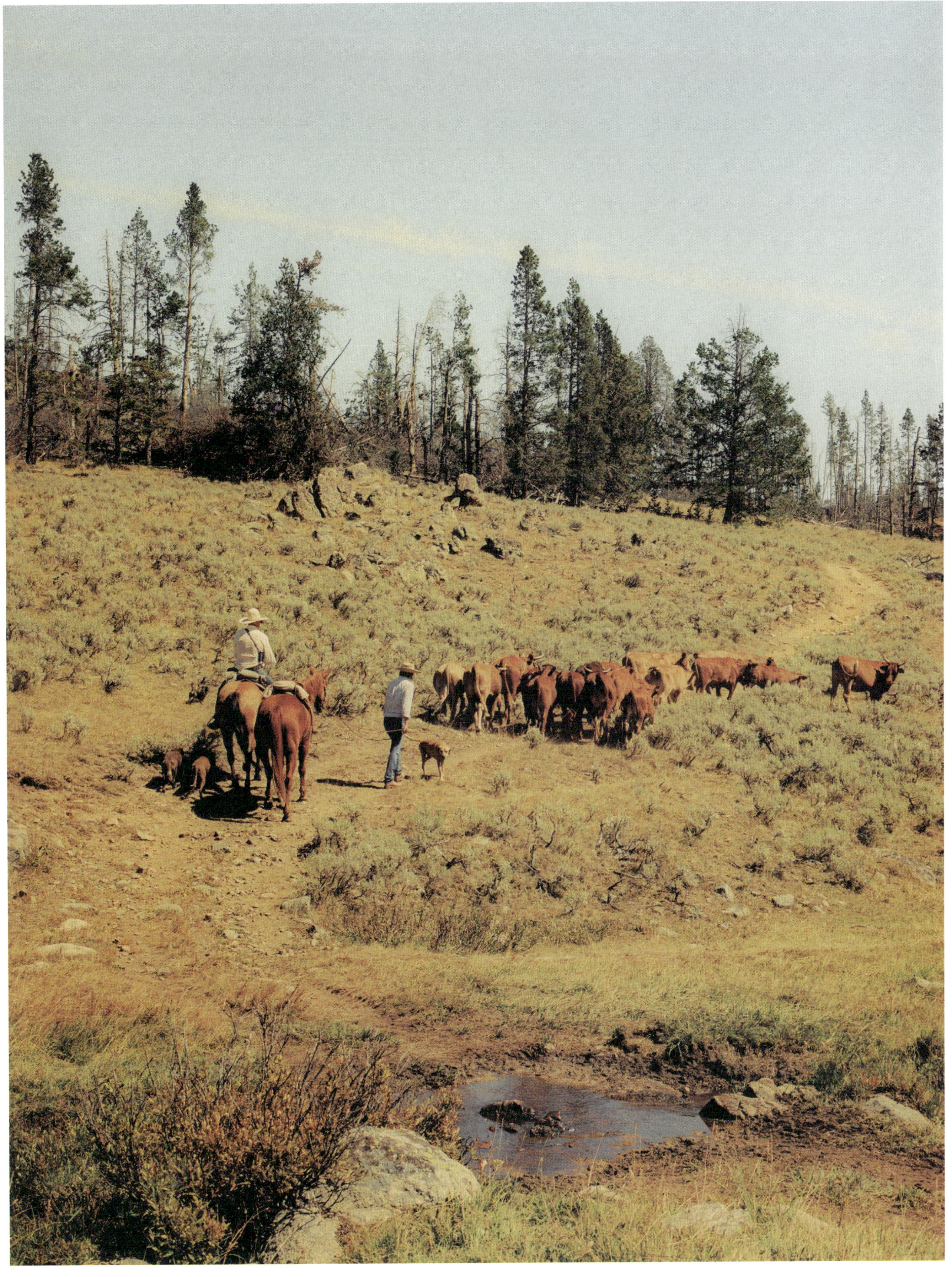

We rotate our cattle through pastures to mimic the migratory grazing patterns of native ungulates.

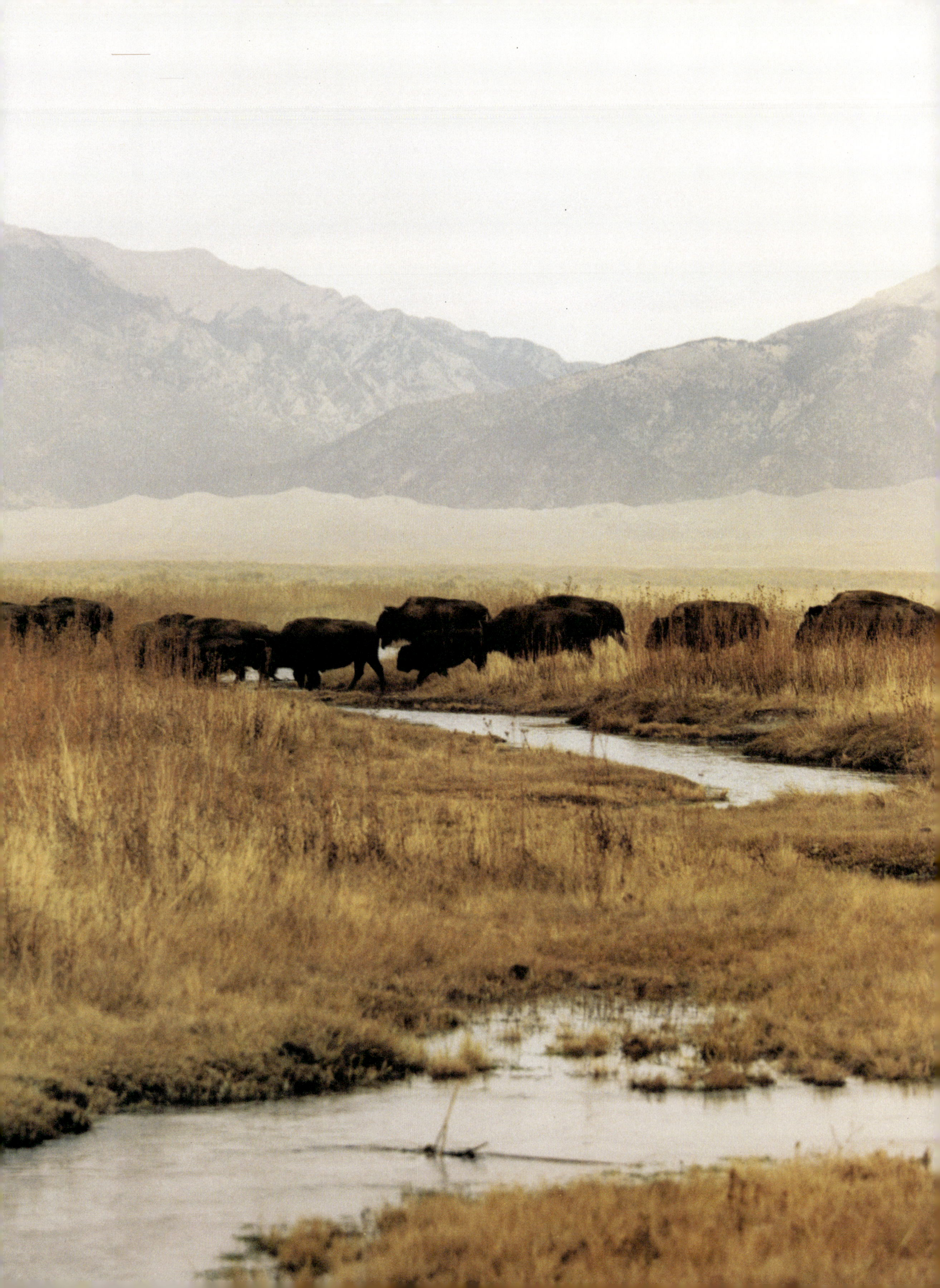

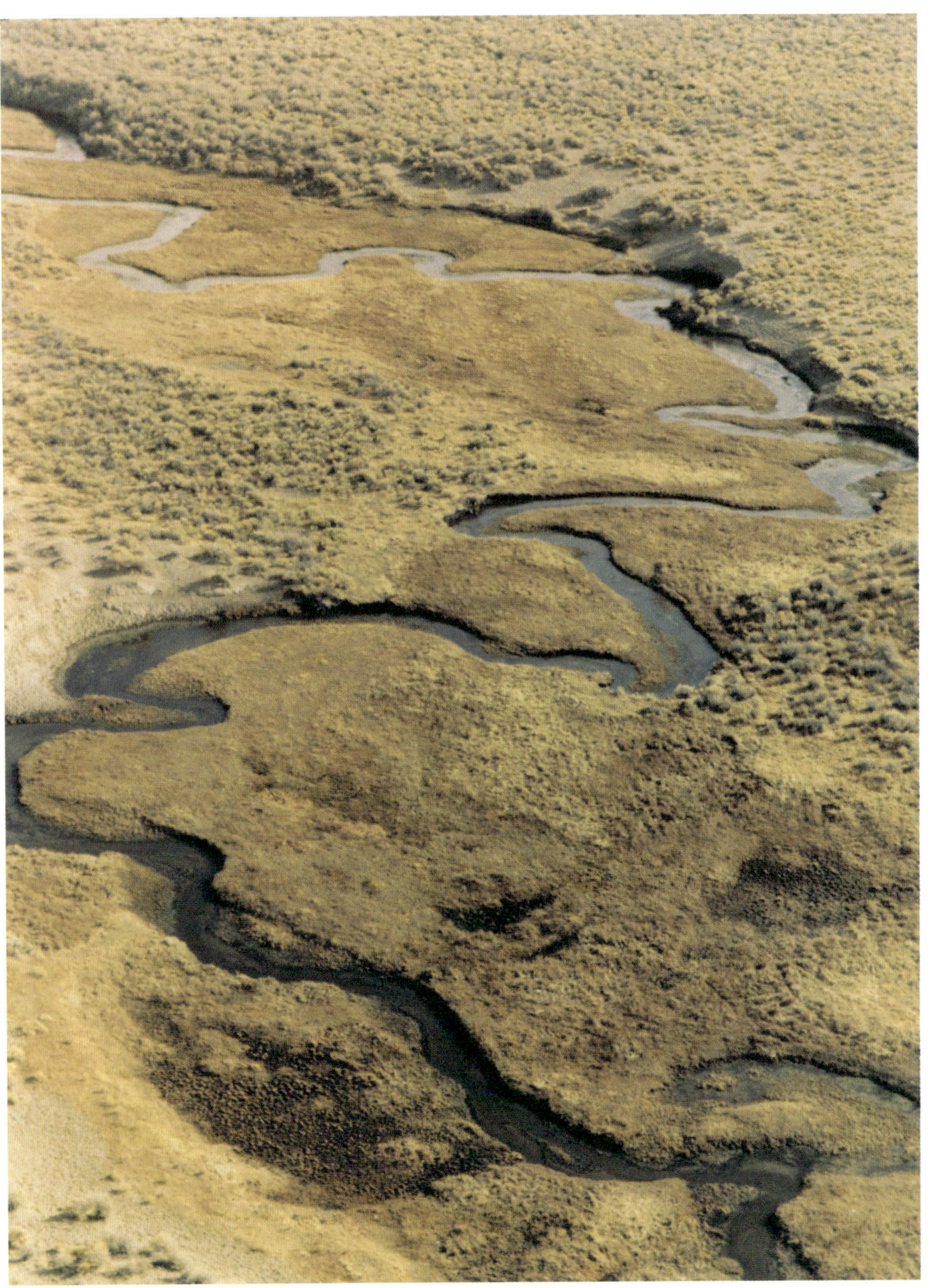

The grasslands of North America evolved with the presence of large herds of migratory bison that moved across the landscape based on factors like the seasons, the growth of grass, and the presence of predators, including human hunters. We manage our cattle to mimic the transience of grazing bison, allowing them enough time in each pasture to prune the grasses but not enough time to overgraze.

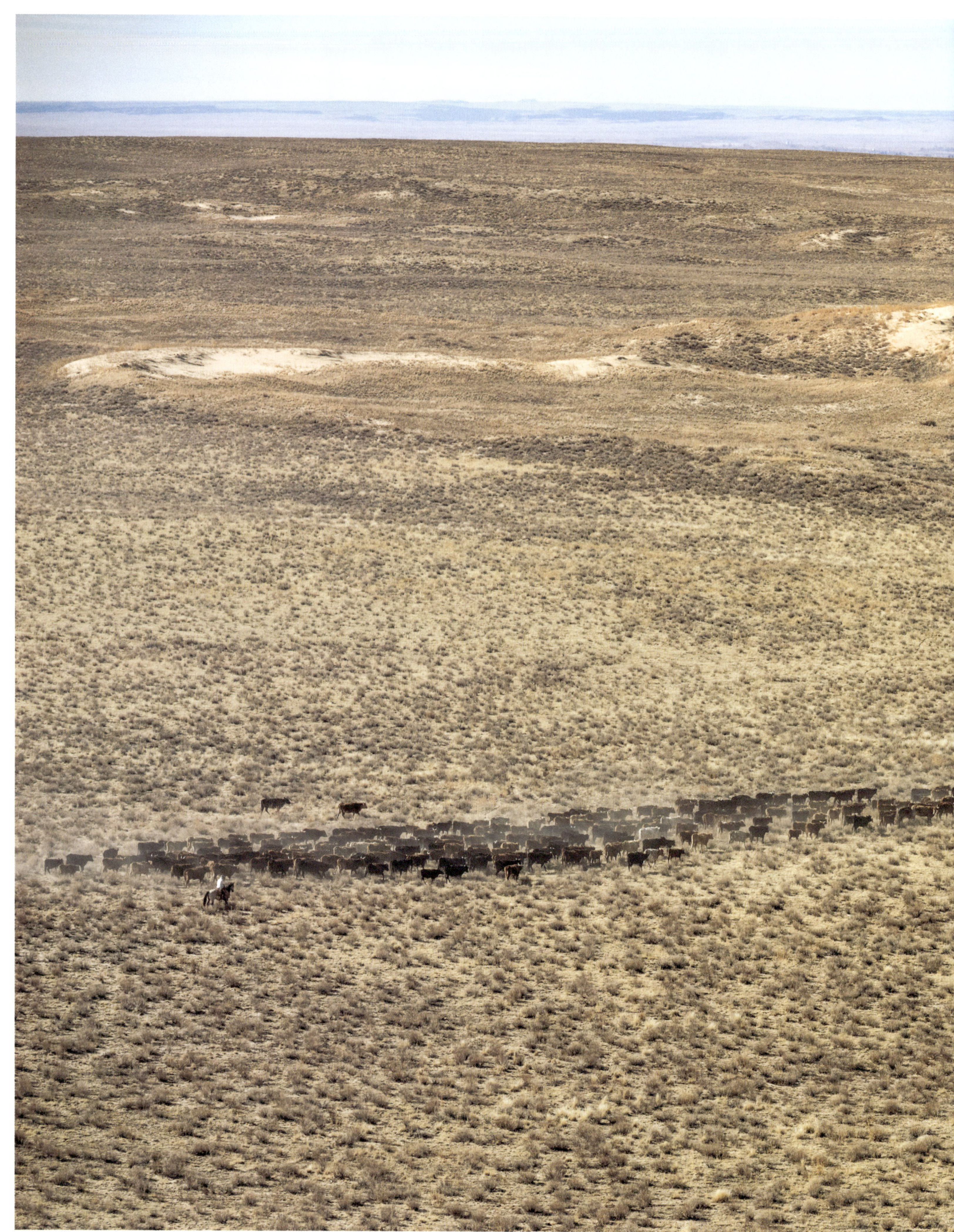

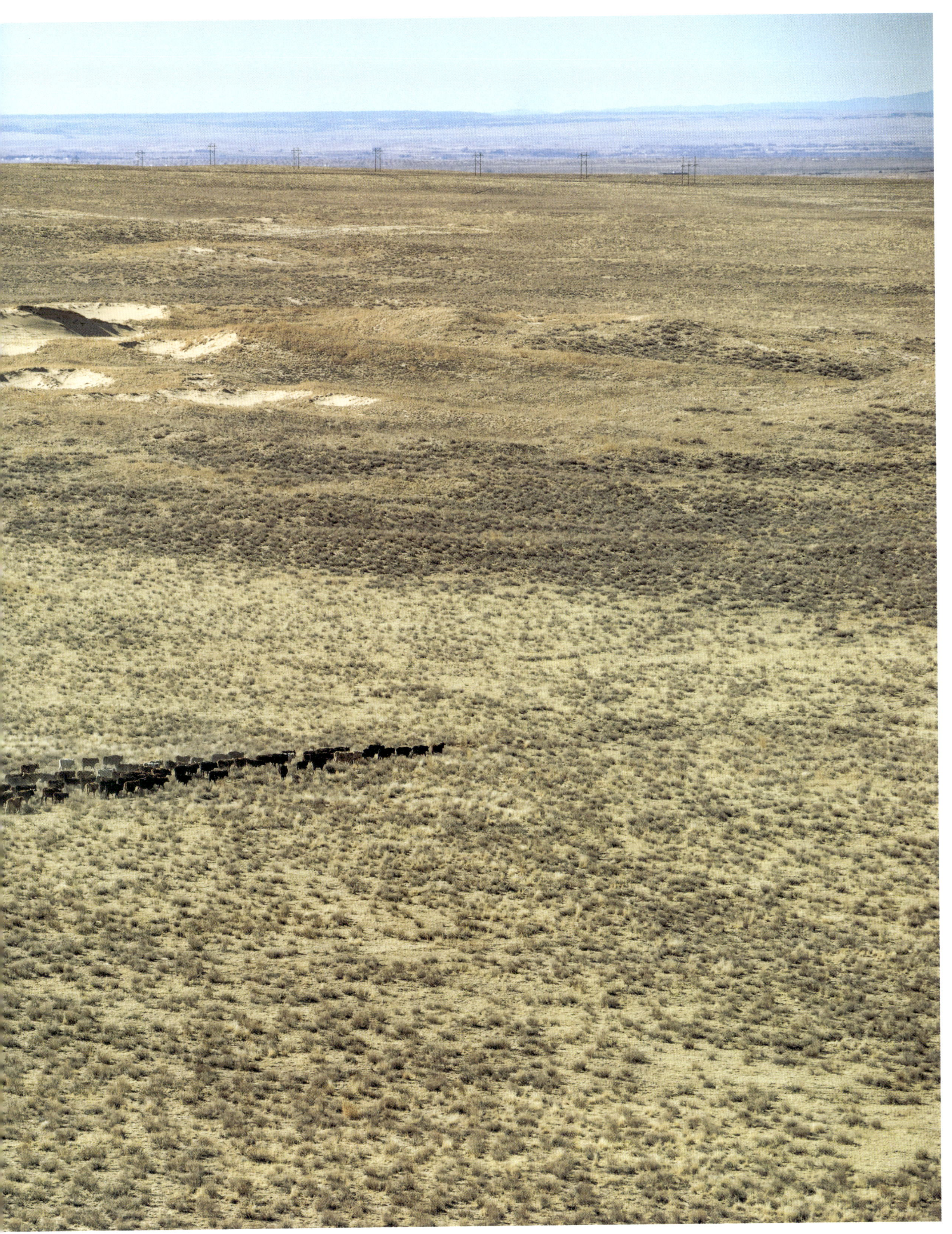

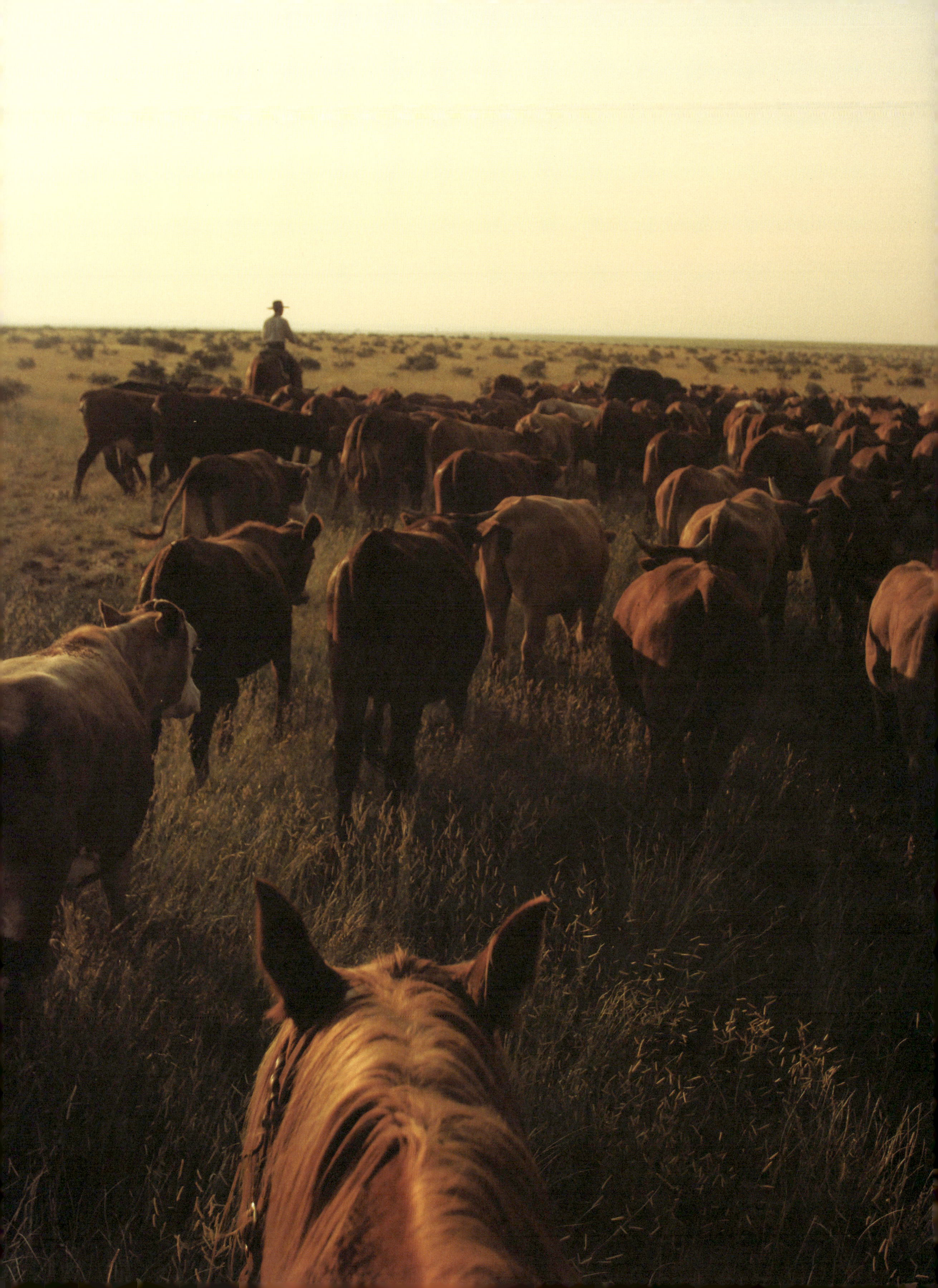

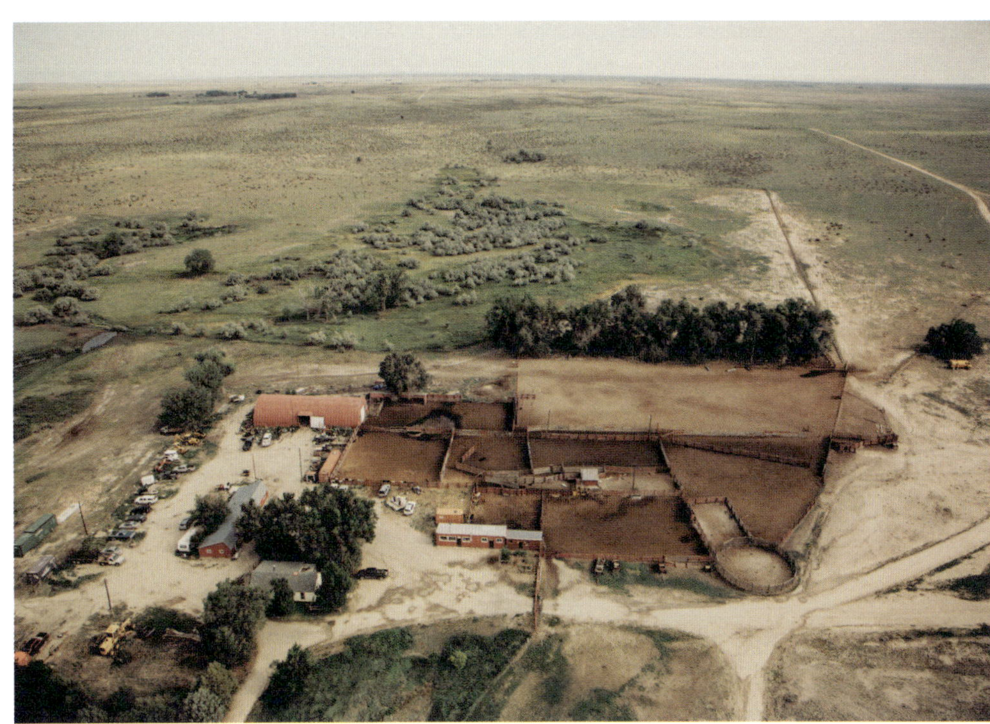

MOVING CATTLE

Our stewardship of the land is dependent upon our ability to move cattle, which is an art that can never be mastered. Cattle are wily, their daily routine so apparently simple that it obscures their true intelligence. Some days the cattle will work with you (and you'll feel very proud of yourself), and other days the cattle will defy you (and you'll likely be cursing yourself). No matter how many times you move the same herd of cattle between the same exact two points, it never unfolds quite the same way. They are very much alive to the environment around them, and how that environment is structured in turn shapes the way they move about and interact with those surroundings. With a handful of cattle in a controlled environment, like a narrow alley in the corrals, the way one animal might move in response to pressure is more predictable. But out in the pasture, in an unrestricted sea of grass and brush, there are many more variables at play.

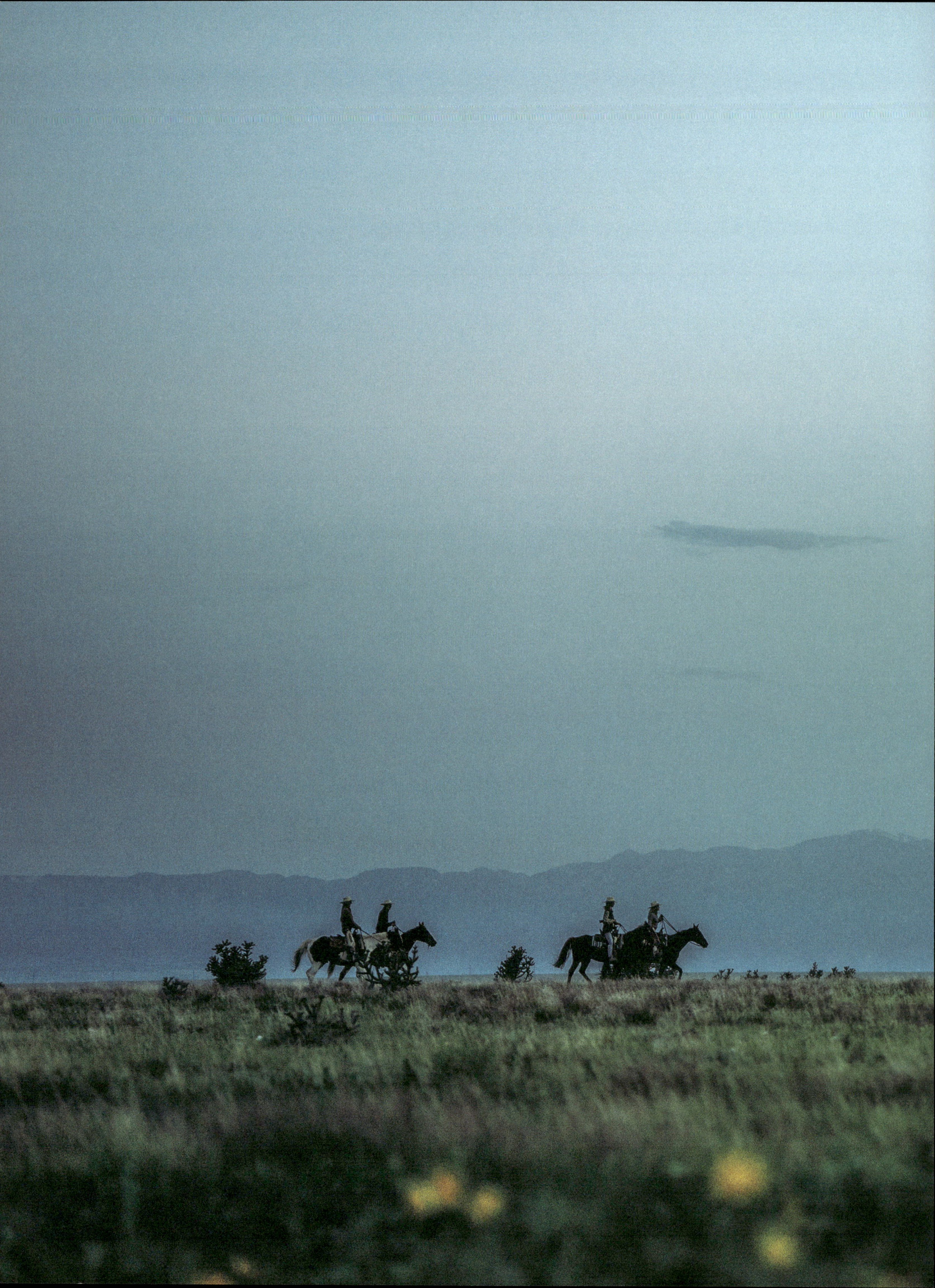

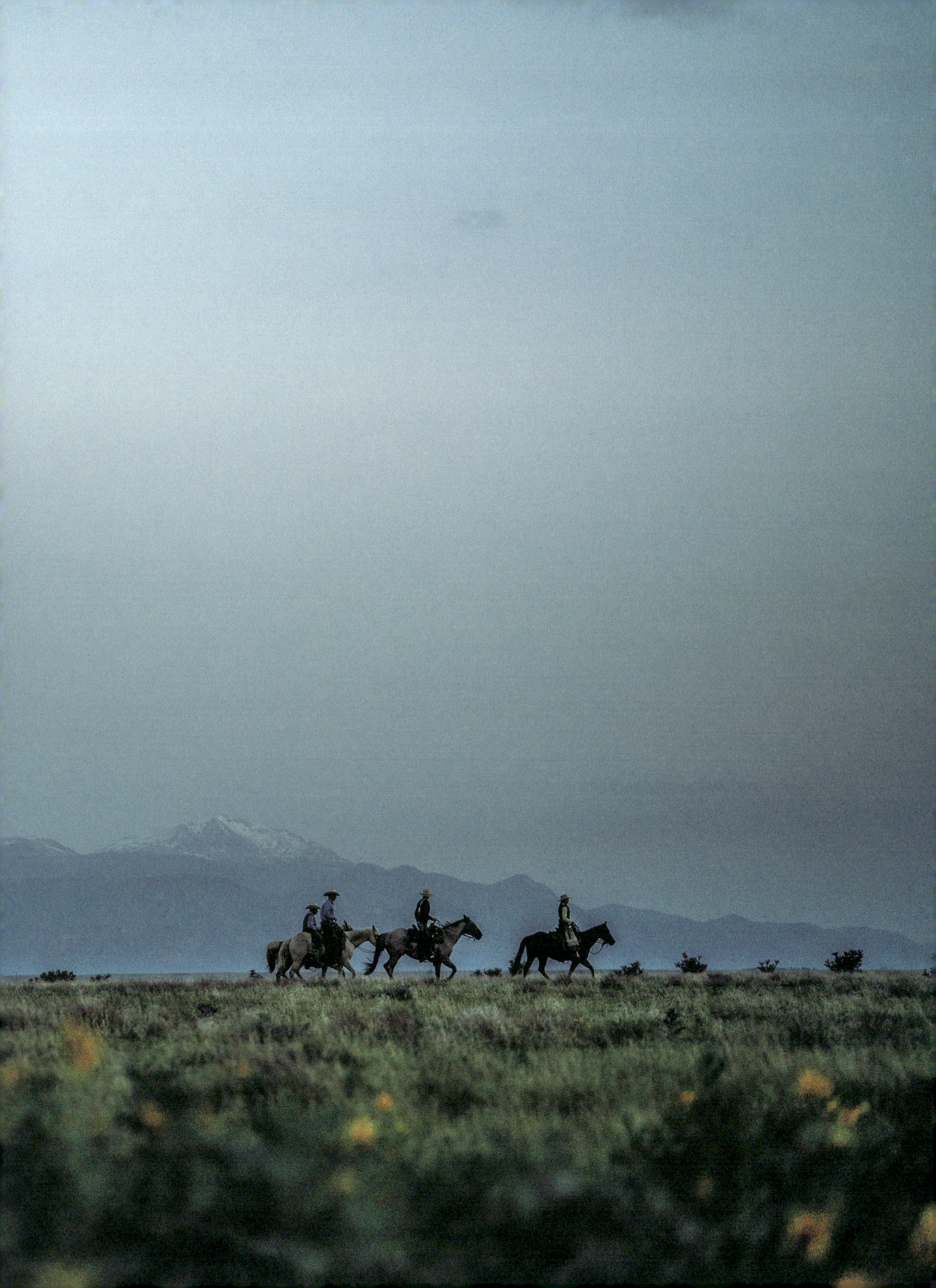

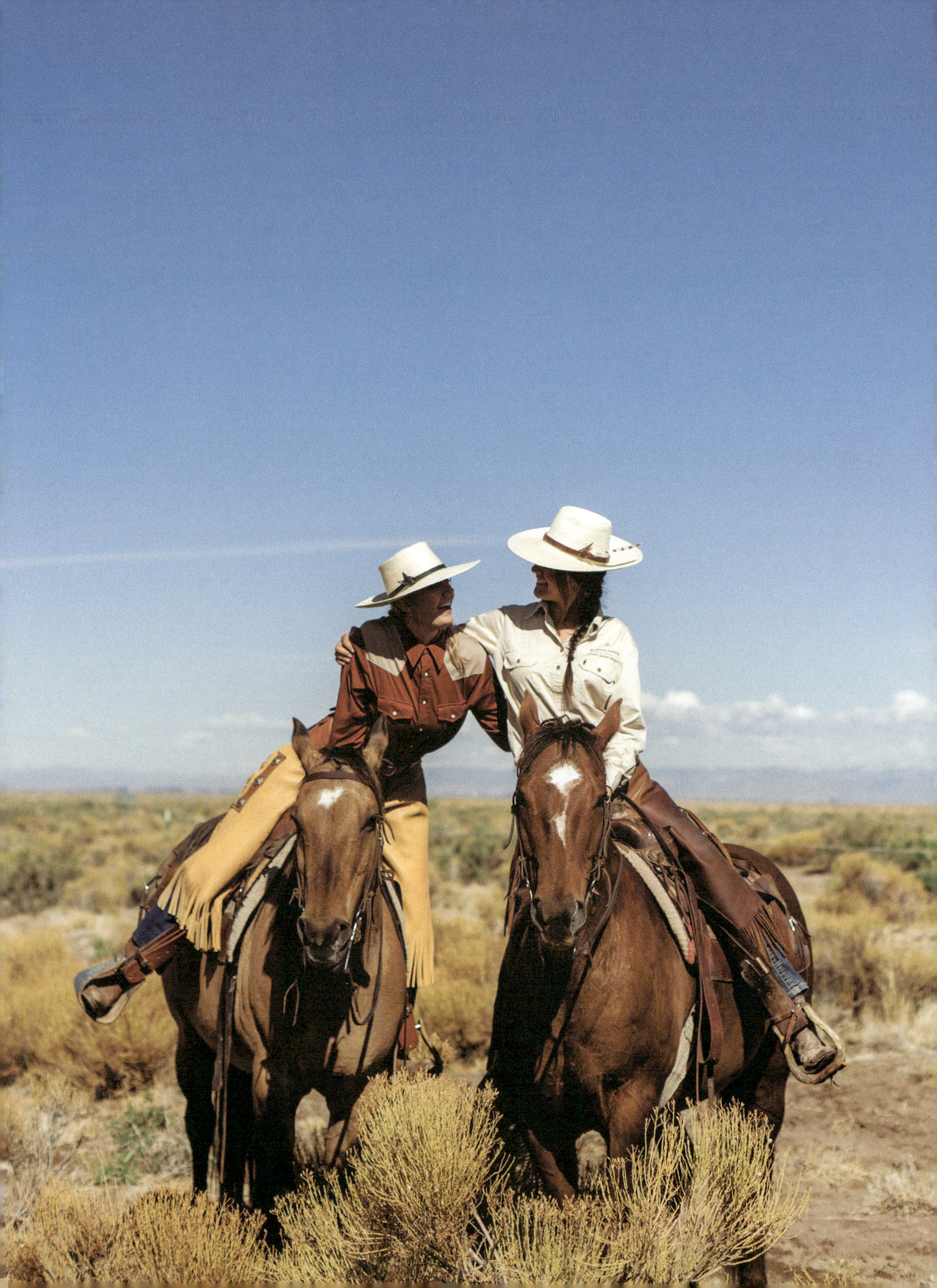

THE

PEOPLE

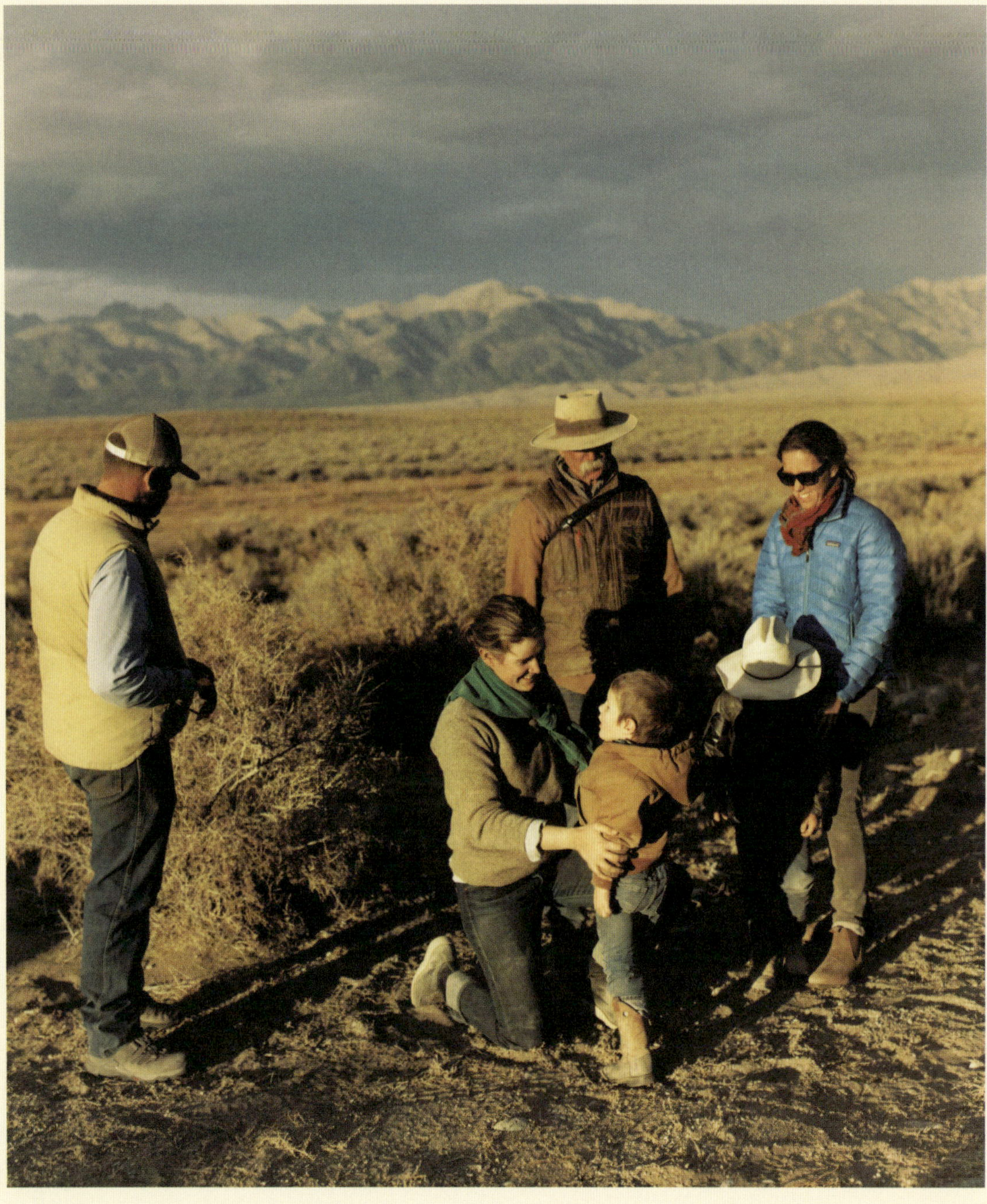

Ranchlands is about so much more than just raising cattle. It has become what it is today because of the people who come from places far beyond these ranches. They often arrive with no knowledge of ranching, stumbling across this place and becoming tied to the land, the animals, and the people. With all of their different backgrounds, they are the ones who will ultimately help carry the torch. They're the future of ranching: guests, birders, school kids, college kids, hikers, leatherworkers, art lovers, concert goers, people attending clinics and workshops, and especially apprentices and interns.

The apprentices and interns are a special group of men and women who help run the ranches, learning to manage the land alongside us. Not only do they figure out how to handle cattle and horses, which takes years of repetition, but they also learn to host guests and communicate with outside partners like the Bird Conservancy of the Rockies, the Audubon Society, and local cattle associations and agencies. They must learn to be highly organized and communicate well. They have to understand ecology, how to use cattle as a tool to put conservation strategies into effect, how to identify and treat sick or injured livestock, how to fix anything that breaks, and how to weld, build fences, and construct and repair extensive water systems. And it all must be done at scale.

The goal of the program is to create leaders who will take these ranch-management strategies and use them well into the future. We want to impart in them a vision of how ranching can be used as a catalyst for change. These men and women are often young, in their 20s, coming mostly from urban backgrounds and from all corners of the United States and even Europe. They arrive at the ranch with degrees from top liberal arts colleges, specializing in English, outdoor education, and conservation—studies that are not typical of normal ranch hands. They are highly intelligent and share common interests in the outdoors, working with animals, preserving the natural resources on the ranch, and learning to be leaders.

One of our first apprentices, Nick Baefsky, remembers the learning curve he faced upon arriving at the ranch: "The first time I moved cows on the Chico Basin Ranch, I wondered why no one was working hard except for me. I was used to a conventional style of moving cattle where everyone stays at the back pushing hard, the calves stressed and running, whips popping, and people yelling. On this particular day, I was at the back, trotting aggressively in a zigzag pattern. And there were Duke and Michael Moon (ranch manager at the time) riding along on the side, way up front, just talking and visiting. How lazy they were! Yet the cows were going along at a good pace, better than I'd ever seen. I gradually learned that when cattle need to be trailed someplace, you should try to use the cattle's innate desire to follow other cattle in order to keep the herd in a long, sinuous column. I learned to become proficient at moving cattle in this low-stress, efficient manner by making about every mistake a person can make. I let them string out too much, and then the next day, I kept them wadded up too tight. I pushed too hard and from the wrong side in an effort to get the cows rolling out of a corner and did the same thing moving cows off a drinker, stopping the flow of cattle completely and even turning them back the way we had come. I was in the wrong position for the gather, or I was too fast, then too slow. I put cows through the wrong gate, and I let a large group of curious yearlings follow my dog in a big circle and missed the gate altogether. Duke and Michael made it look so easy!"

What Nick didn't say in his story is that we condone these kinds of mistakes at first. While we don't let anyone hurt themselves or do something crazy or expensive, we let mistakes happen as important learning opportunities. Today, Nick is a master of his trade. Through the experience of doing the wrong things, he gained a deeper understanding of how to do the right thing.

This means that leaders of Ranchlands have to behave and set an example. It forces us all to walk the talk together, leveling the playing field. Sharing ideas and learning together has become a central pillar of the Ranchlands ethos, not just for the apprentices and interns but for all of us. The wider Ranchlands community is made up of people who are interested in learning more about our ranching legacy and how it catalyzes ecological transformation.

While our shared work is to transform these landscapes that we call home for the better, a side effect of learning how to do just that is often that we transform ourselves, and we learn who we are as people along the way. The work we do is emotional. The timeline that we use to look at land is generational. With our kids and their kids hoping to have a place to come back to, our work extends beyond ourselves. We are all intertwined in these processes of growth, discovery, and transformation—not only of land and animals but also of thoughts, abilities, dreams, and possibilities of imagination. This in turn binds us together as people who have all been changed and marked in different ways by the ranches we have worked on and the people we have worked with.

It makes us all teachers—apprentices teaching interns; interns teaching guests, visitors, and school kids; the land teaching all of us. The longer we spend getting to know a place, the more it starts to feel like a friend we've known for so long. The land itself becomes a protagonist in our everyday lives. This culture of learning creates a level of commitment to one another and to our mission that permeates everything out here.

The attitude generated by a learning atmosphere is contagious and creates a "can't wait for tomorrow to get here" feeling that permeates daily life. Yes, the work days can be hard and long, but the most common feeling is that we're all having fun. There is little ego or negative attitude among us, which also helps create the welcoming atmosphere that attracts so many people from beyond our border fences.

The Ranchlands circle includes those who are interested in learning more about ranching by simply coming to be part of the crew for a week or attending a concert or grass-monitoring workshop. These people make the dimension of Ranchlands bigger. Ultimately, if we have the courage to dream big enough, it could very well become a cascading collective of people who become strong enough to do something really cool, like change the world.

RUTH REES PHILLIPS

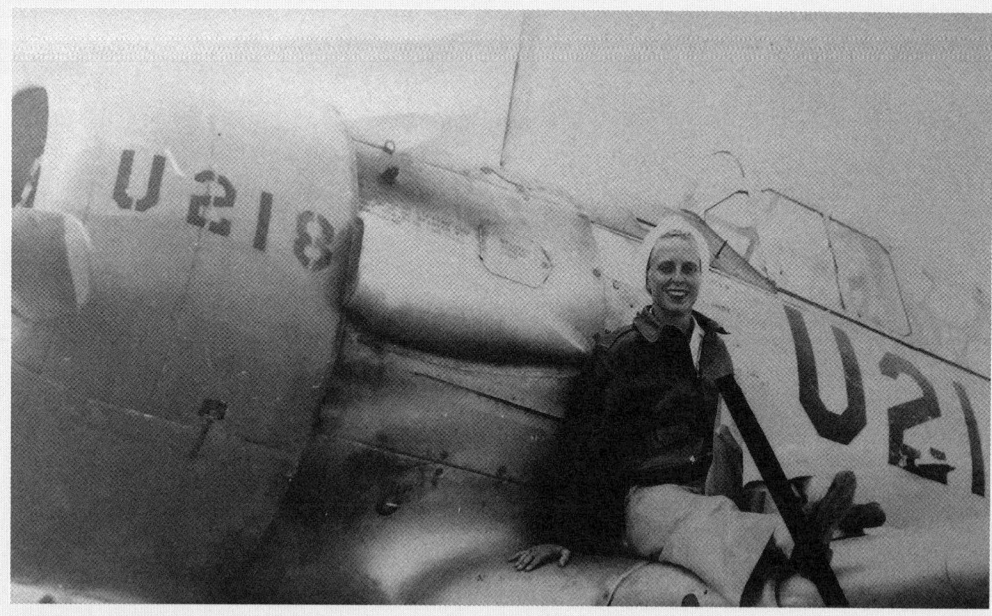

Ruth Rees Phillips was my mother. She was raised in San Antonio, Texas, where she met and married my father, Duke Phillips II. Together they began a ranching career that led them to Big Bend, Texas, the isolated llanos in Venezuela, and northern Mexico to the Hacienda Sierra Hermosa Ranch, where they lived for 23 years raising me and my three siblings.

She flew airplanes, something women didn't do very much back then. Having learned to fly after joining the Women Airforce Service Pilots (WASPs) in the early 1940s during World War II, she was the chief pilot in our family, doing most of the flying in our red Cherokee Six for town runs (which would have taken five hours by truck), visits to the ranch commissary, and any medical emergencies. There are a lot of stories of flying with her that I vividly remember.

Once, when the weather was a bit dodgy, she dumped a case of tonic water for a neighbor at the end of their airstrip without even turning the engine off, then took off again to go back home. The neighbor, Laurie Lasater, remembers the scene: "One evening, just minutes before dark, a plane landed on our airstrip as we were sitting on the porch with our friends Tom and Jan Newsome and enjoying a gin and tonic. We watched in awe as it stopped at the south end, the engine still running. A female pilot jumped out, unloaded a box, jumped back in, and waved to us as she took off again. It was Ruth, dropping off a case of tonic water while hurrying to get home before dark. Of course, we told Tom and Jan that we had a charter flight to deliver our tonic water every month."

Another time, I remember our neighbor Maxi called to tell us on the two-way radio that we needed to come get our mother because she had landed at his ranch due to the inclement weather. When we arrived, she had a big smile on her face and was on her second scotch and soda. She said we were going to have to fly the plane home since she and Maxi were just having too good a time and she'd had a long day.

Yet another time, she organized and participated in an air race in Mexico and talked the neighbor into going with her. I remember the big splash she made in the local paper and my father telling the story afterward. "Not only was she a gringa (a white woman), a foreigner, and a woman in a male-dominated culture, but she was also in an airplane, waving at everyone with a great big smile as they taxied in at the end of the race," he said. "She was flying the damn thing." He said the photographers all ran over to her plane, abandoning the rest of the participants and surrounding her.

Ruth was always up for anything. Always positive, she was the most empathetic person you could ever meet. To the other children on the ranch—there were eight families with a dozen kids per family—she was their second mother. She was the doctor, the family counselor, and the woman who everyone went to with any kind of problem. I remember after a fishhook went through the middle of my thumb, she gave me a tumbler of whiskey and a piece of wood to bite down on, and as the cook held me, she cut it out. I remember her repeating to me as she squinted and worked, "Hold on, sweetheart, it's almost over."

DUKE II

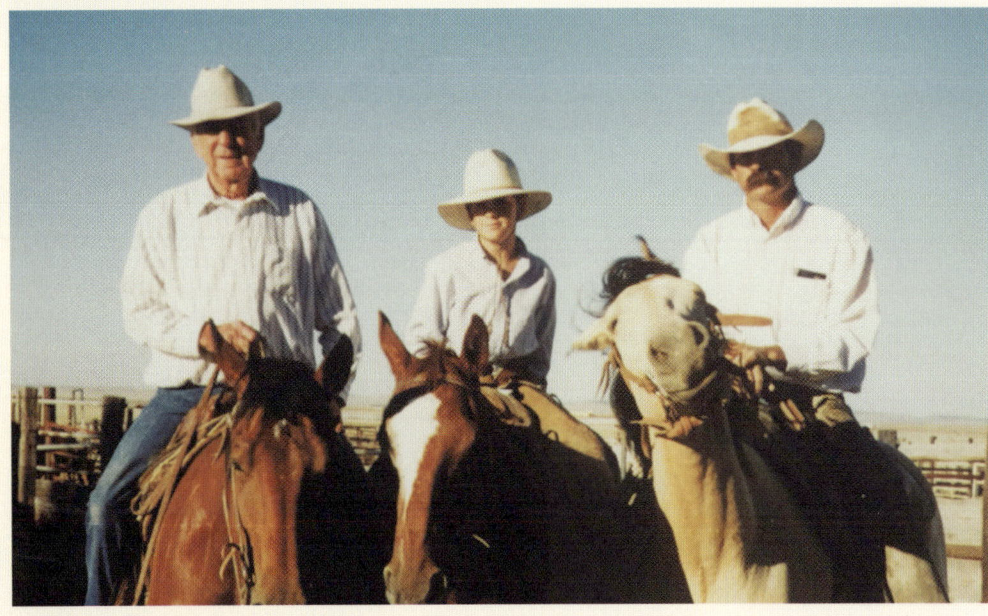

Duke II, Duke IV, Duke III

Someone once called my father a rugged individualist, which comes pretty close to the truth. Growing up as the son of a wildcatter and rancher, he knew from an early age that he too would one day be a rancher. As a grown man, that's how he lived his life—on a big piece of land far away from people. After serving two tours as a squadron commander flying P-51 Mustangs in World War II, where he was shot down by Zeros, he met and married my mother, Ruth, in San Antonio, Texas. After he graduated from Texas A&M, they moved to Big Bend, where they were droughted out. I remember my mother saying, "The only thing left was our suitcases." And my dad nodded his head.

Aviation would become a mainstay in our family because my mother was also a pilot in the war and because the places where my father took us were so remote. I remember birthday parties at the ranches in our neighborhood. There would be a dozen planes parked on the dirt airstrips and a piñata swinging from the shade tree. We lived on extremely isolated ranches in the llanos of Venezuela, Mexico, and Hells Canyon in Oregon.

When I was a boy, it seemed like my father was afraid I'd grow up not knowing how to work. I'd be up at 3:00 a.m. wrangling the horses, then milking the cows in time to start homeschool with my mother at 6:00 a.m. At noon, my brother and I would grab horses that were left in the corrals and finish the day riding with the men out prowling the pastures. Because my father placed me right in the middle of life on the ranch, I developed a deep fondness for traveling on horseback, for wild places where very few people had been before, and for caring for these places as if they were my own home. I worked six-and-a-half-day weeks alongside men who stayed busy those long days without even thinking about it.

As hard as he could be, my father was also gentle and thoughtful, and we were extremely close. As a teenager, I remember his hand on my neck or arm when I sat on his chair or stood close to him. I always felt that he was proud of me and that he listened no matter what wild idea I had—always supporting and always encouraging. Having this kind of backup from someone I deeply respected gave me the self-confidence to feel comfortable taking risks and to not be afraid to think differently than those around me.

My dream growing up was to work and create a large ranching business with him, which never happened. But because of his influence, I have been able to think and work independently, which has led to creating a ranching business that I run with my son, Duke IV, and daughter, Tess. My dream has ultimately come true but with a twist. I am always surprised to see how closely my son resembles my dad, even though he did not spend a great deal of time around him. He is charismatic like his grandfather, who always had a flock of kids hanging around him when I was in high school, the other fathers looking on wistfully. Duke IV's sensitivity and ability to lead people and make sound decisions are a complete remake of my father. People are drawn to him because he is genuinely interested in them, the same as his grandfather—both have been known to spend hours talking with people they've just met. Duke IV still wears the short-sleeve shirts that Duke II handed down to him more than 15 years ago and the same kind of Velcro-fastened tennis shoes that his grandfather used to wear. And his piloting skills are on the same level—I've trusted no one more than either of them to fly with.

In his later years, my dad came to live a few miles from the Chico Basin Ranch, so he could be close. He would come over when we were branding or working cattle in the corrals and sit in his chair as close as he could without getting in the way, intently watching and listening. Without even looking at him, I could feel his pride—the same feeling I often felt as a boy, which created the self-assurance that has informed my life as a rancher. It's the same pride that I feel for Duke IV now when he is roping, flying, and talking with his team or a friend. And it's the same pride I feel when I see my grandkids running around the ranch in their hats and boots without a care in the world, waving swords or fishing poles around. If it were not for my father's vision of working and living on a dusty, lonesome, old ranch, we would not be working as a family raising cattle and kids as we are today.

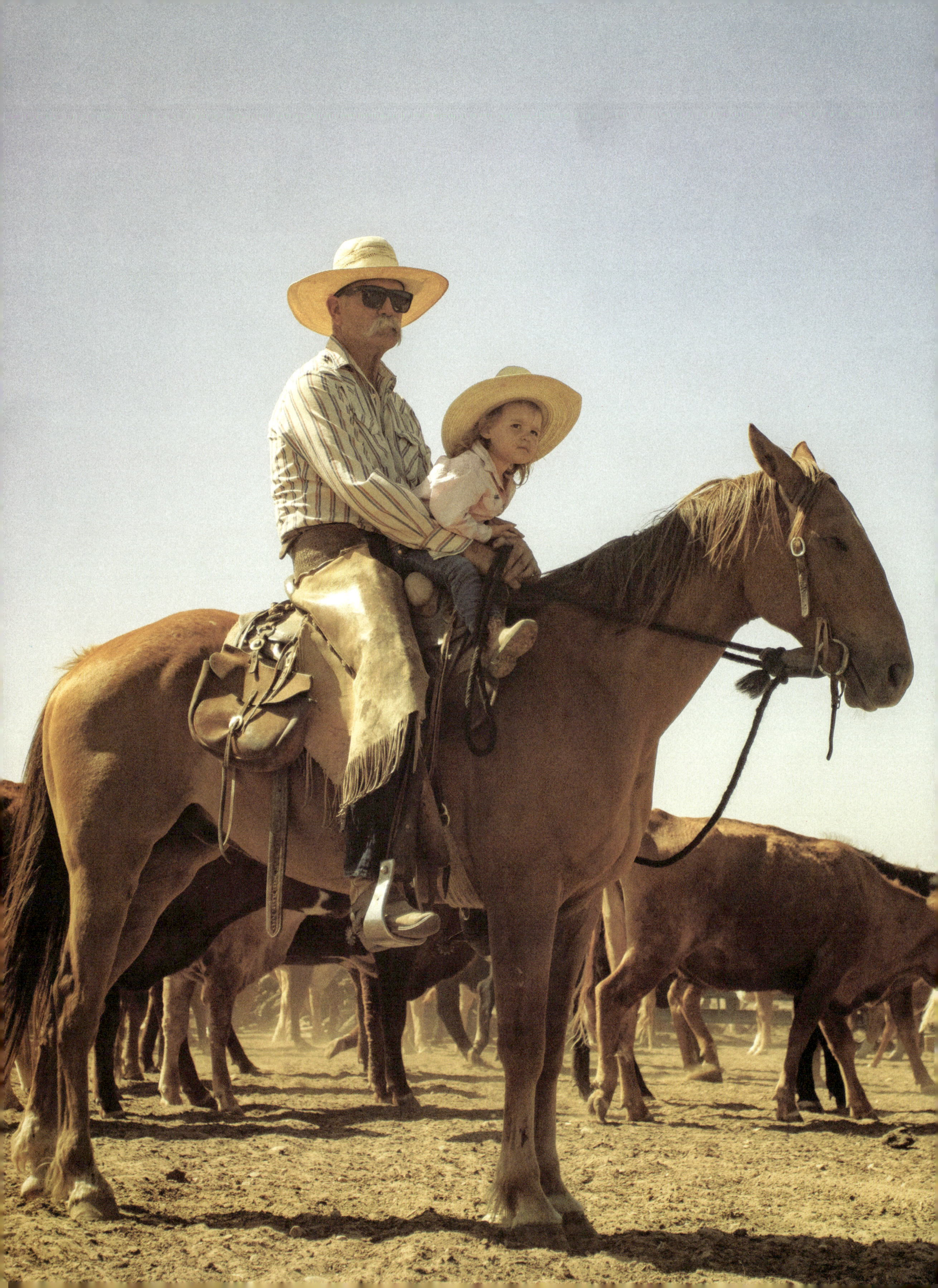

MADI PHILLIPS
Leather Shop Manager, Ranchlands

"We raise our children here on the ranch, and they're involved in everything, even Bison Works—the long days, the wind, the sand in your eyes and mouth, the branding outside with Duke preg-checking bison in freezing temperatures. You see life through a new lens when you include your children."

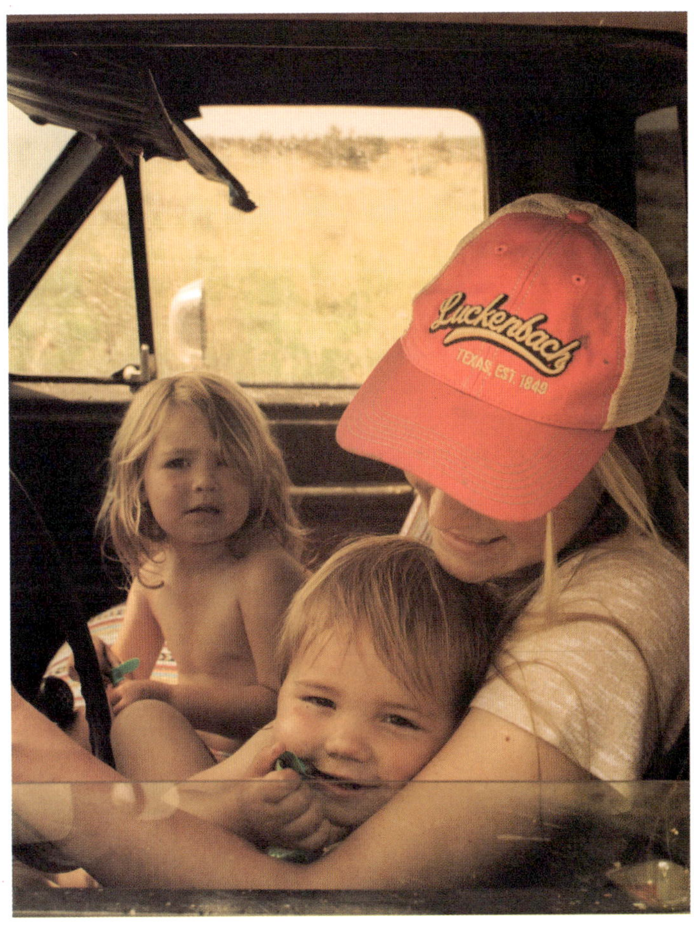

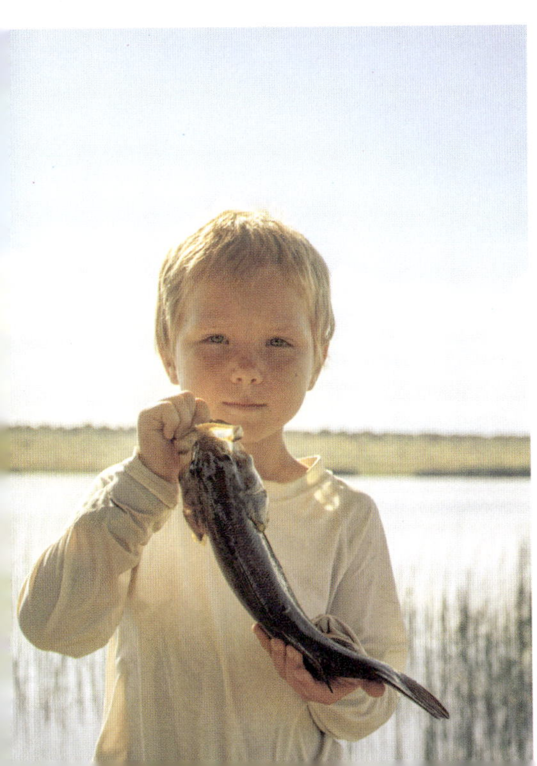

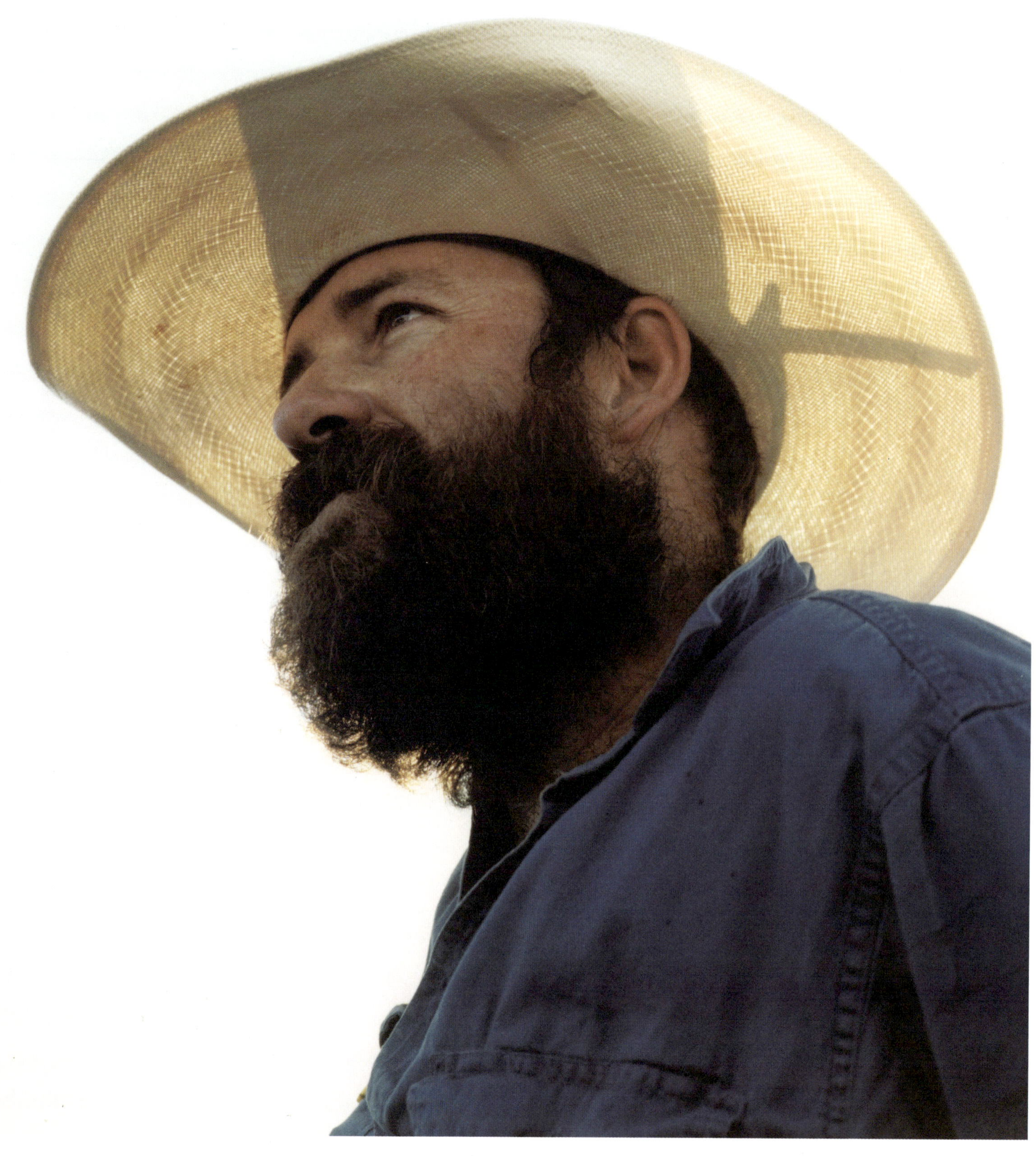

DUKE IV
Co-owner and COO, Ranchlands

"Where else can kids grow up learning all the things that ranch kids are exposed to? It's not just about riding horses but knowing how to work, manage lands and animals, and understand how to fit into a diverse group of people who are engrossed in a project that's much larger than themselves."

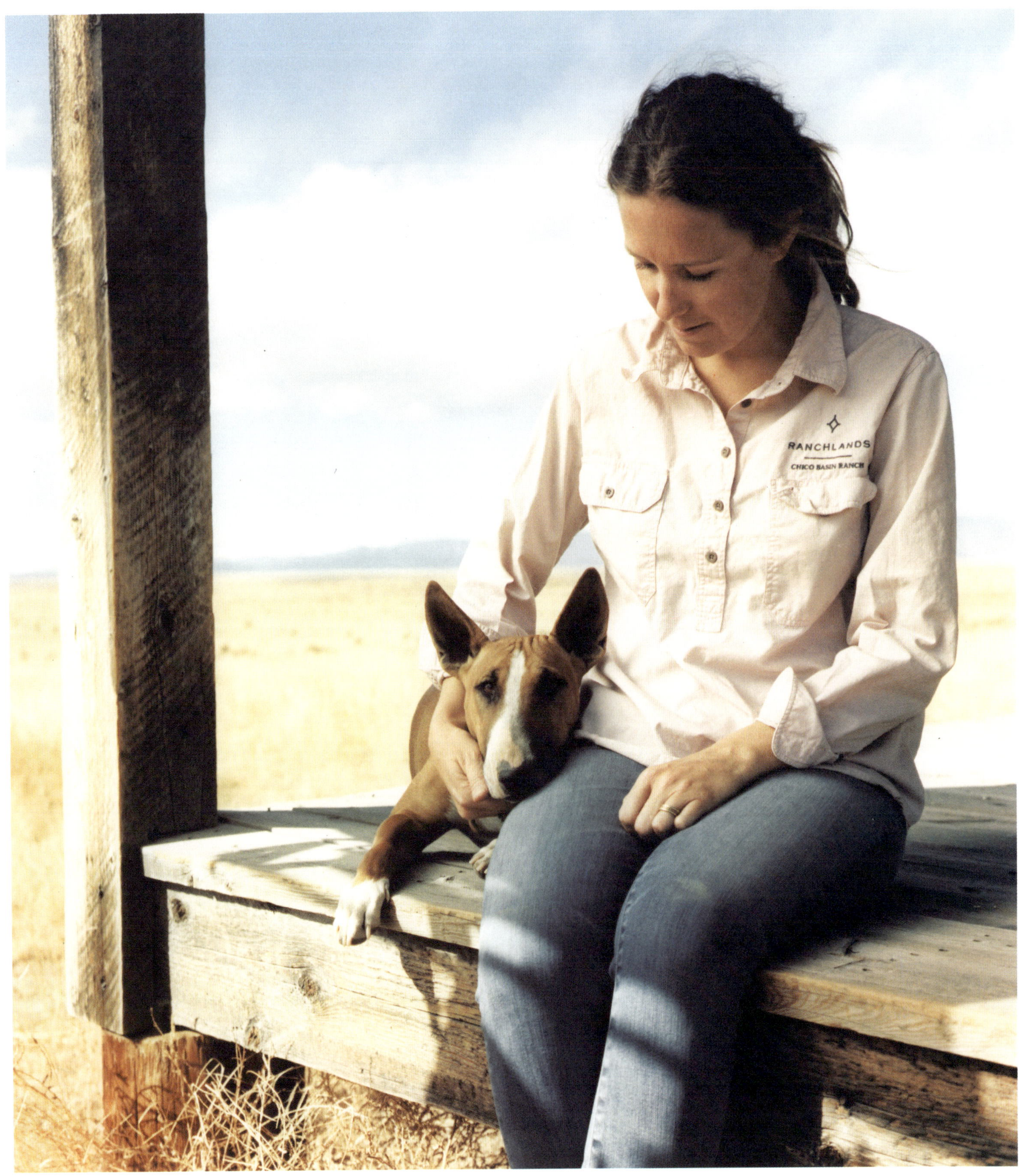

TESS LEACH
Co-owner and Head of Business Development, Ranchlands

"People often ask if I feel lonely living on the ranch, or if I worry about my kids not having exposure to different cultures. Ranchlands has evolved into more than just a place; it a community of people and a close-knit family. Our team is from all over, with diverse backgrounds and a genuine love and excitement for one another and for what we are doing. We aren't just coworkers; we're a dynamic family that I am so grateful to be a part of and to raise my kids around."

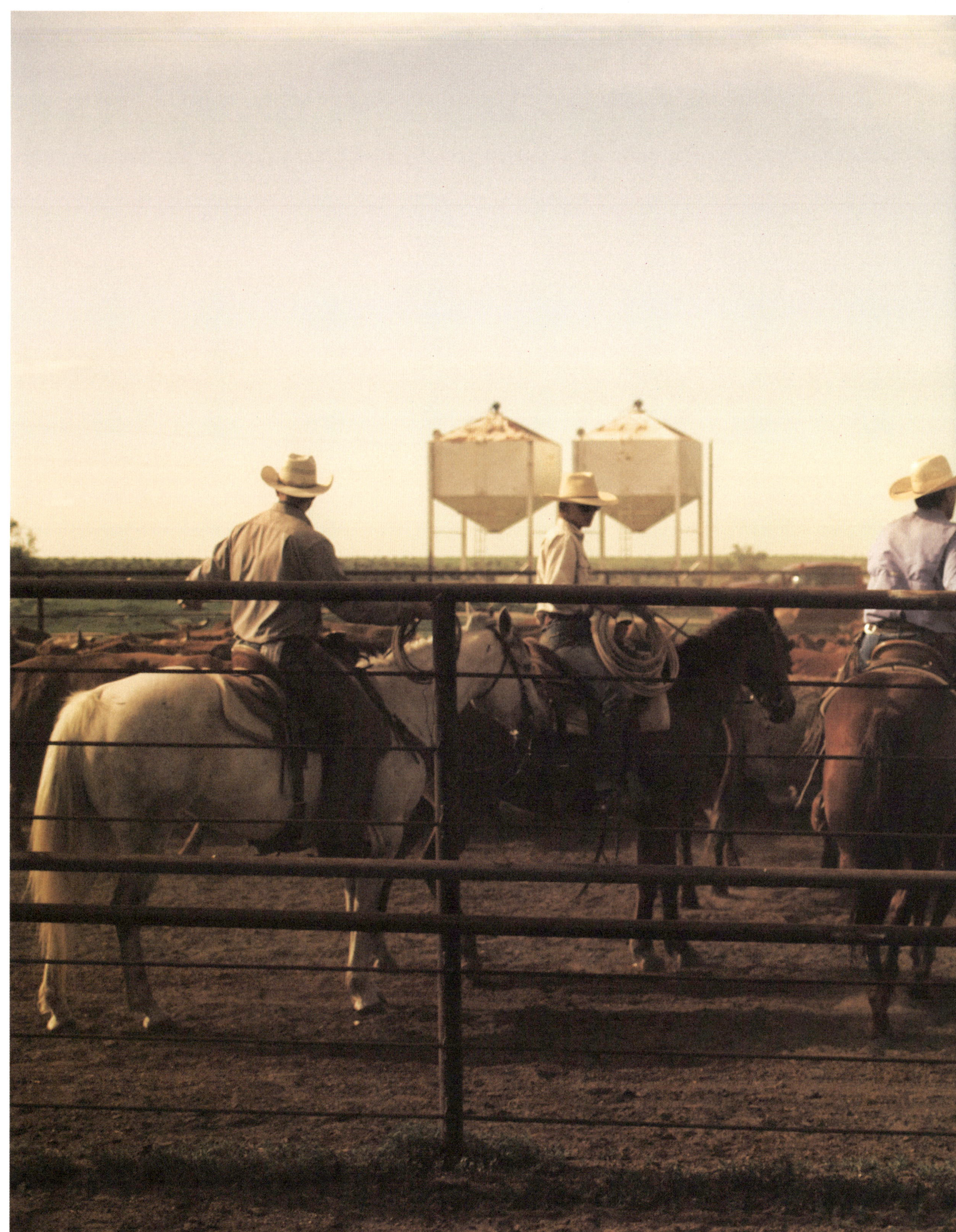

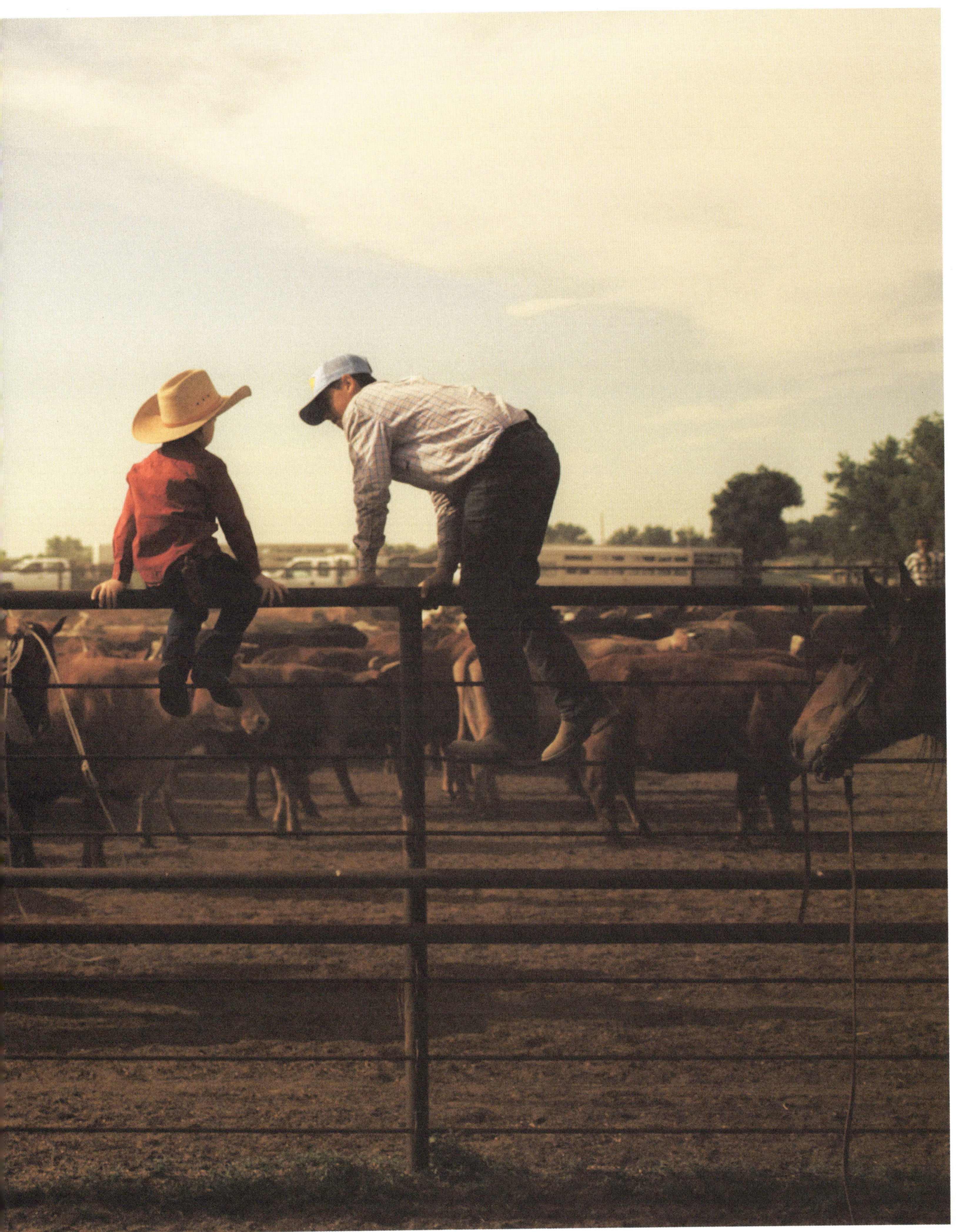

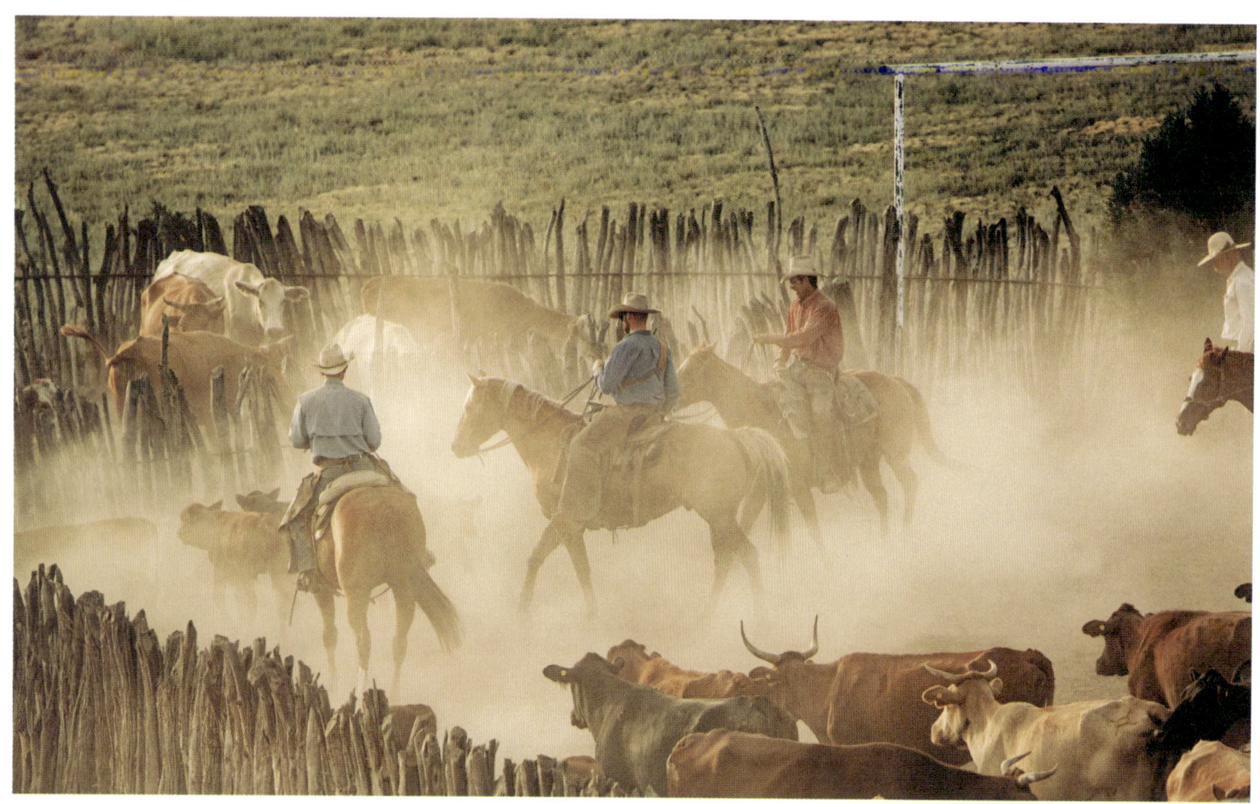

NICK BAEFSKY
Apprenticeship Graduate, Ranchlands

"When I started my apprenticeship at Ranchlands, I had no clear idea of what I aimed to get out of the experience. I did know I wanted to have a career in cattle ranching, and this seemed like a way forward with that."

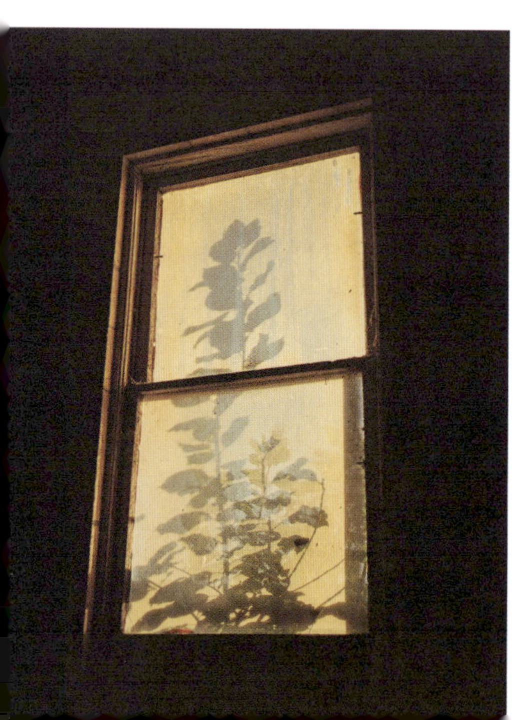

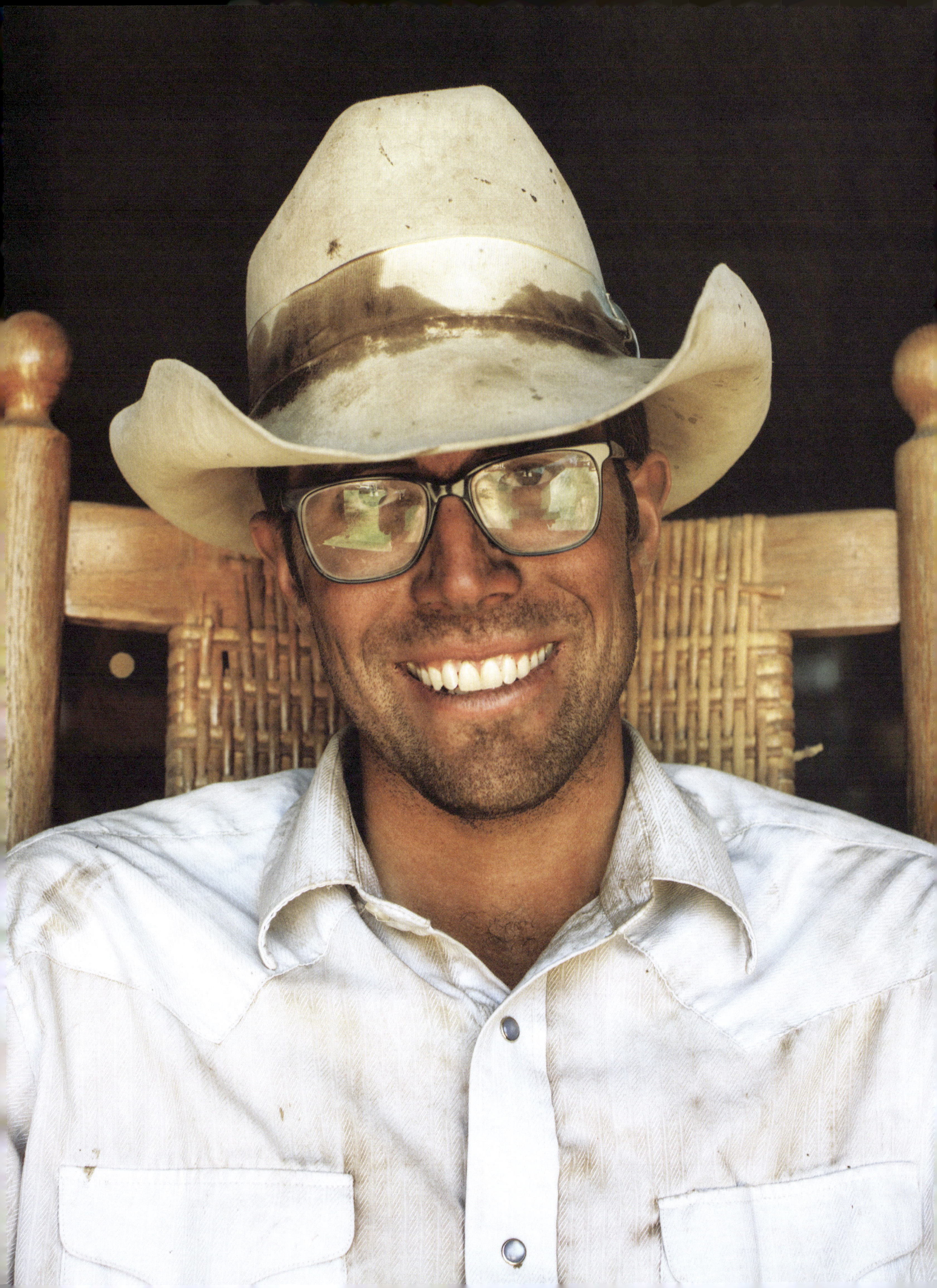

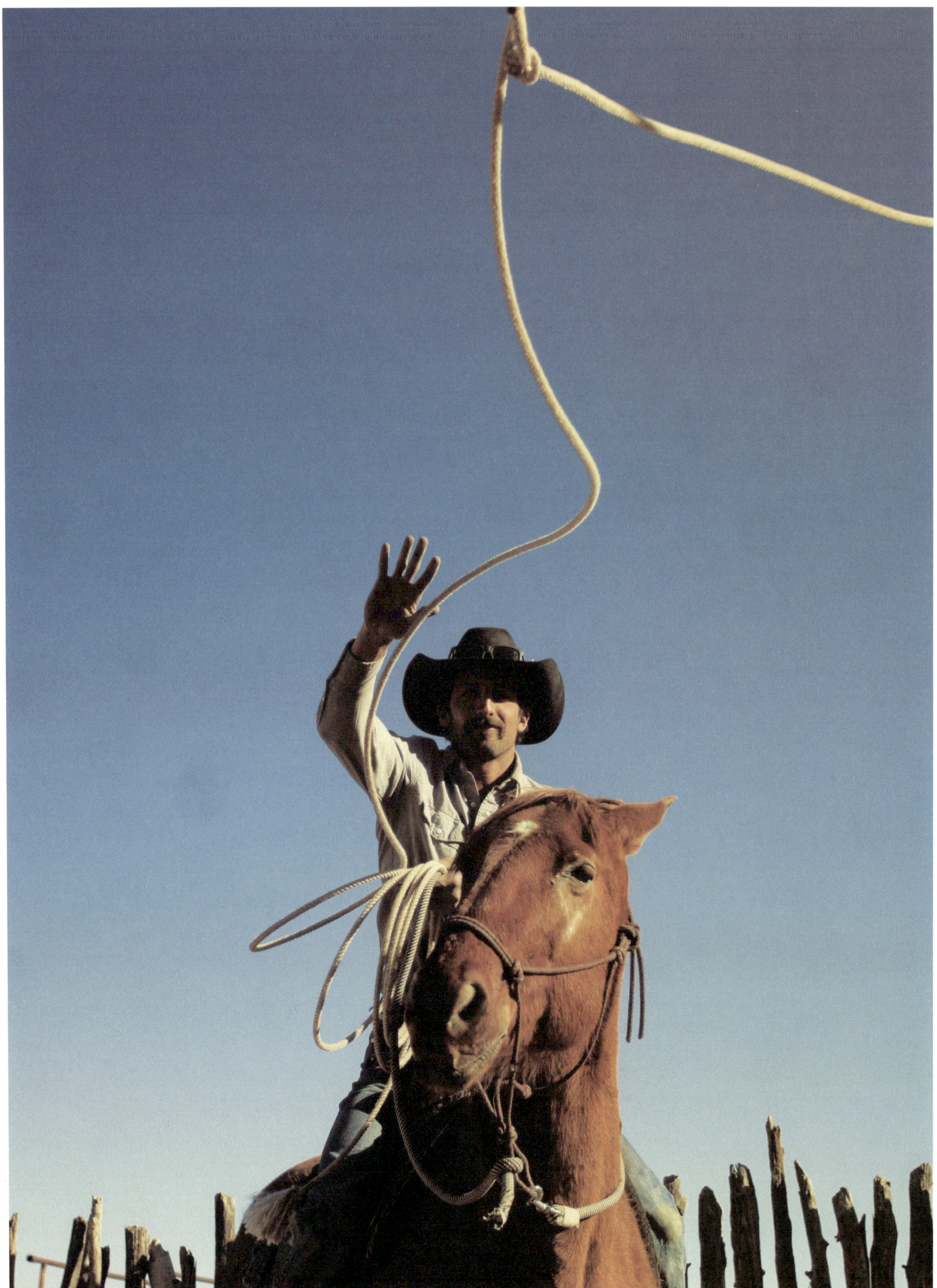

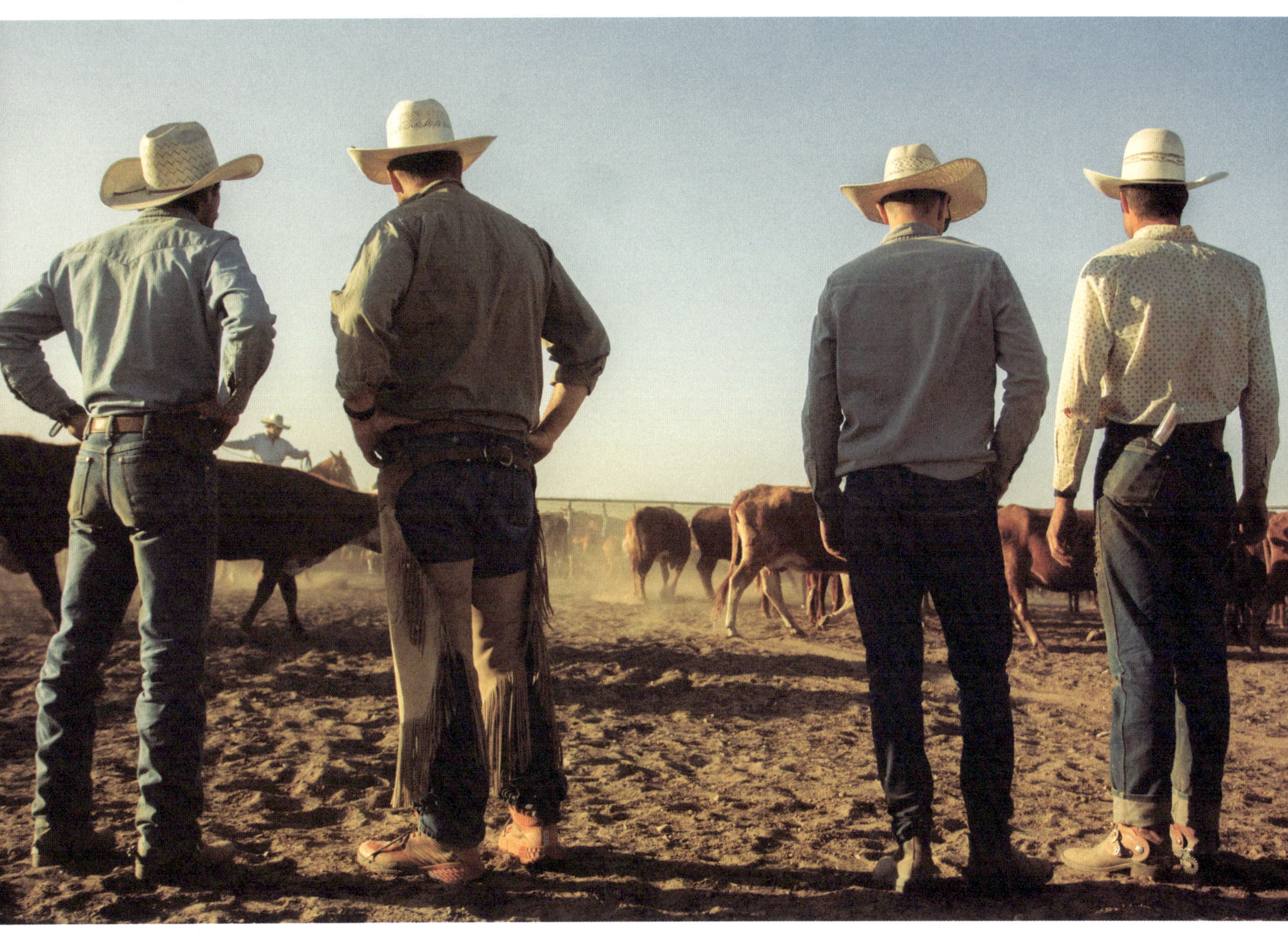

MIKE GIORDANO
Managing Apprentice, Frying Pan Ranch, Texas

"I get to do the thing I love most, which is working cattle across these ranches with some of the best friends I've ever had. That's why we work so well together—we're genuinely having a good time even when it's challenging. Sure, sometimes we're choking on dust, or suffering from extreme heat, or feeling completely exhausted from the physical work of fixing fences for hours. But we're always in it together, which means we're trying our best to be as effective as we can be for one another. Whatever comes our way during a day's work, I'm just happy I get to be along for the ride."

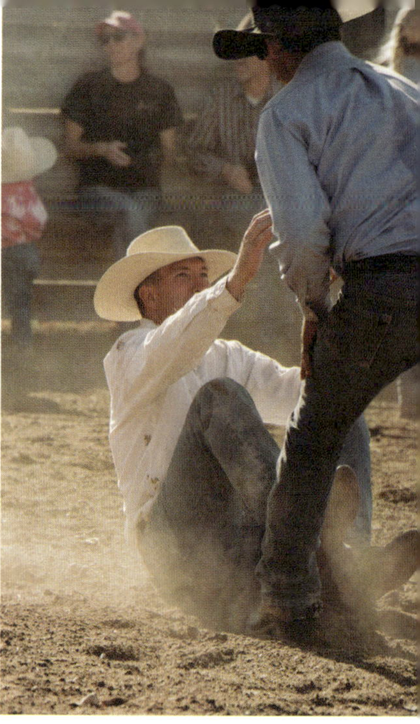

JONATHAN TULLAR
Foreman, Paintrock Canyon Ranch, Wyoming

"When you have an environment where everyone understands that we're all learning, the egos go out the window. It's a lot more fun this way, and it's also an authentic opportunity to become better. It's still serious, and everyone's dedicated to learning, but it feels collaborative and approachable. If there's something you want to bring to the table and share, Duke is always open to that. If there's something you want to learn, you can always ask to be taught."

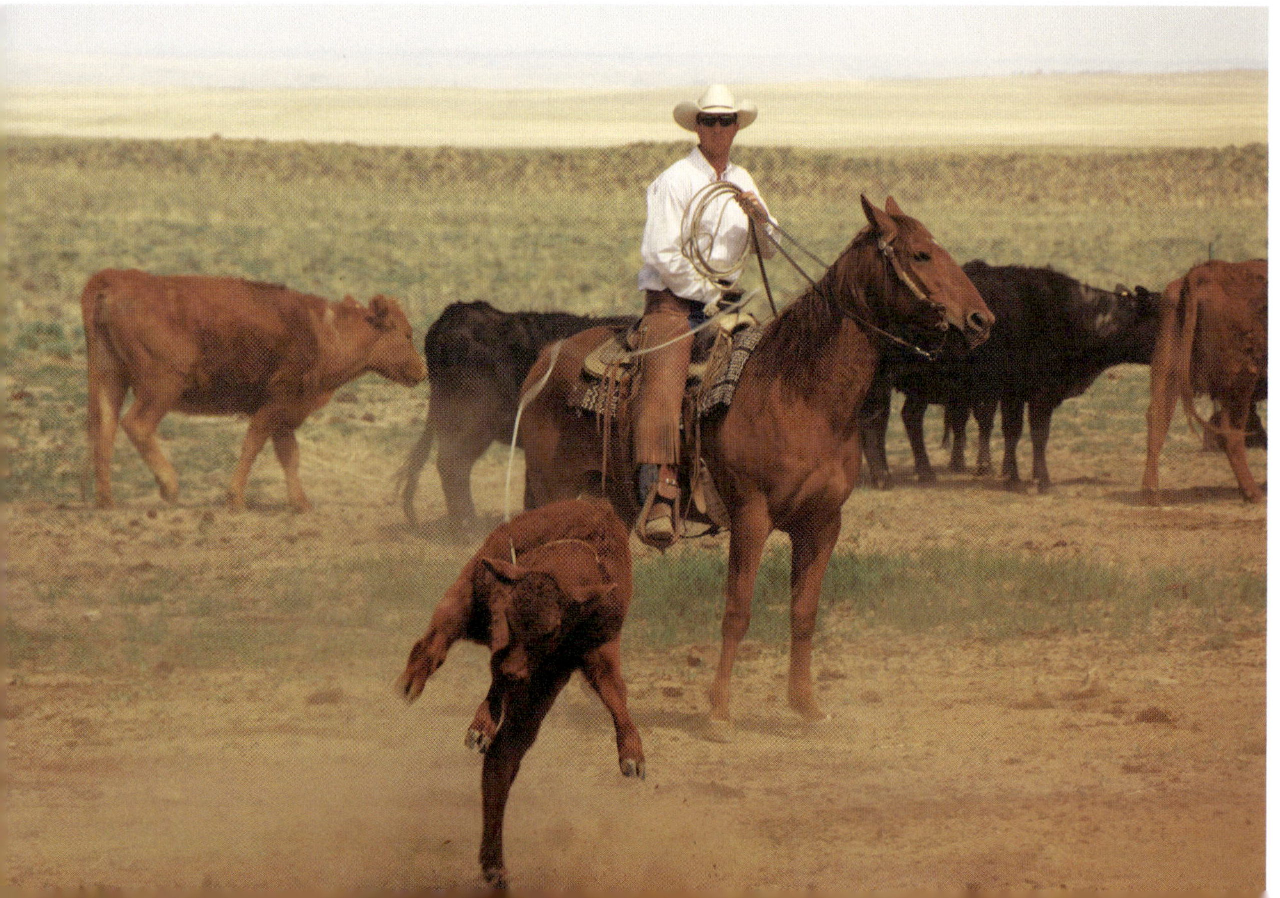

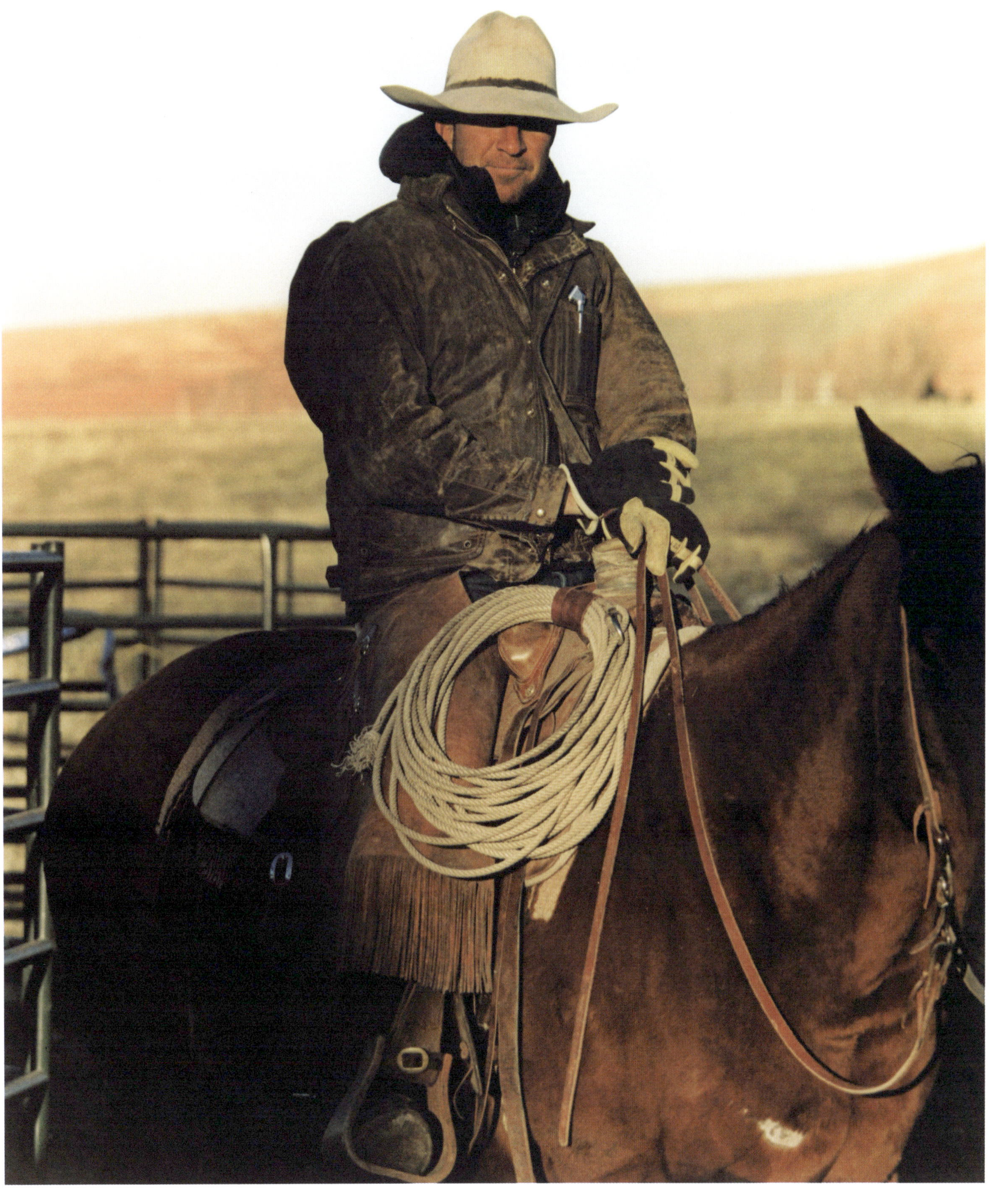

Ranchlands began in the business of grazing animals, but it has become more about growing a group of people who are working together with one mission: to create a better world through the management of land, people, and animals by living our ranching legacy every day.

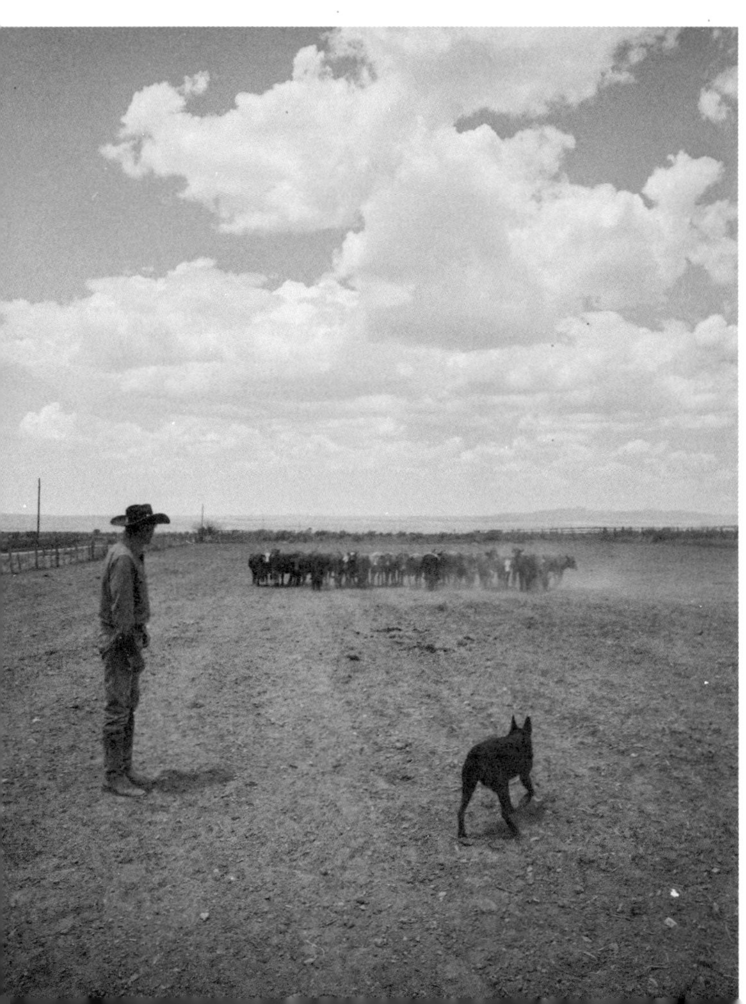

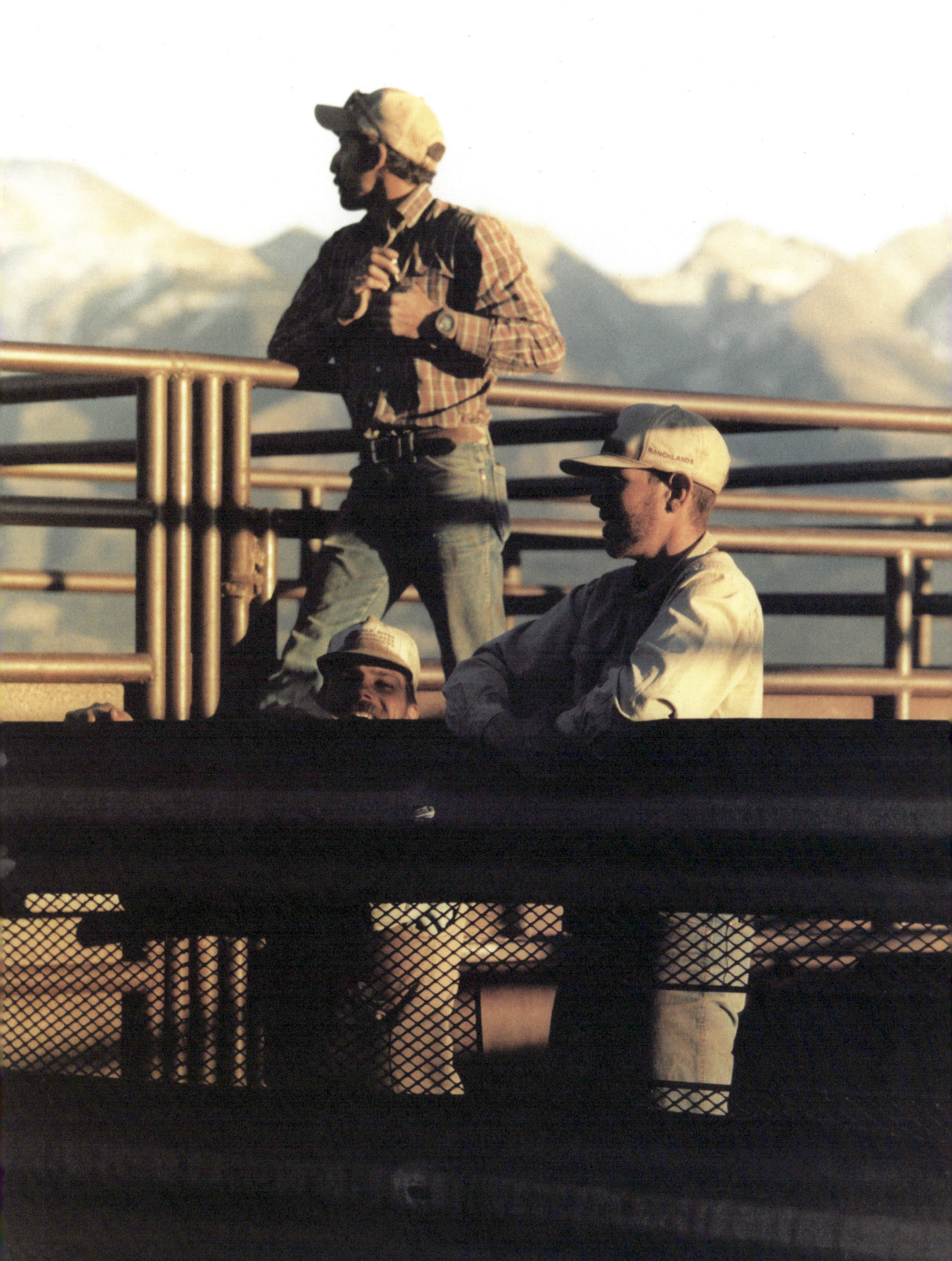
RANCHLANDS

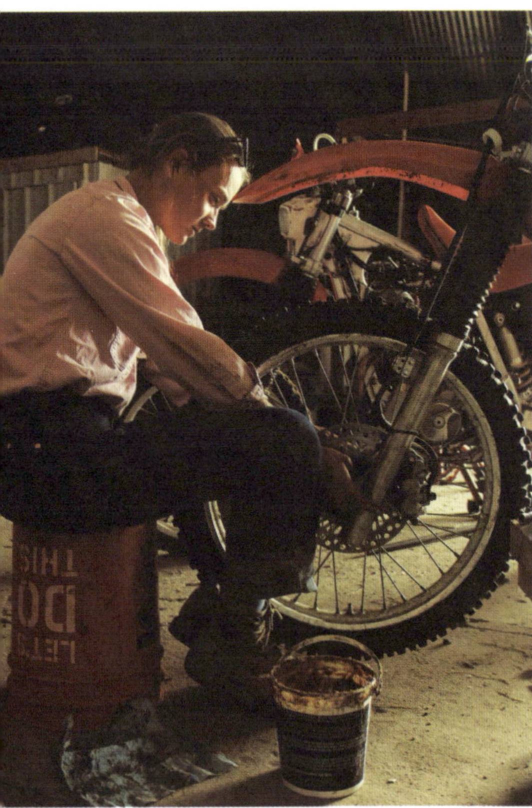

ANJA STOKES
Apprentice, Chico Basin Ranch, Colorado

"Nothing felt natural to me when I started. Even riding horses in this context felt so different, though I'd ridden my entire life. The saddles felt uncomfortable, and I didn't understand the cues these horses knew. I didn't have the first clue about mechanical things, and I lacked common sense on electrical and plumbing issues. But as time went on, I was entrusted to figure things out with less guidance and to grow confidence in my own ability to solve problems and eventually help others who were just starting out. Ranchlands teaches us that how you navigate challenges and process mistakes are a critical part of life here. You're going to mess up, but it's about doing things better or differently the next time."

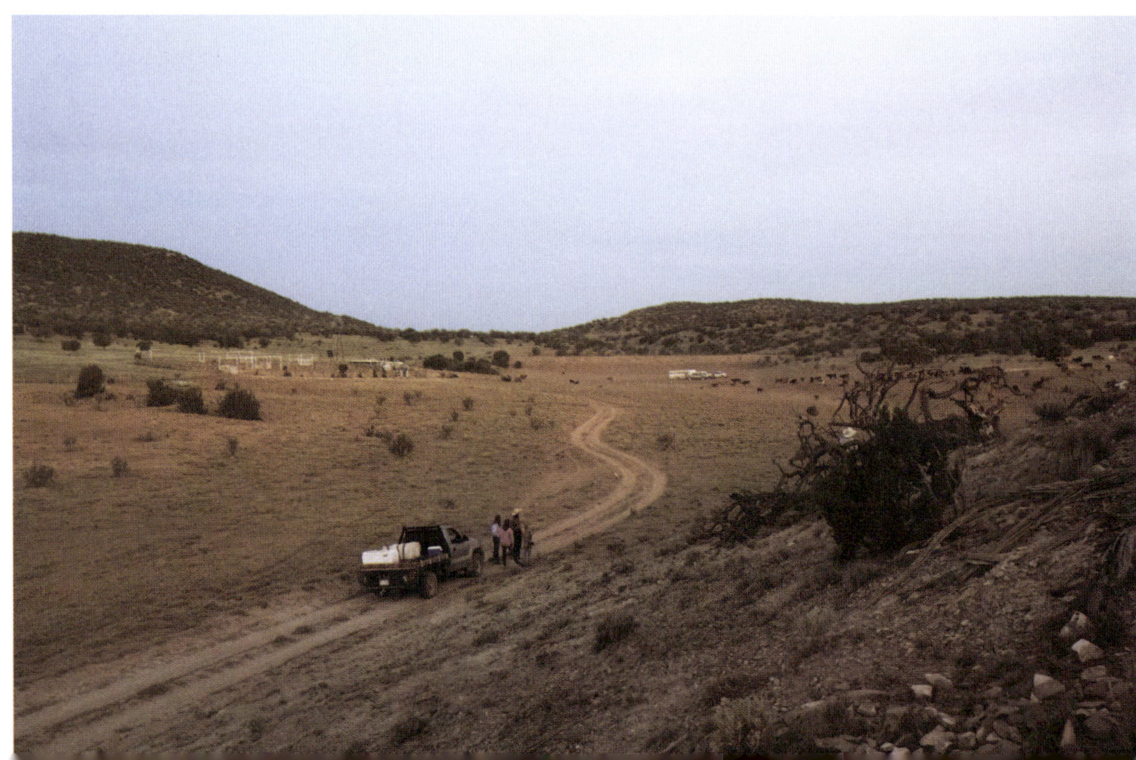

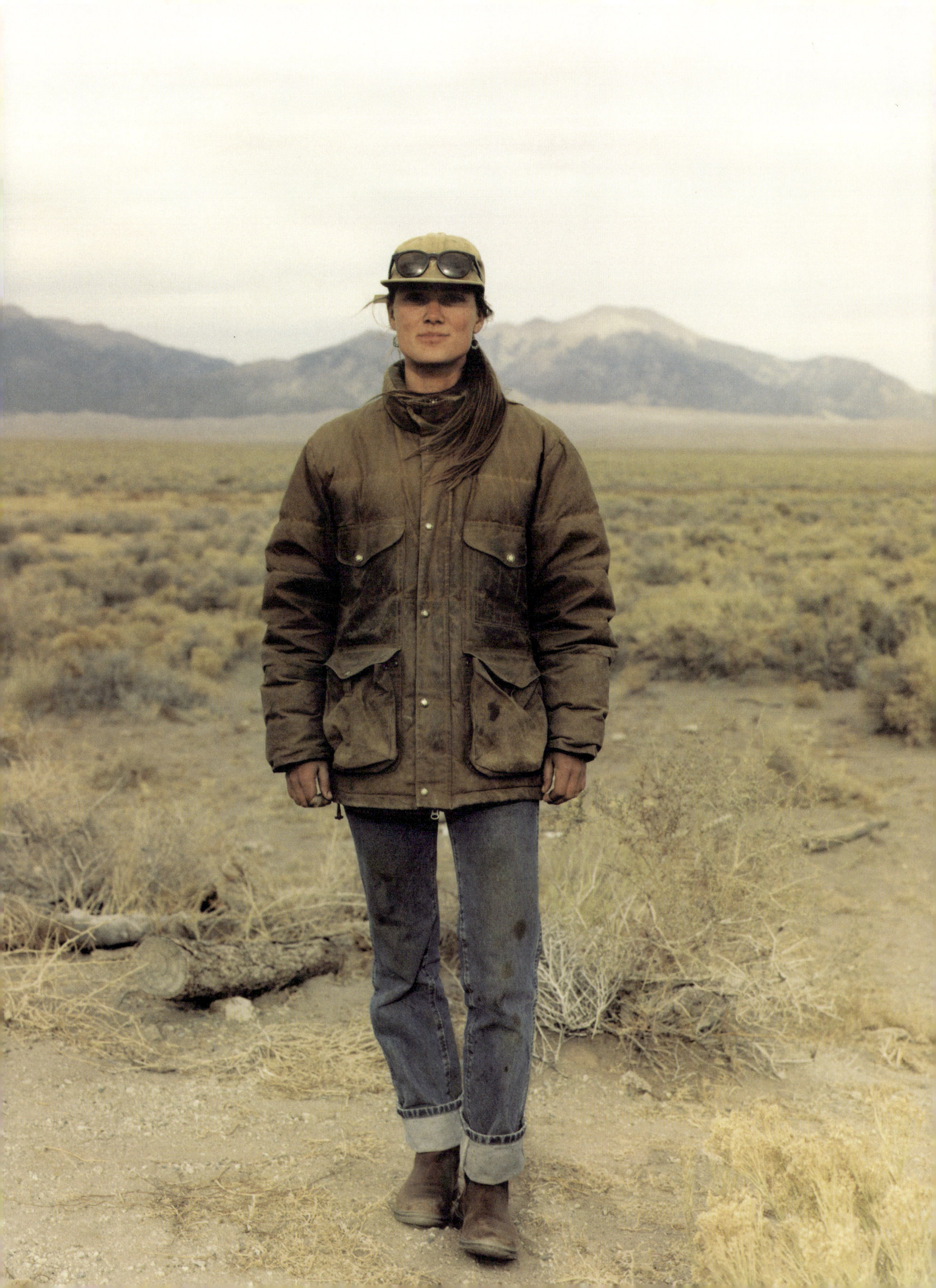

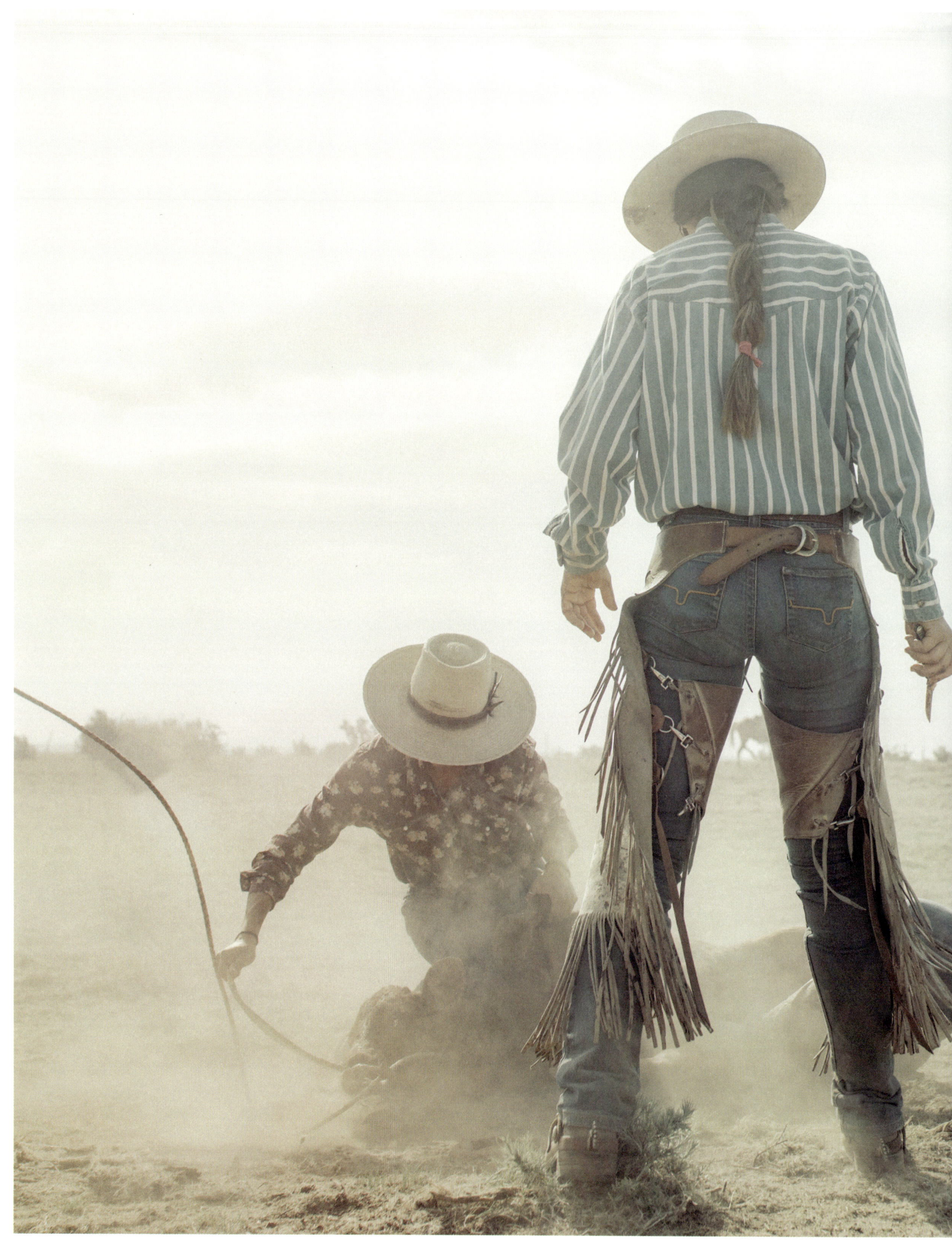

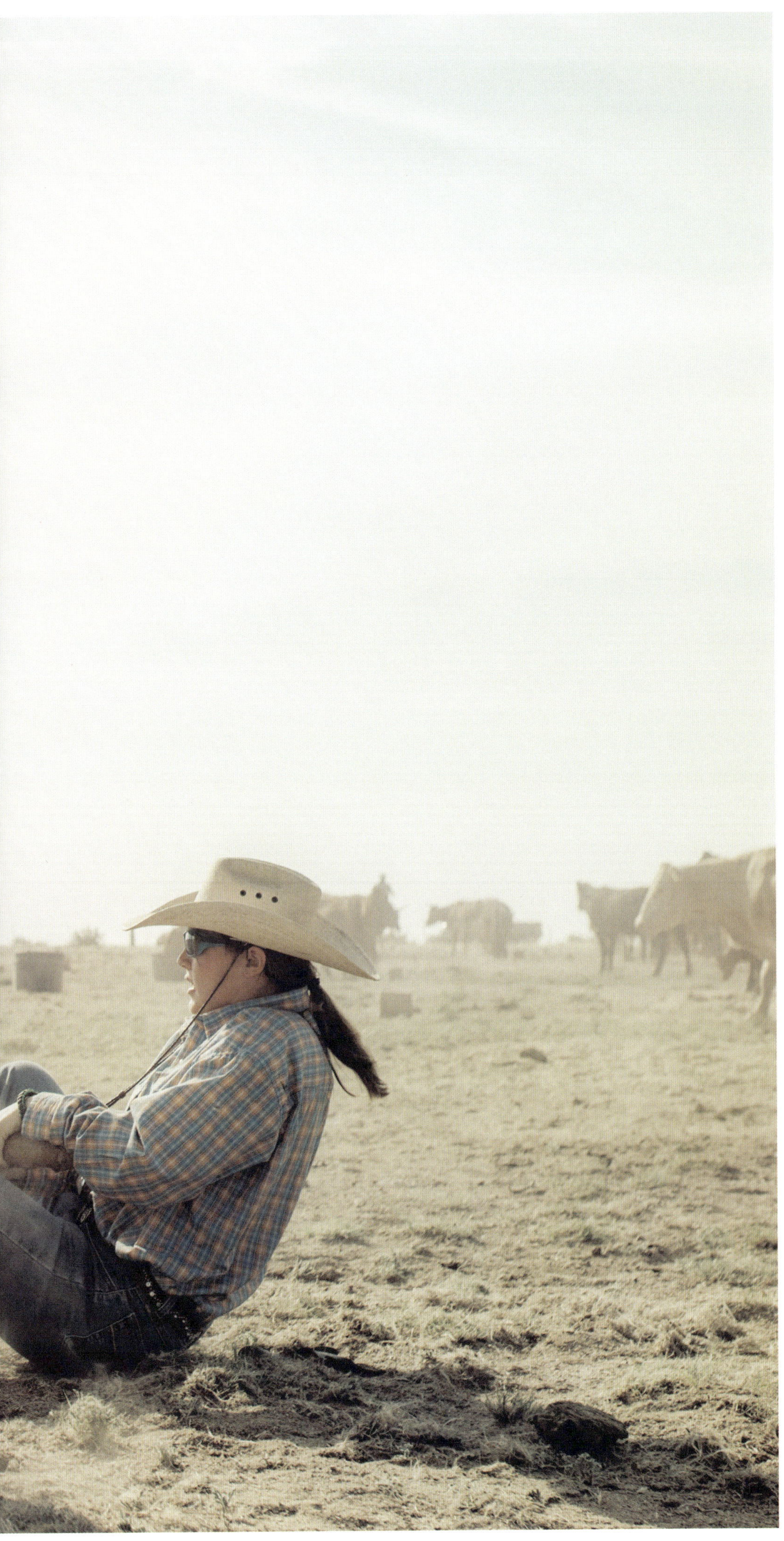

LYNAE RISINGER
Managing Apprentice, MP Ranch, New Mexico

"It starts with your first calloused knuckle or your first solo mission and quickly progresses to your transition from intern to apprentice. In the spring, it might be pulling your first calf. In the summer, your progression could be signified by being asked to rope at a branding. Rites of passage come as natural as the seasons around here. Unexpected moments of pride and triumph are an indication of personal growth in your craft and your place in this profession. A rite is not something that can be artificially produced, signified by a signed certificate, or subdued to a monotonous, structured timeline. It is a euphoria that never ends and will continue for a lifetime as one continues to seek knowledge and gain understanding."

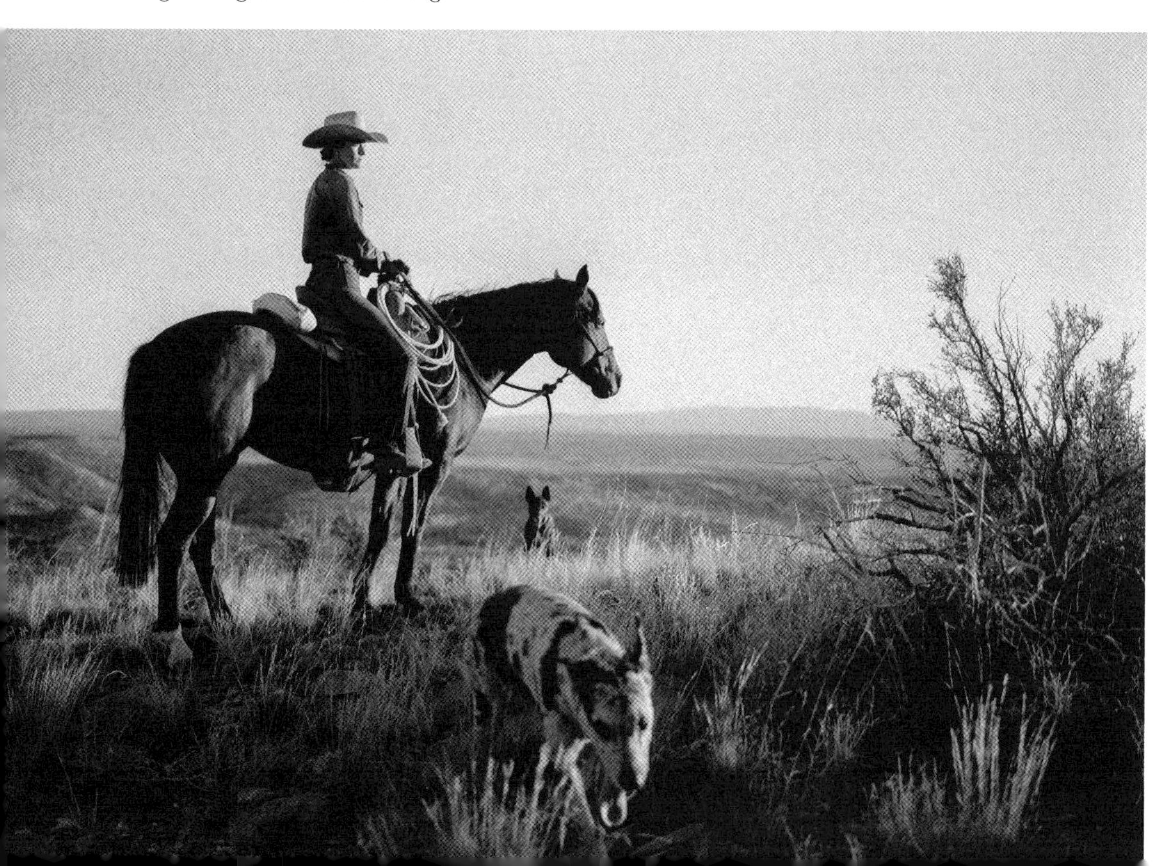

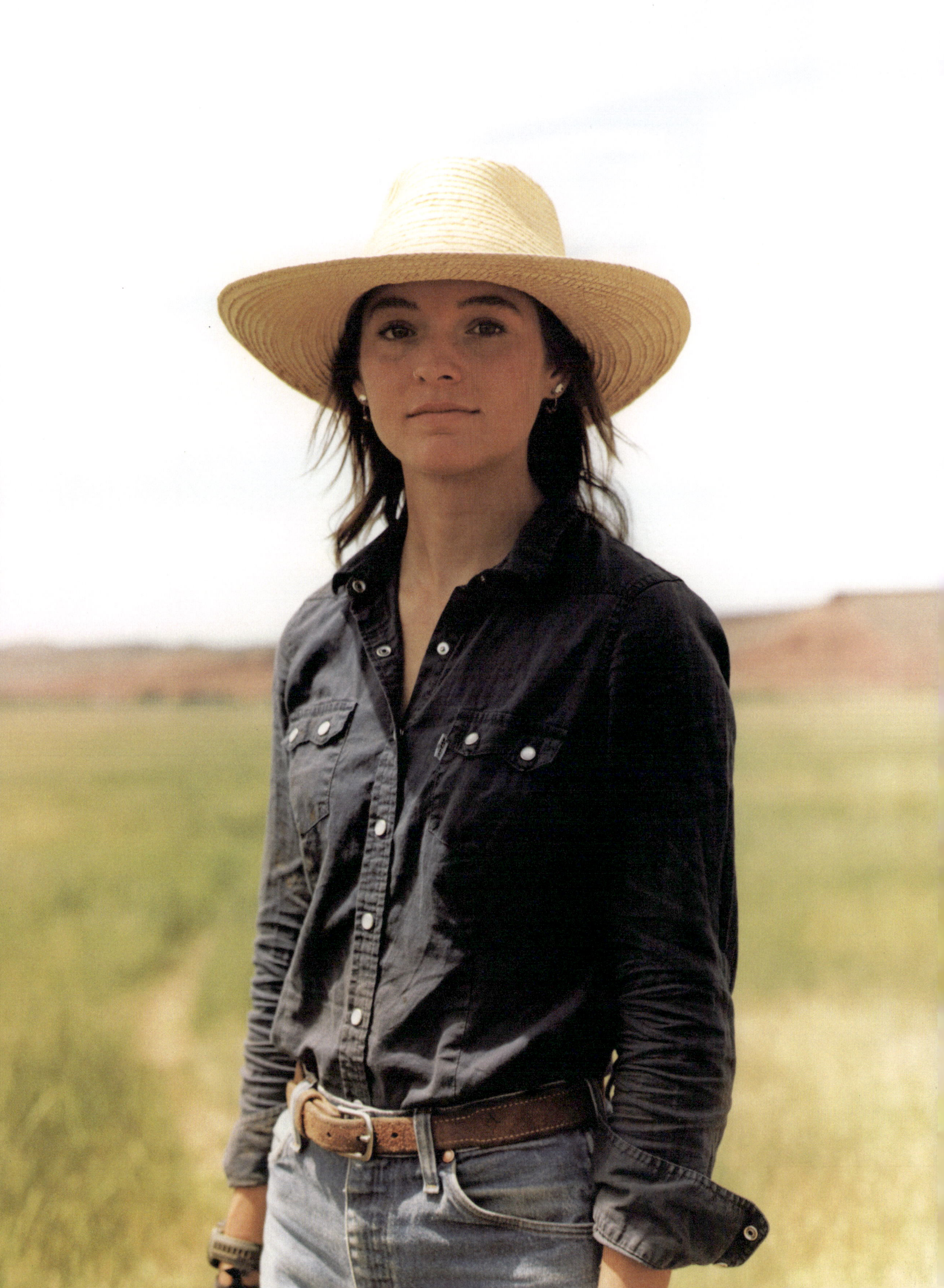

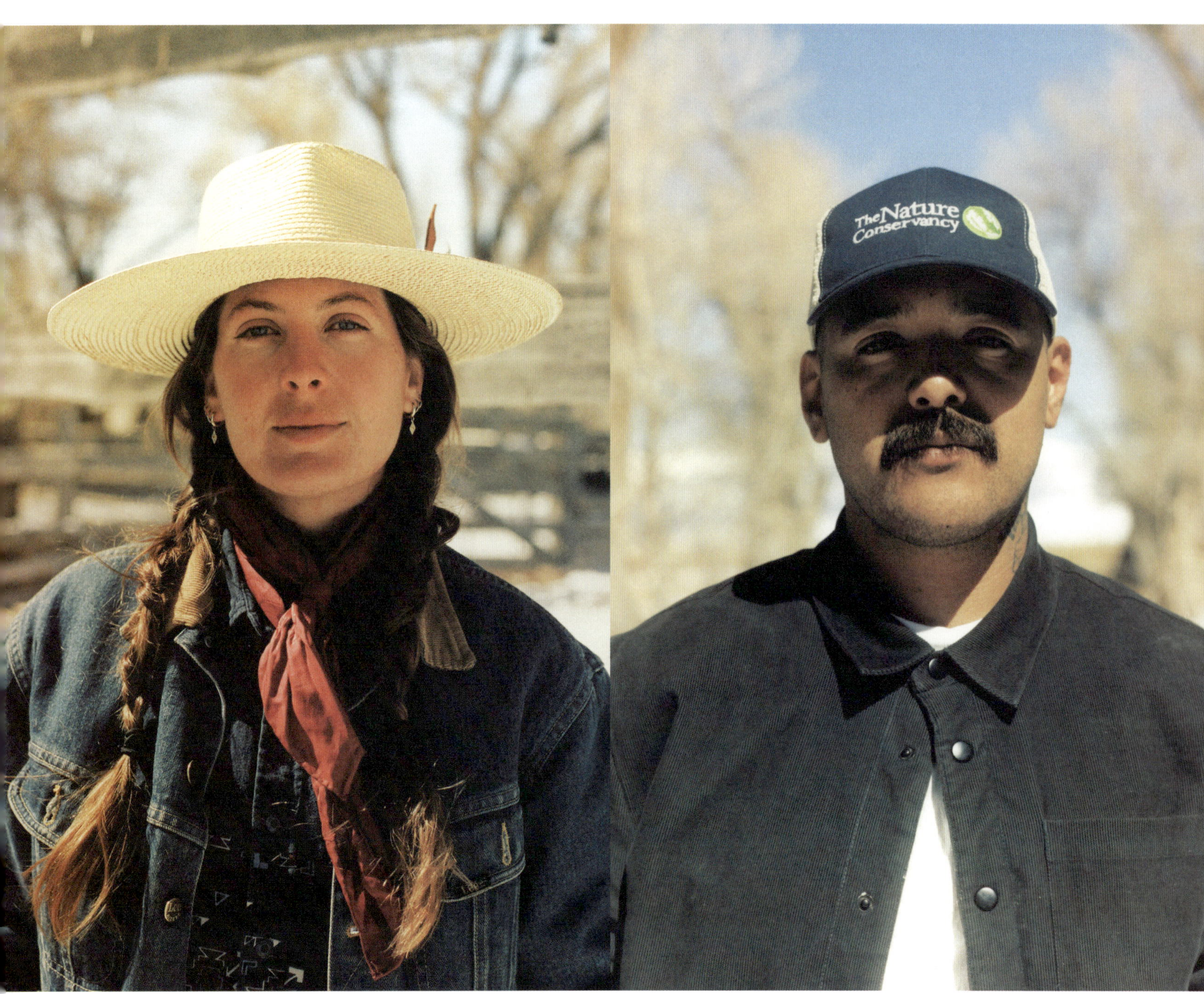

SIERRA MACDONALD
Programs Director
Zapata Ranch, Colorado

IVAN GUILLEN
Paintrock Inn Chef
Paintrock Canyon Ranch, Wyoming

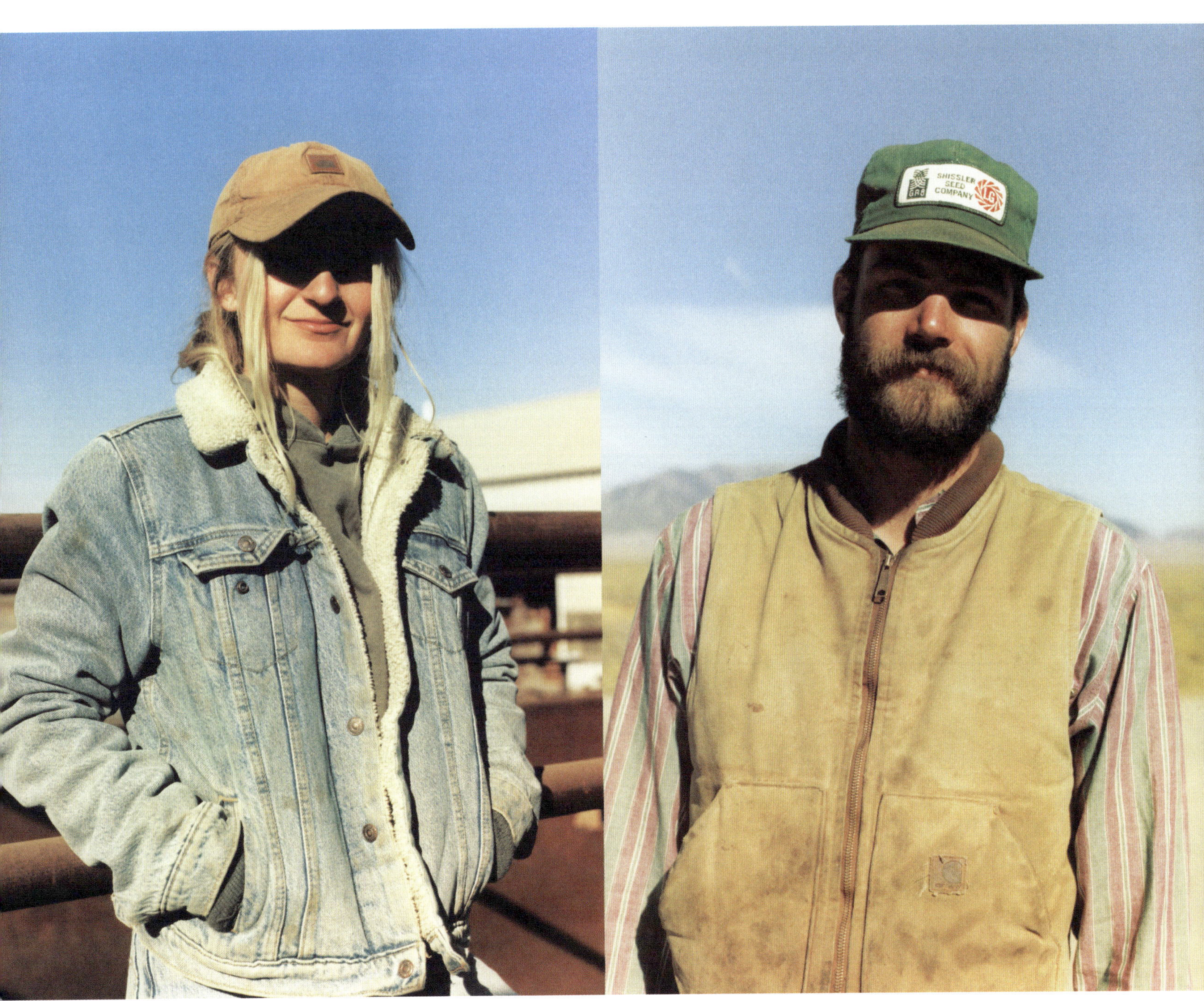

CLAUDIA LANDREVILLE
Ranching and Editorial Content
Frying Pan Ranch, Texas

DYLAN TAYLOR
Apprentice
Chico Basin Ranch, Colorado

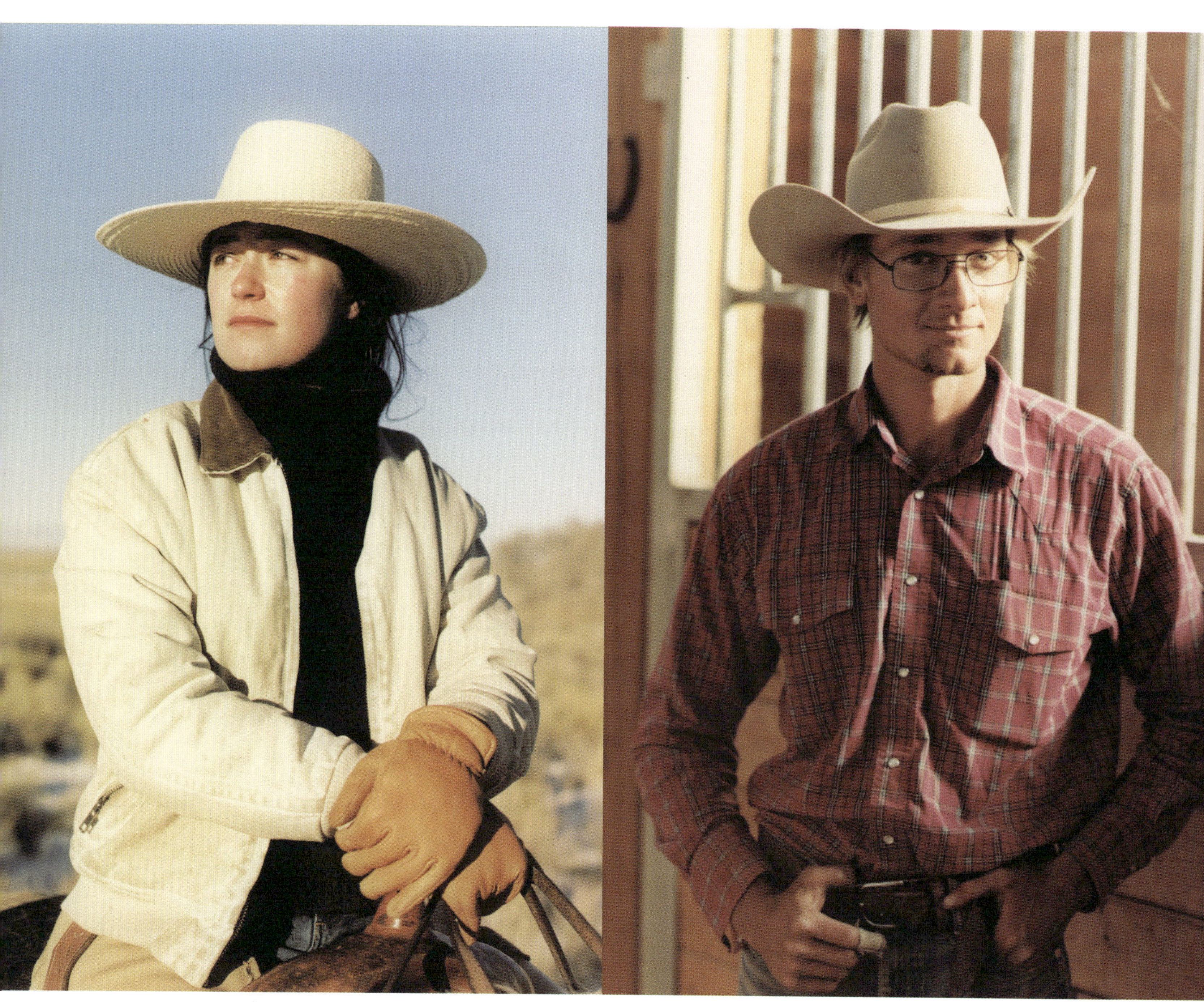

RUBY DUNBAR
Apprentice
Chico Basin Ranch, Colorado

MAX VANHORN
Facilities Manager
Zapata Ranch, Colorado

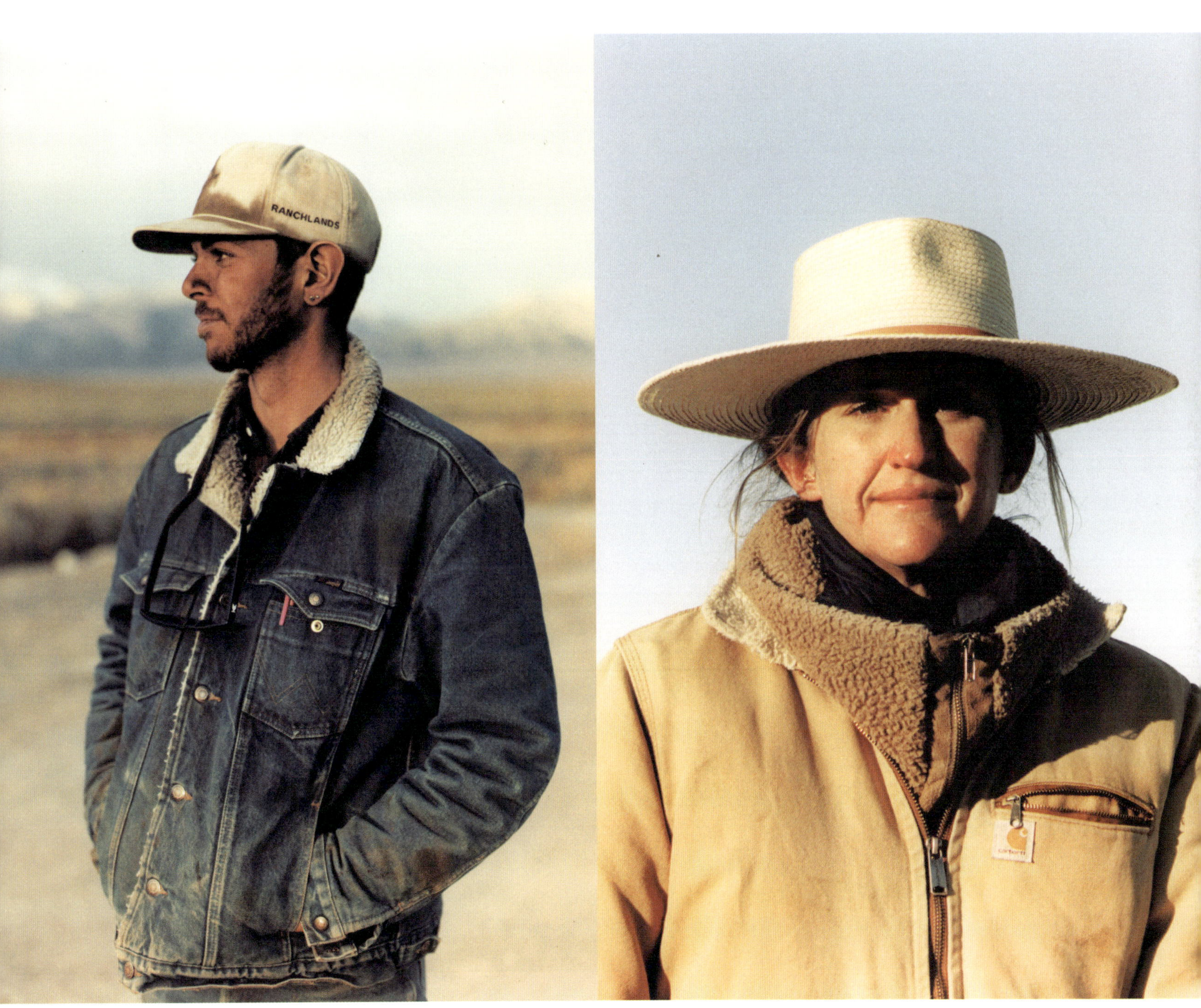

BRANDON SICKEL
Managing Apprentice
Chico Basin Ranch, Colorado

LAUREN O'TOOLE
Ranchlands Stays Program Director
Paintrock Canyon Ranch, Wyoming

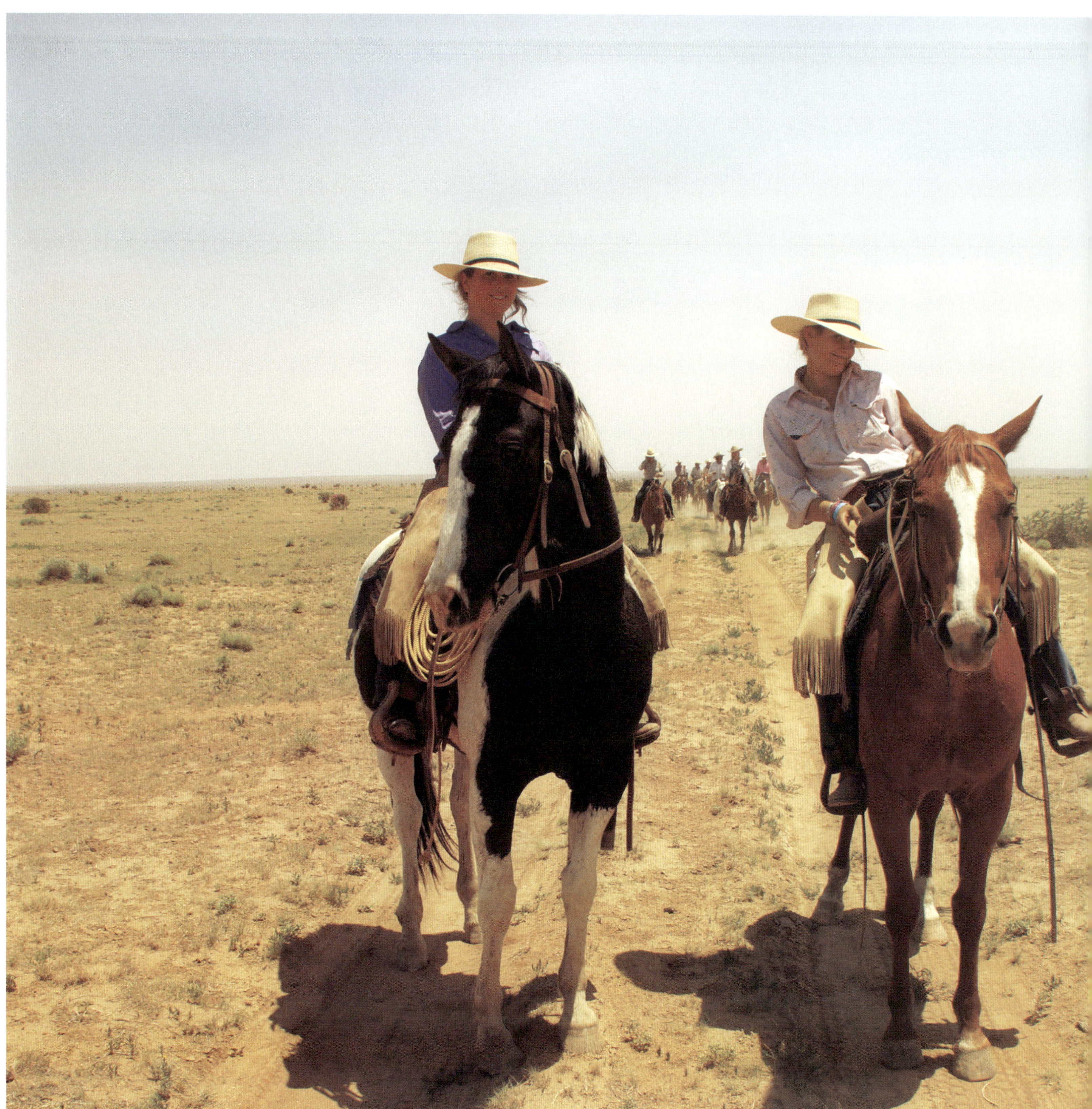

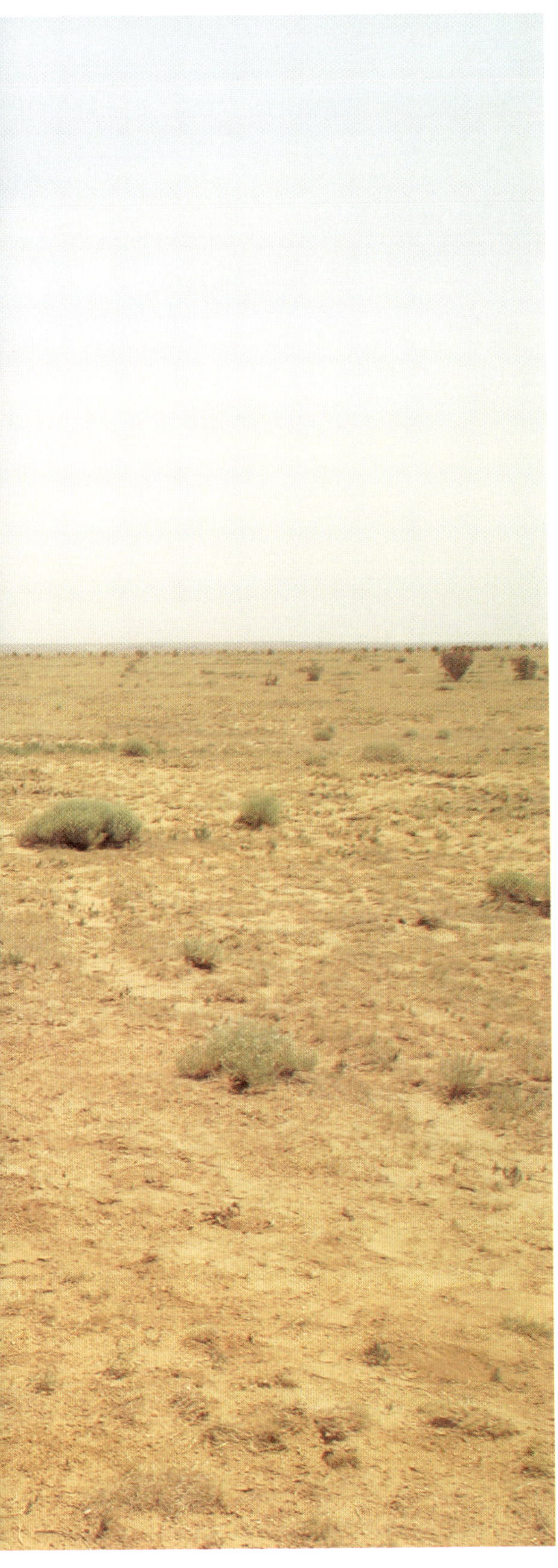

If we are lucky, and if enough people come through our open gates to hear our stories, these lands may sing forever.

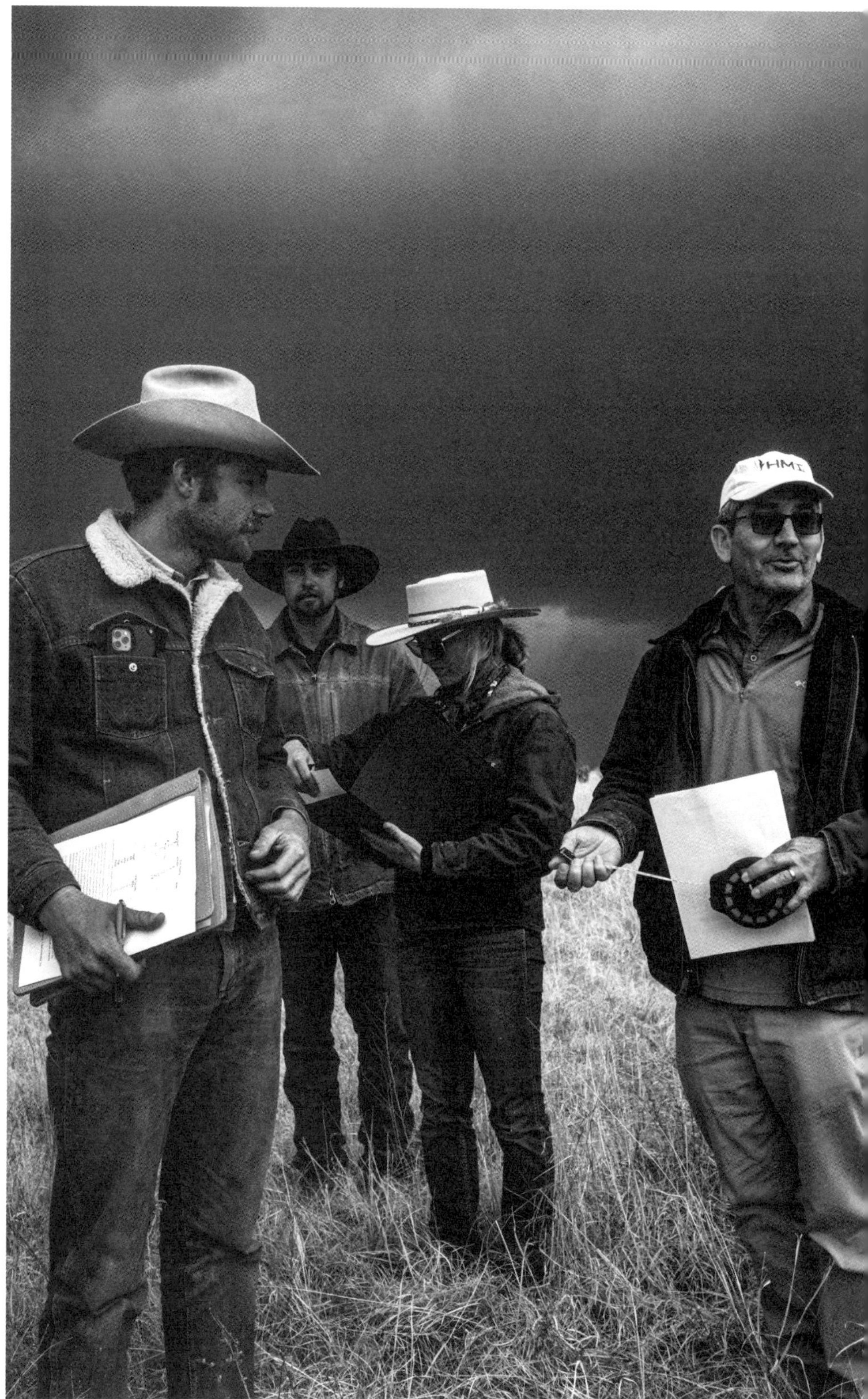

OPEN-GATE POLICY

Ranchlands partners with a variety of conservation and ecological groups—such as The Nature Conservancy, state wildlife agencies, Bird Conservancy of the Rockies, Holistic Management International, and the Savory Institute—to host workshops, field days, scientific projects, and other educational events. At the Chico Basin Ranch, our open-gate policy welcomes birders and other wildlife enthusiasts who visit the ranch to look for rare dragonflies, mountain plovers, and swift foxes.

We work to encourage people from outside our community to build a relationship with us and with the land, and hopefully to understand what would be at stake if ranching landscapes were lost. Our collaborations with scientists and ecologists are two-way streets: we provide them with natural laboratories to conduct their research, which provides us with data about the effects of our conservation programs on the land and wildlife.

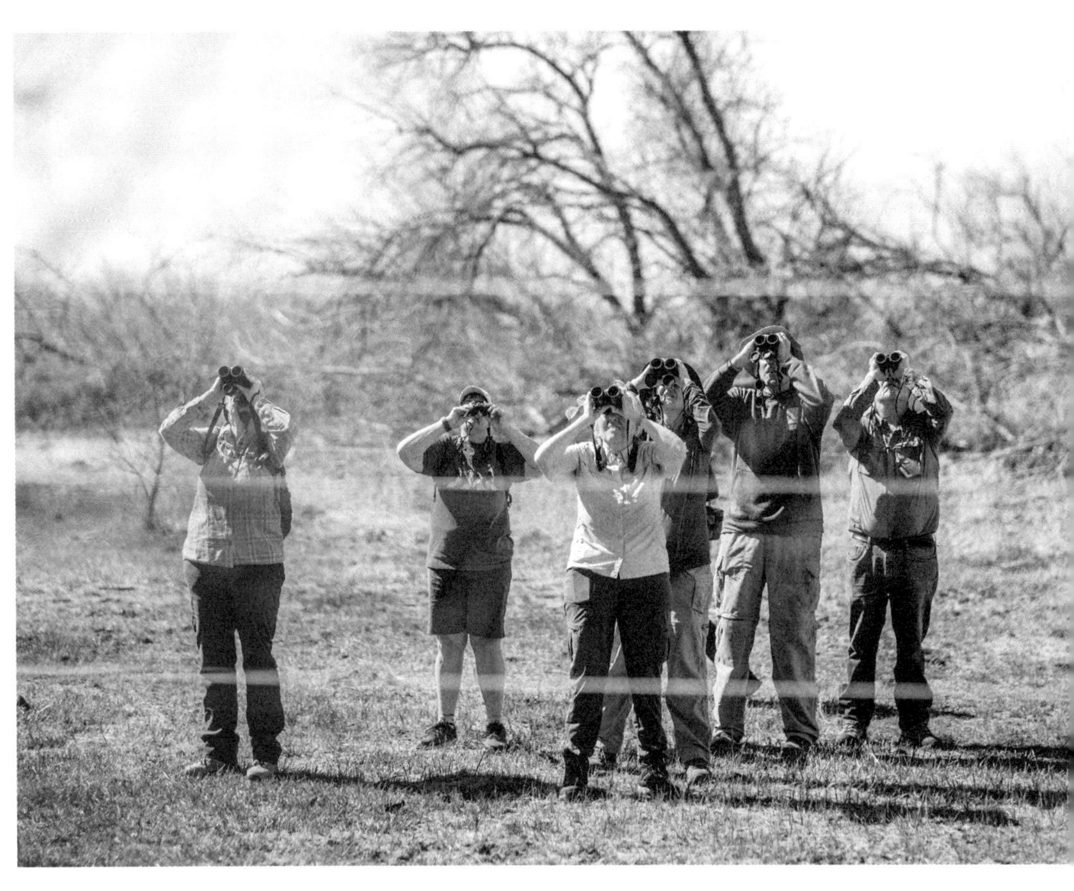

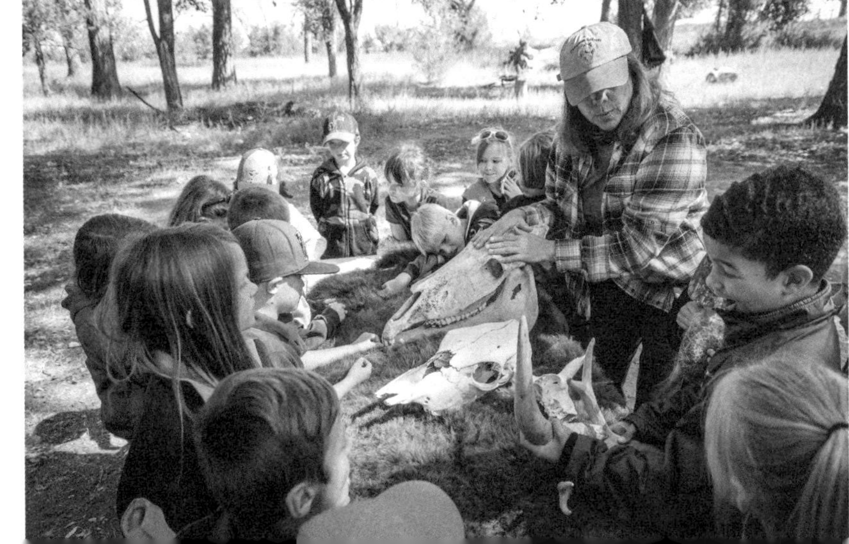

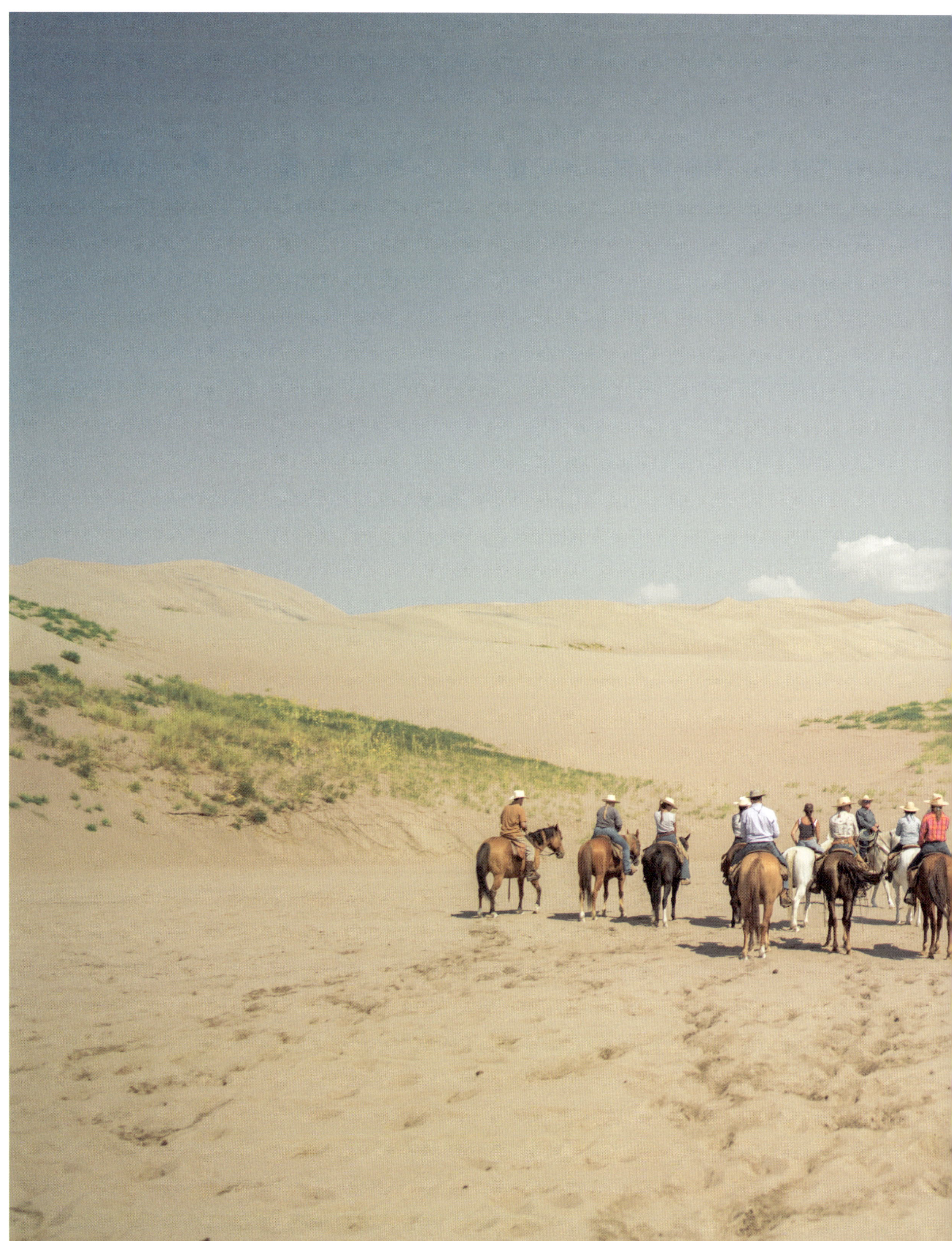

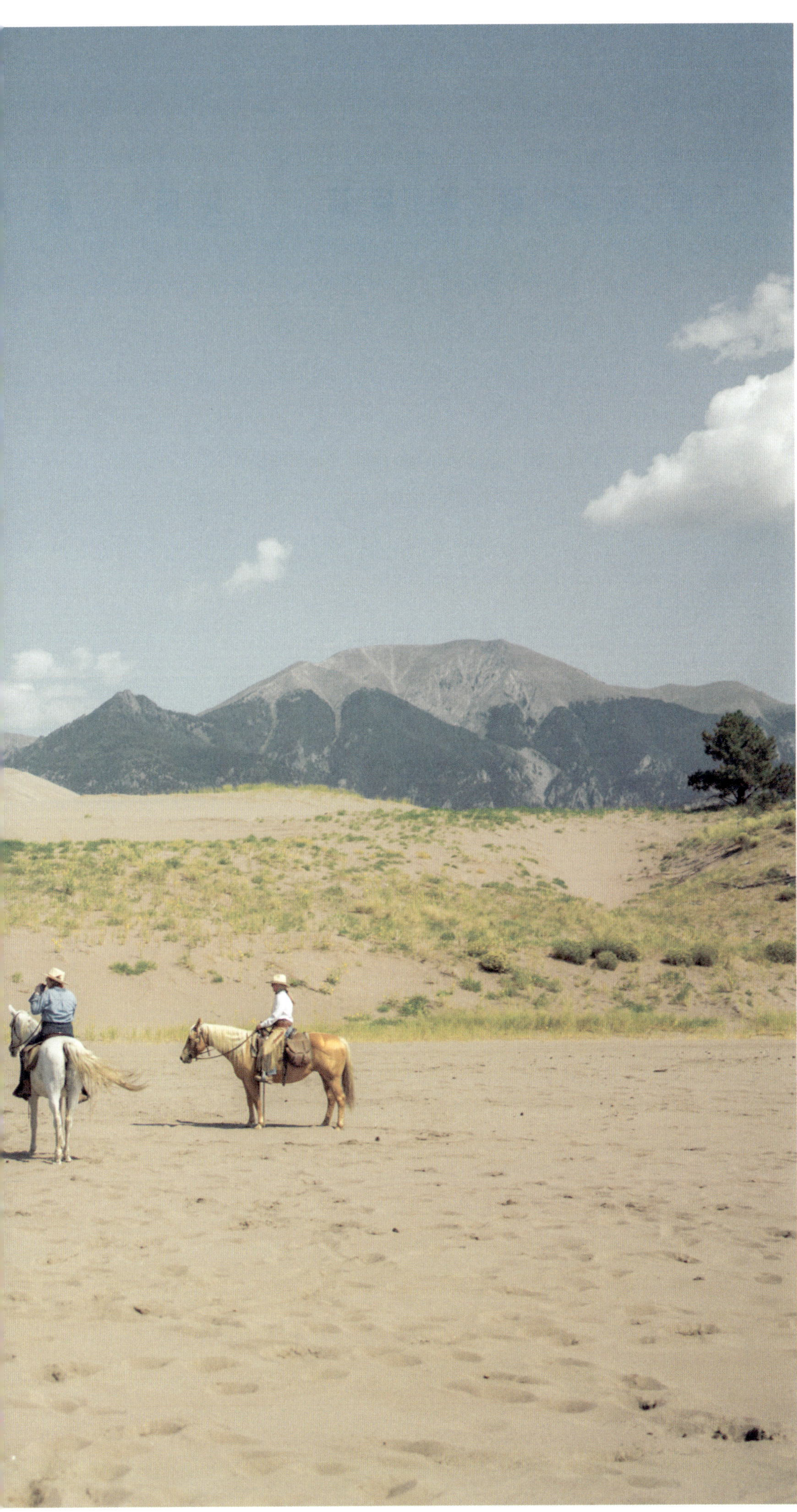

Ranch vacations are another avenue for people from beyond our border fences to visit, learn, and develop relationships with the land. Unique experiences like horseback riding in the largest inland dunes in North America or packing horses into the Bighorn Mountains of Wyoming are paired with opportunities to learn firsthand about ranching, conservation, and topics like art, literature, and horsemanship.

Ranching is an endeavor that culminates in healthy meat, and as food producers ourselves, food is often at the core of the experiences we present to guests or share within our own staff and community. We aim for the meals served to our guests to be an extension of everything else we do at Ranchlands—ethically produced, environmentally conscious, and investing in a network of food producers who steward working lands and protect them as open spaces.

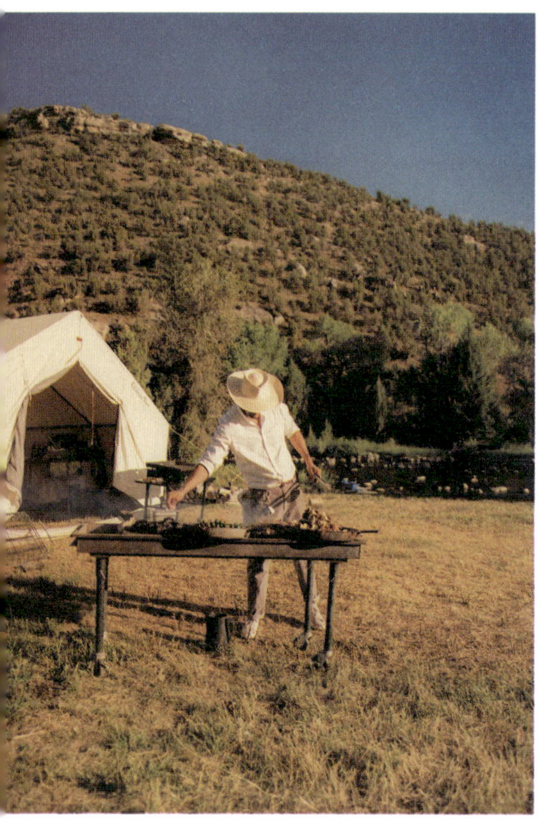

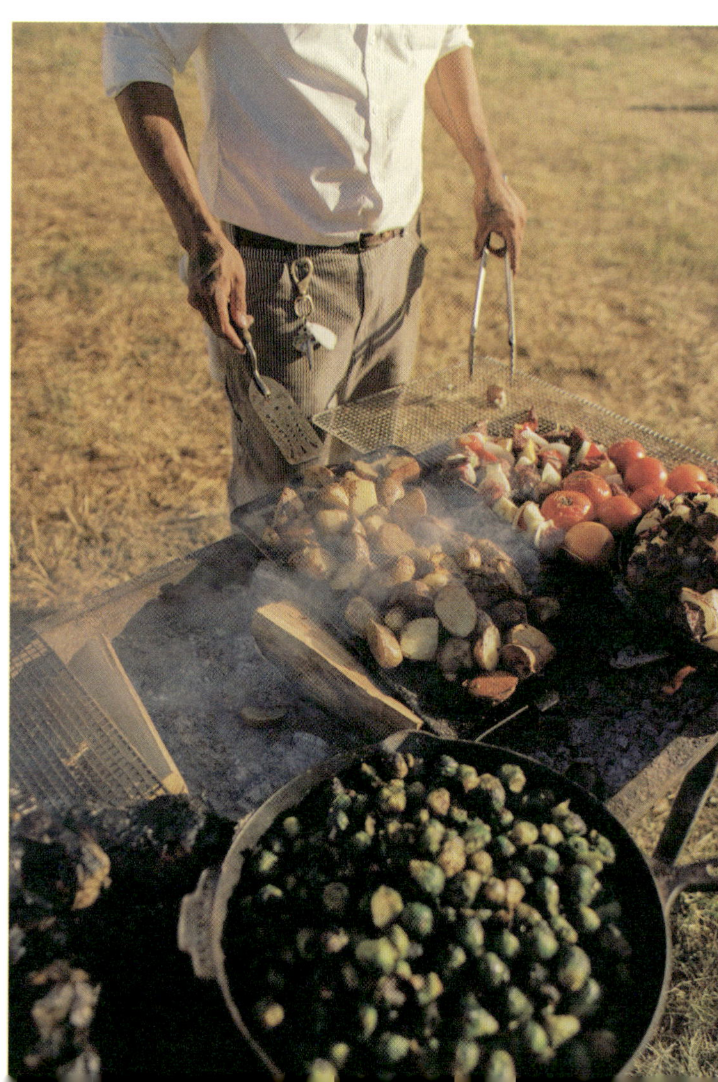

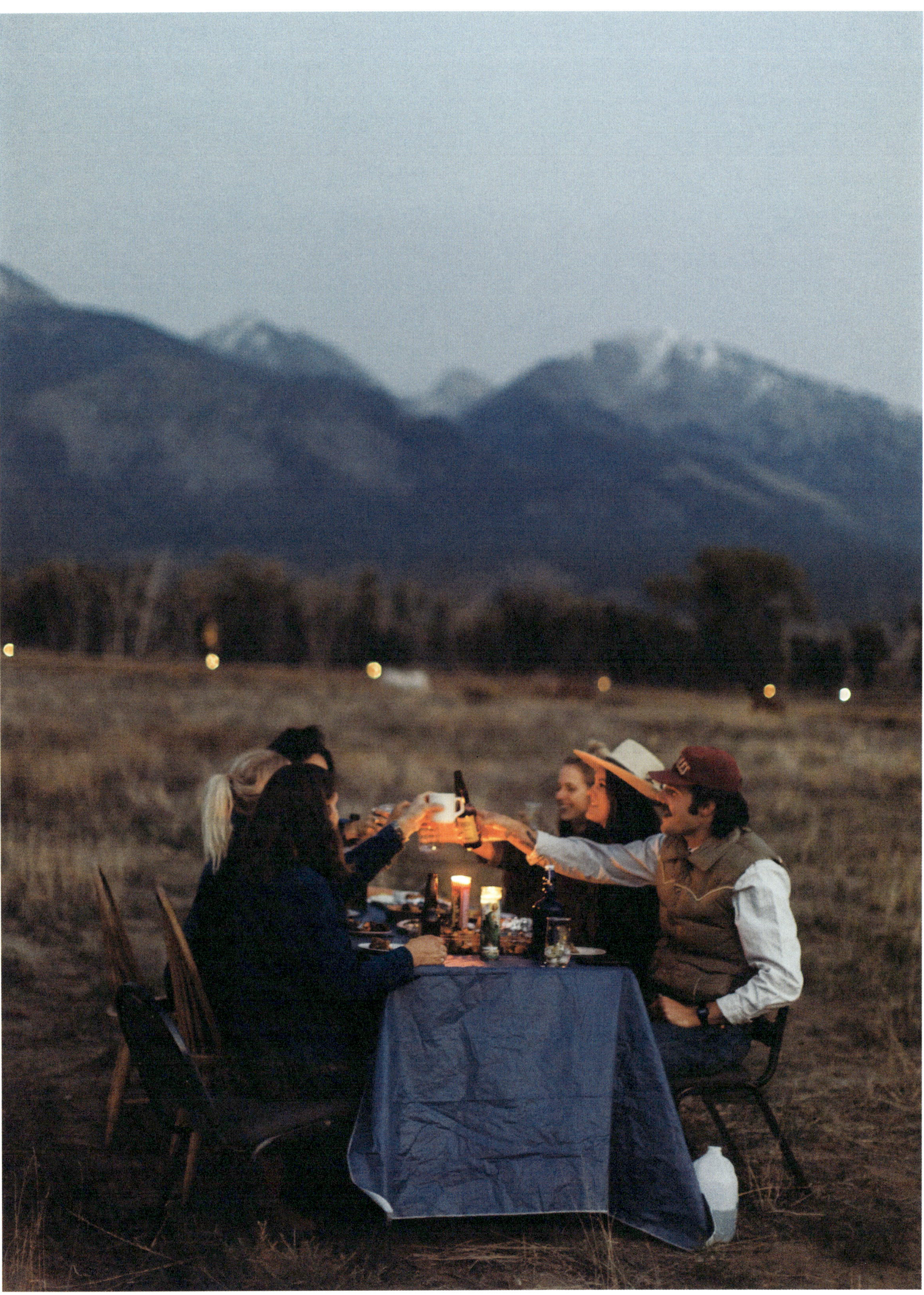

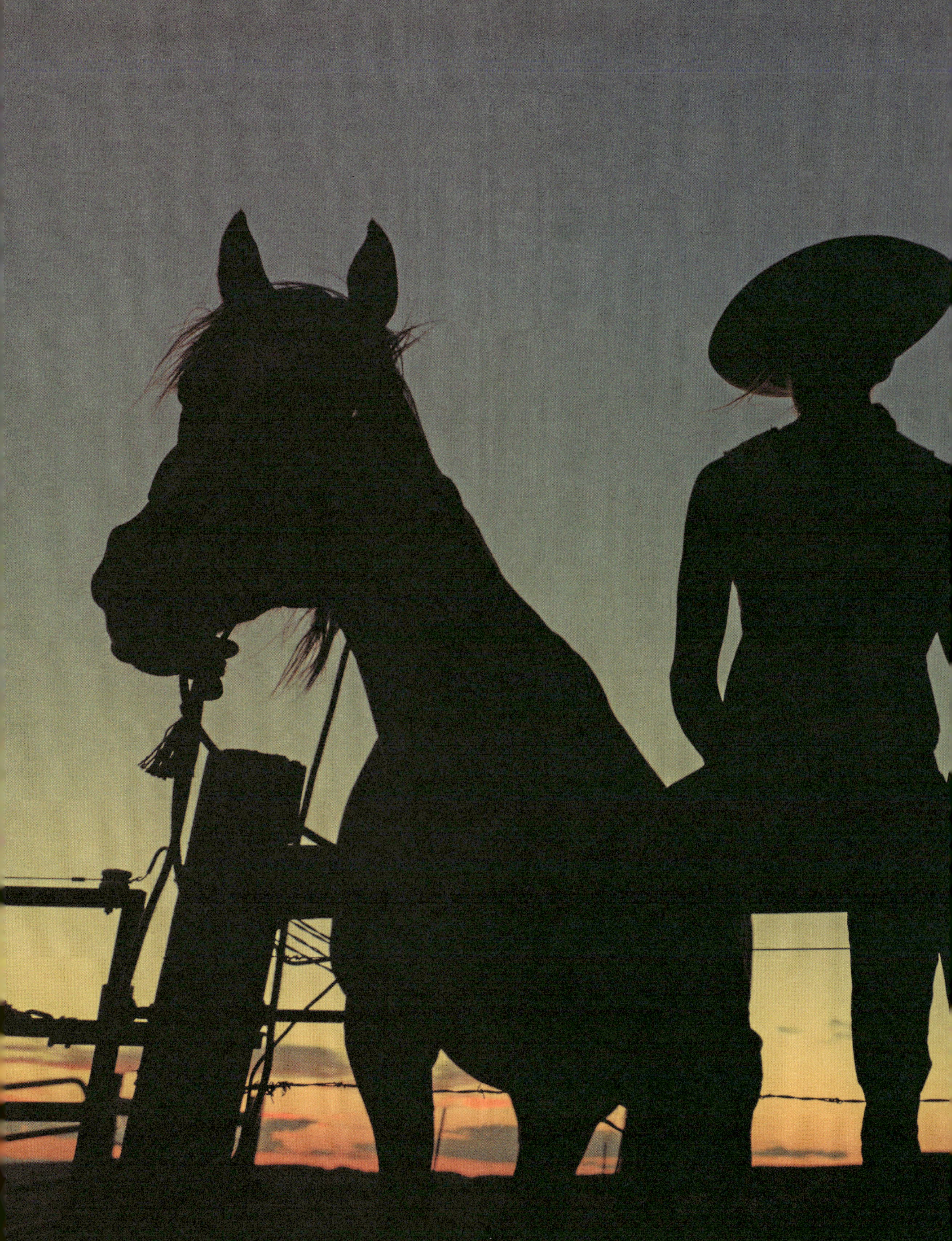

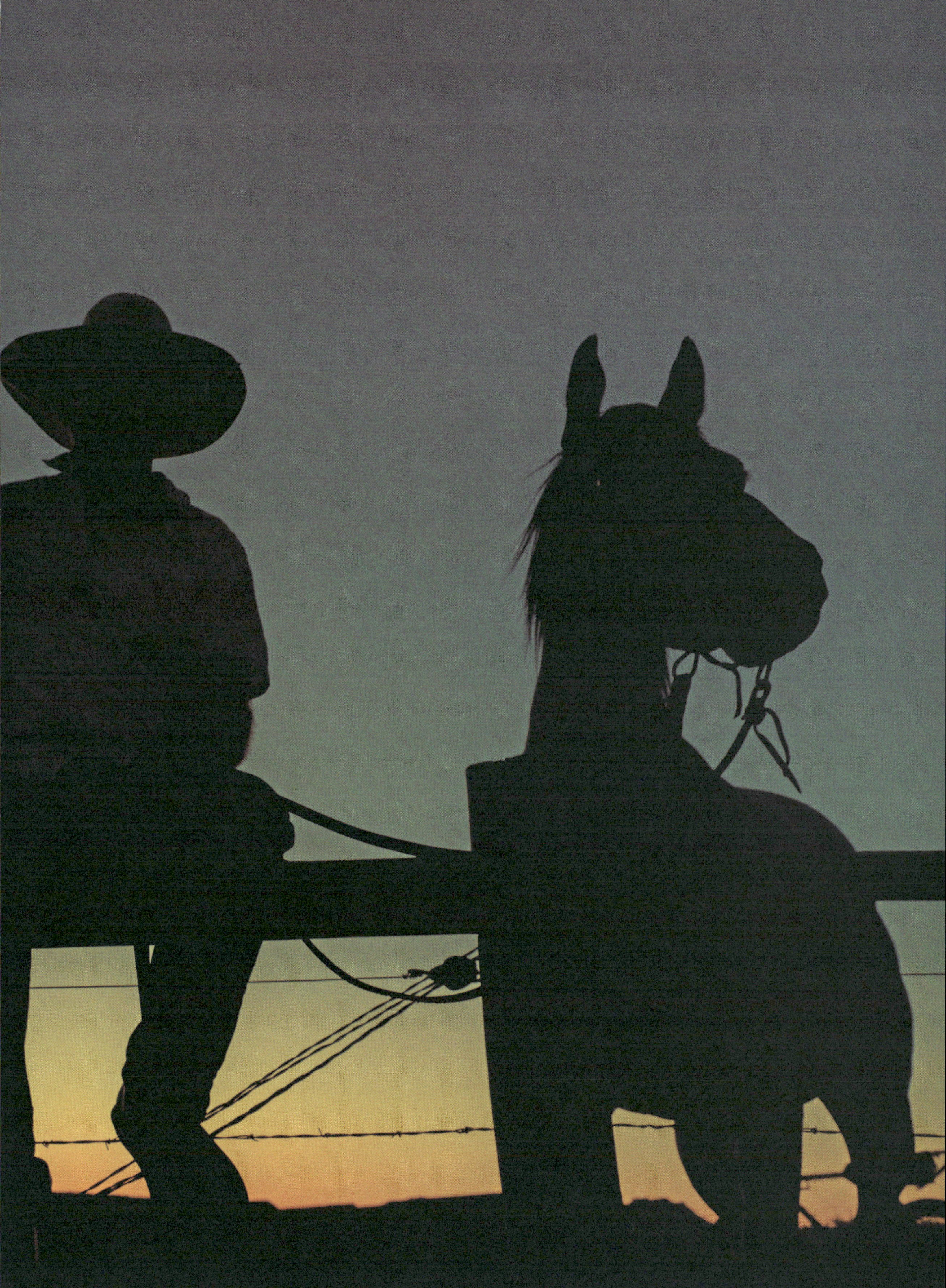

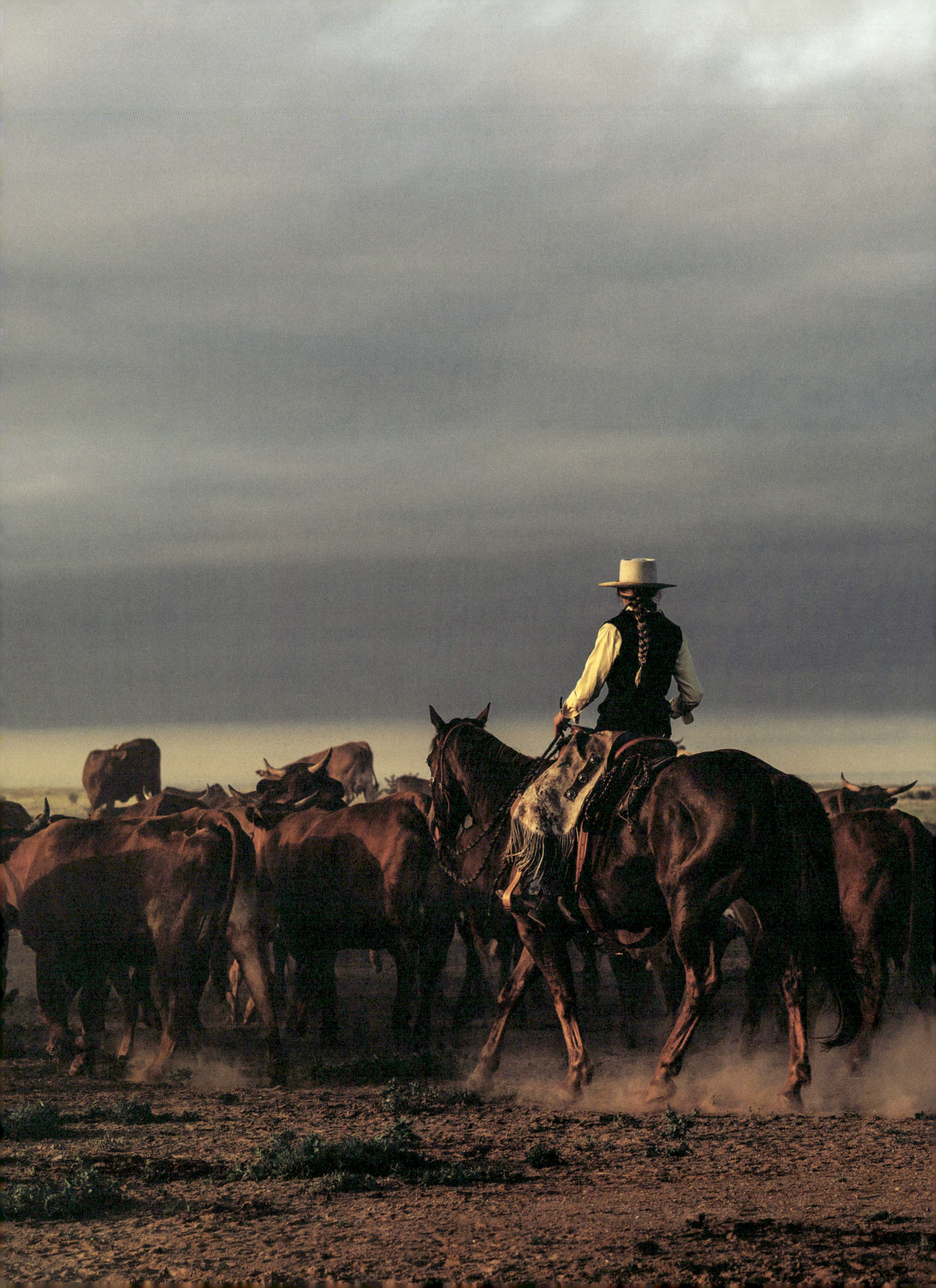

THE

ANIMALS

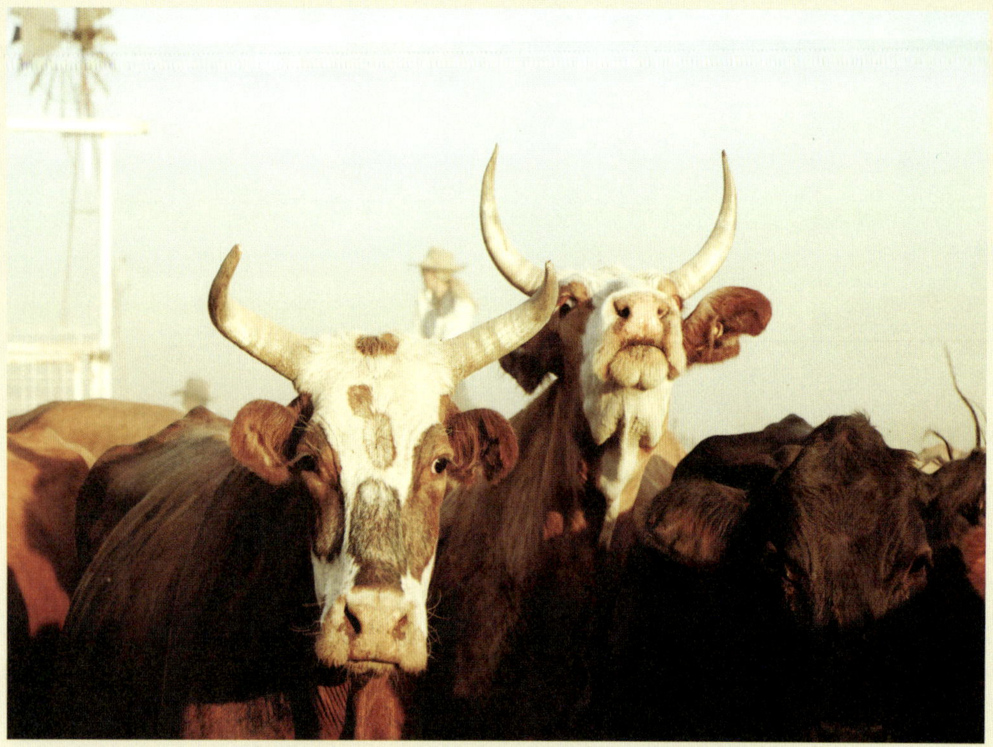

in the big pastures and selecting the ones that excel empowers them to become efficient at finding ways of living within the array of nature's forces. It gives them power to grow, to be in sync with the natural cycles, and to become what they are meant to be, the same as wildlife. Humans often think we can overcome nature, that we can move faster and make things better by taking shortcuts, such as breeding cattle without horns, or mass treating them with pesticides, hormones, and antibiotics for maximum weight gains. In the end, this attitude creates a sterile environment that props up the animals instead of allowing them to evolve with nature as all living creatures have done over millennia. The human problem is thinking in short time frames instead of projecting into the future. We have to understand that time, and lots of it, is part of the movement of life.

Animals are everywhere on a ranch. Beyond our cattle and horses, all manner of wildlife abound. Migrating songbirds, badgers, bugs, bobcats, coyotes, owls, ungulates, and even mountain lions and bears live in every bush and pool of water across the entire ranch. In addition, underground and invisible are trillions of microscopic organisms.

Domesticated animals—our cattle, horses, dogs, cats, and chickens—support and live alongside us. This community of animals is part of our identity. Tending to their needs dictates the list of priorities that govern our daily work. We depend on them for our sustenance, and their presence and dependence on the same ground that we have formed our entire lives around gives us a deep sense of belonging to the places where we live.

CATTLE

Ranchlands' core business is raising cattle, measured each year in pounds of beef shipped off the pastures. As with everything in our industry, the majority of cattle-raising modes are dictated by convention. Things such as the breed, color, size, shape, and presence (or lack) of horns influence the market price we receive. However, Ranchlands' philosophy of raising cattle is predicated by our philosophy of how we must live and work in harmony with nature, both ethically and economically.

A favorite pastime is going to a watering point at noon, when the lounging cattle are bulging with water, and spending hours sitting in the herd, listening to their noises, watching them, smelling them, and playing with the curious calves that always come up to investigate us.

We manage the genetic makeup of our entire herd to ensure that our cattle are as adapted to their natural environment as possible. We produce the bulls that are used on all of our ranches from a special seedstock herd of Beefmaster cattle, one of only three American breeds. Dating back to the 1930s, the breeding program has always selected animals to be naturally adapted to their environments. Our herd has not been sprayed with pesticides or given antibiotics or hormones in more than 90 years. Almost everything about the cows is contrary to the contemporary status quo. We sell a cow if she loses her calf to a predator or to parasite infestation, rather than shooting the predator or treating the herd with pesticides. As a result, the cows are all colors, have horns, are moderate in size, and roam the range similar to elk, deer, and antelope.

Horns are there for a reason. They help the cattle protect themselves from predators. Treating an entire herd for parasites like flies makes them weaker and more vulnerable as a whole. Leaving them alone to graze where they choose

HORSES

Horses always amaze us: their level of sensitivity, their athleticism, and their desire to do whatever is asked of them if we are sensitive enough and don't look at them as something that needs to be dominated. Every time they are in the pasture or corral and we walk around visiting with them, it feels as if it is for the first time. Their noses and hairy lips are their fingers, used to touch our hands and faces, moving around our ears and eyes, breathing deep and smelling our hair, exploring like a blind person's hands wanting to know the contours of our faces.

Horses are one of the main elements that connect us to our past. They are also what keep us connected directly to the land. Spending time with these animals is what we enjoy most. There is no feeling in the world like arriving at the saddle house at 3:00 a.m., turning on the dim light bulb, grabbing your bridle, and heading out the door into the corrals to catch your horse and saddle it.

On these mornings, mostly everyone is quiet, going all directions at once. That is, unless some joker is having a happy morning, whistling, singing, and talking to whomever will listen. Everyone is busy, walking around one another in the limited space of the saddle house, carrying saddles and blankets, putting on their chaps, tying jackets onto saddles, combing their horse's mane, and packing water and lunch into their saddlebags while the horses stand with their reins looped over the rail. Soon the moment comes that everyone is waiting for—the leader for the day (not always the boss) grabs the reins and walks his or her horse into the darkness toward the gate, leading it out into the horse trap. Everyone follows through and mounts, waiting for the last person—usually the one who has been doing all the whistling and talking—to catch up, close the gate, and mount. With everyone up, the leader turns his or her horse toward the open pasture at a walk, then slowly moves into a trot. Everyone turns and, as one body, trots off into the darkness.

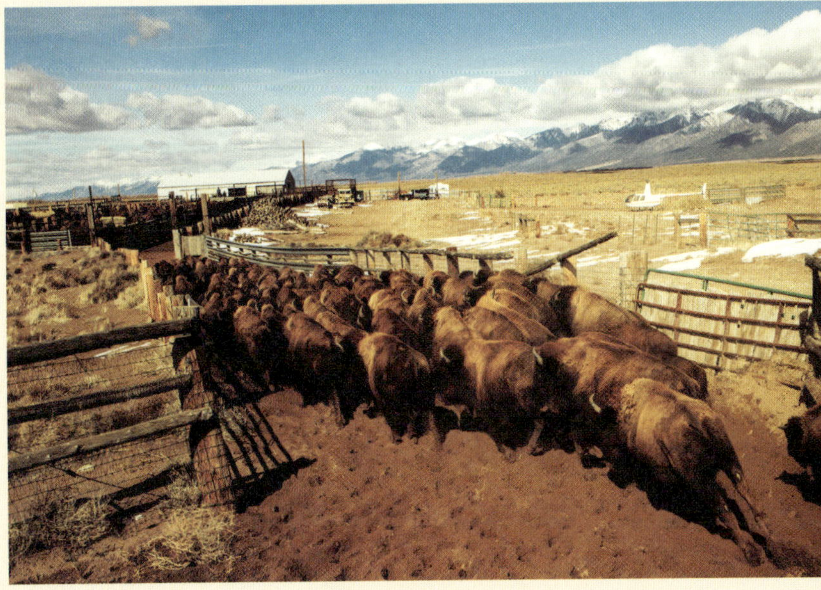

The ride can last for an hour or two at a steady trotting pace. When arriving at a gate, the protocol calls for someone in the group to open it but not the leader. It's a matter of respect for those who have gone through to turn and face the rider who is closing the gate and wait until he or she remounts before riding off again. Traveling together on horseback in the dark is like nothing else—settling into the rhythm of your horse and riding into the day that's coming, watching it slowly grow from the very first light of dawn, spreading into the sky. This is when the bonding happens, the closeness that comes between horse and rider, and among one another. Connections are built from riding long distances together, covering a lot of ground and running across wildlife: a deer or an antelope, or an owl flying low along the bottom of a creek bed.

Across all of our ranches, there are more than 150 horses. In the beginning, there were seven, but over time, more and more were brought in until a friend gave us a bunch of broodmares. Then we found a stallion and began breeding our own horses. It took a while, but the progeny from these mares are now used on all the ranches. And what a change that has been to have good home-raised and trained horses.

The horses are like our own family members, who we have grown to know through hundreds, if not thousands, of miles together—traveling long circles to gather pastures, trailing cattle, roping in branding season, sorting, loading trucks during fall cattle work in the corral, racing each other, riding bareback for a hot afternoon swim in the lake, trimming hooves, and tending to sores and wounds. Horses take us far away from the roads and into the domain that makes a ranch a ranch.

BISON

The bison herd that we manage is a conservation herd, meaning no weaning, fences, feed, or vaccinations, except against brucellosis. The bison roam a 50,000-acre pasture as one big herd, grazing at will in family groups of related females with multiple generations of progeny. We had never been around bison before we assumed management of the Zapata Ranch on the Medano in southern Colorado. Over time, as we applied animal-handling principles that we learned with cattle, we discovered that bison are an entirely different creature, one that we have grown to respect and admire. They remind us of where we came from.

They're also very misunderstood. Bison have a reputation for tearing things up and turning trucks over, but they're not like that. Most people don't know this, but cattle can be far meaner. Bison are also superathletic. They can walk very slowly, or they can outrun a fast horse. After years of gathering and spending time with bison, we joke that they generally smell like goats, sound like pigs, and move like a school of fish when you're trying to gather them.

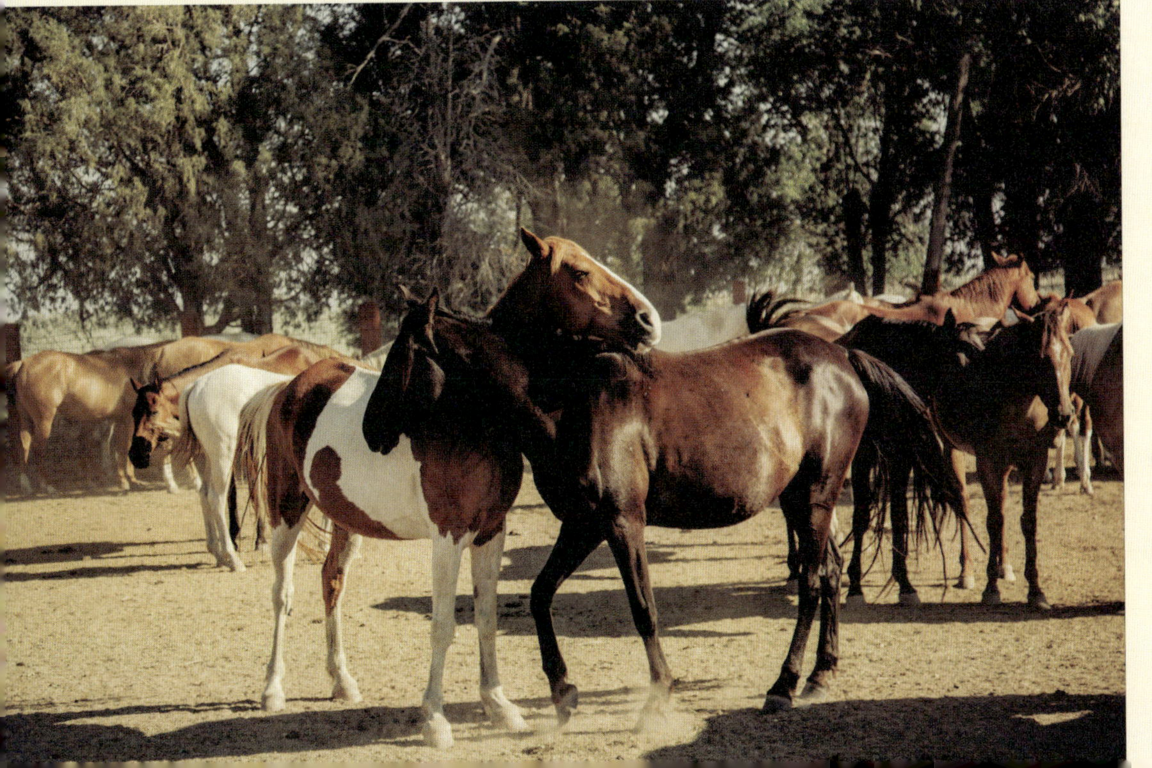

People also think that all bison have to do is walk across the landscape and everything flowers behind them. That might have been true 200 years ago when they had the entirety of the Great Plains to roam, but in today's world of fences and property lines, ecological health is mostly a function of grazing management, no matter which species is eating the grass. Another misconception is that bison meat is much healthier than beef. The quality of any meat is a function of how the animal was raised and treated, what it ate, and if supplements and antibiotics were administered by humans. Meat from bison fattened in feedlots is essentially the same as meat from cattle fattened in feedlots. But meat from grass-raised cattle is more like grass-raised bison than feedlot cattle.

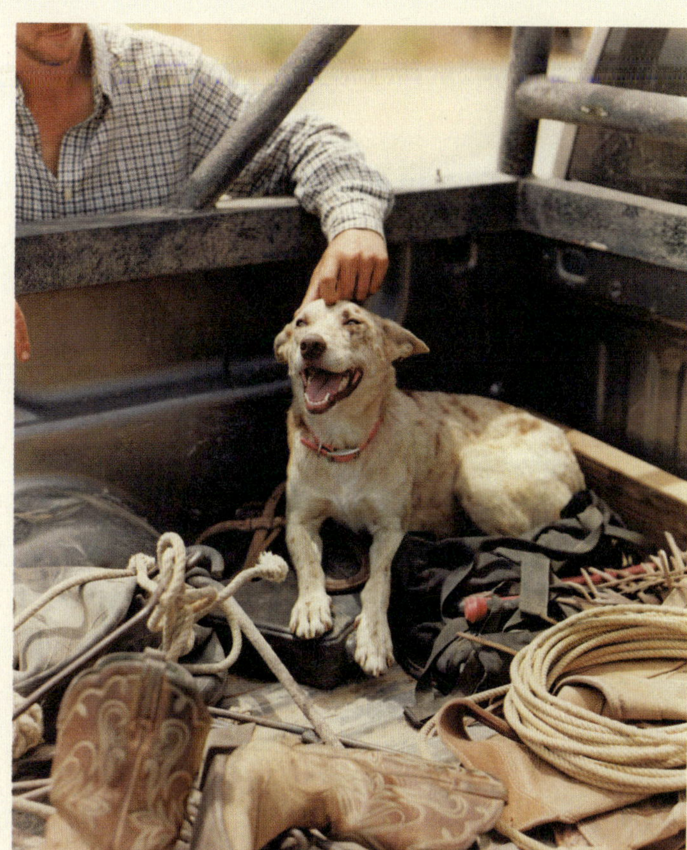

It is awe-inspiring to sit in a herd of 1,000 or more bison, just watching and listening to them. They have the ability to move so slowly and with such grace, giving us a sense of how long Earth has been turning on its axis and of the internal source of life coming from the oldest of things—nature. Our time with bison has enhanced our understanding of working with nature, moving us closer to creating a beef animal that's as wild as possible.

DOGS

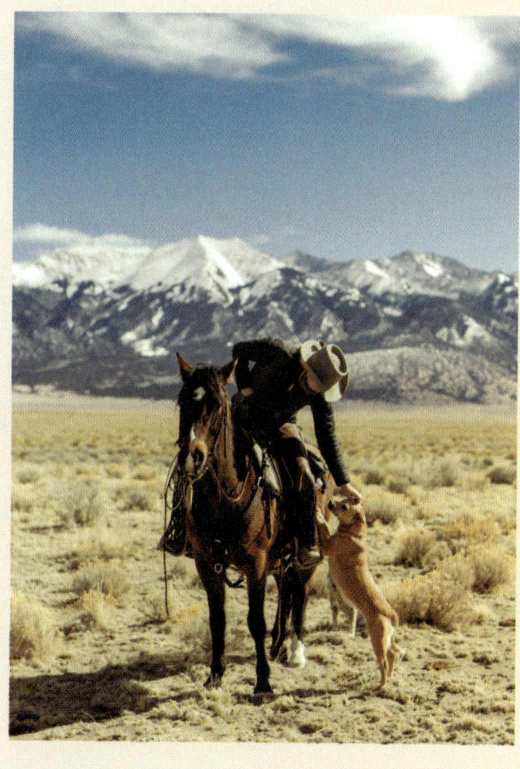

Dogs can change everything when working cattle. They can create the biggest disaster, with cattle and people on horseback running everywhere like a tornado tearing a house to pieces. Or one person can go out with his dogs and handle a good-sized herd of cattle without missing the five riders that would be needed without the dogs.

We have many breeds roaming around the ranches: border collies, hanging trees, Australian shepherds, and crossbreeds. Even our little wiener dog, Punch, rides on horseback behind Duke and thinks he is a true cow dog—until a cow turns and faces him! Some of the dogs are very good; others not so much. Some of them are young; others are old and hang around like retired kings and queens. It's a herd of dogs living and working on the ranches, much like us humans. We often carry them around in our trucks or on the front of our motorcycles, and they always tag along with us on horseback when we go out.

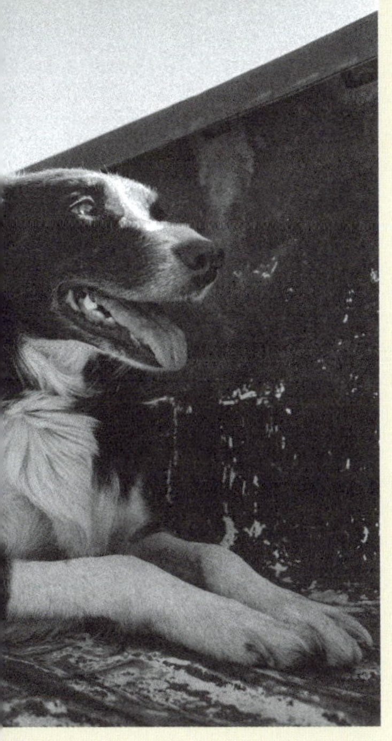

At times, a particular dog has been trained to ride behind his human partner on his horse. Sometimes they live with us in our houses in their own beds, just like city dogs do. Some of us prefer to have them outside in their own pad. They are our work colleagues and, next to our partners, our most treasured companions.

WILDLIFE

Ranchers are surrounded by wildlife. We know which species are the most common and where different individuals live. We pay attention to their needs, such as giving extra space around fawning grounds in the spring. Firsthand contact is a common occurrence. For a long time, we had a coyote that would ride alongside us, close enough that we could see the color of his eyes. He would follow us when we left headquarters, keeping up with us, stopping when we stopped. A bobcat walked over someone's legs once when he was sitting outside in the yard. Another time, we saw crows teaming up with a coyote to hunt a rabbit. The wild animals on the ranch are not only important to the land and its health but also to us, the same as neighbors or friends. Animals are the best indicators of the change of seasons or weather.

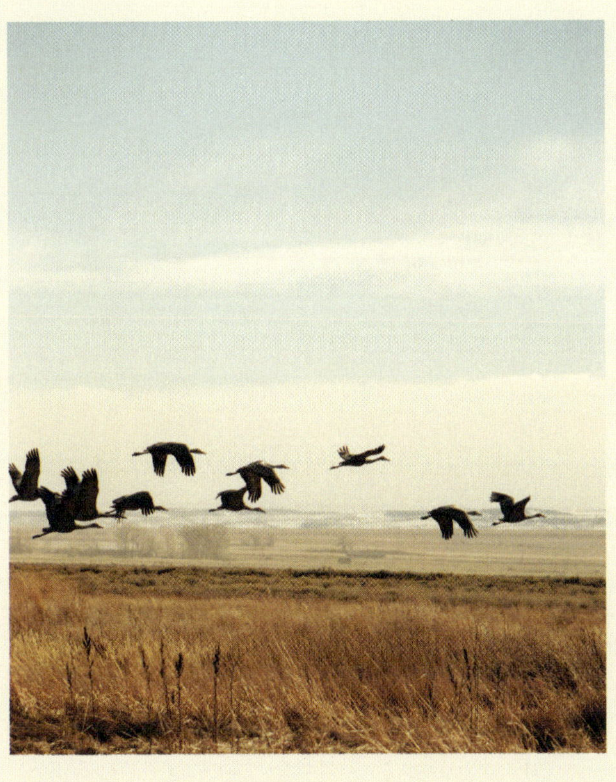

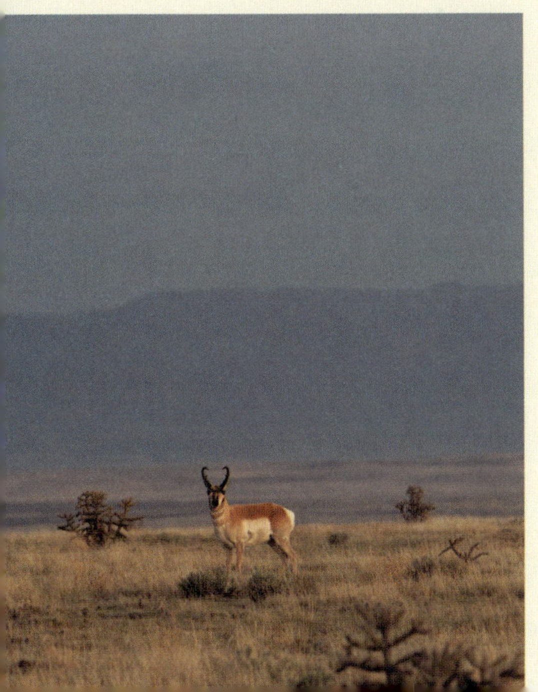

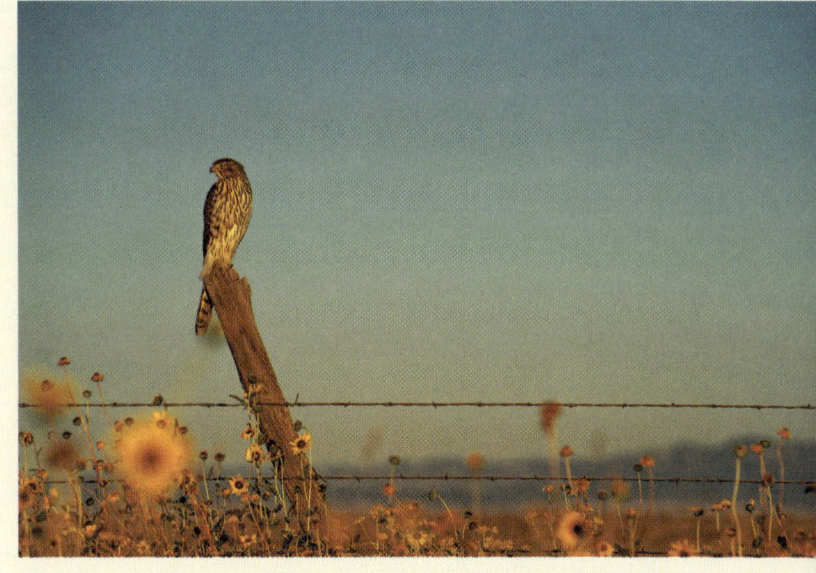

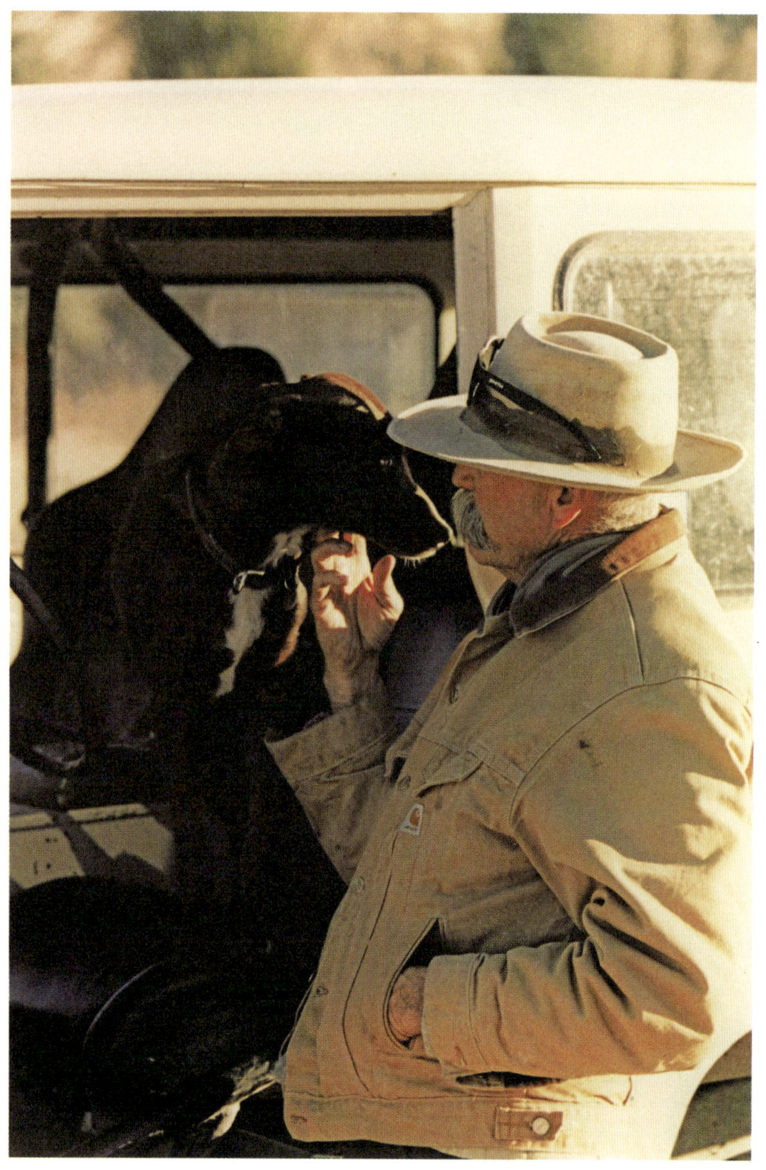

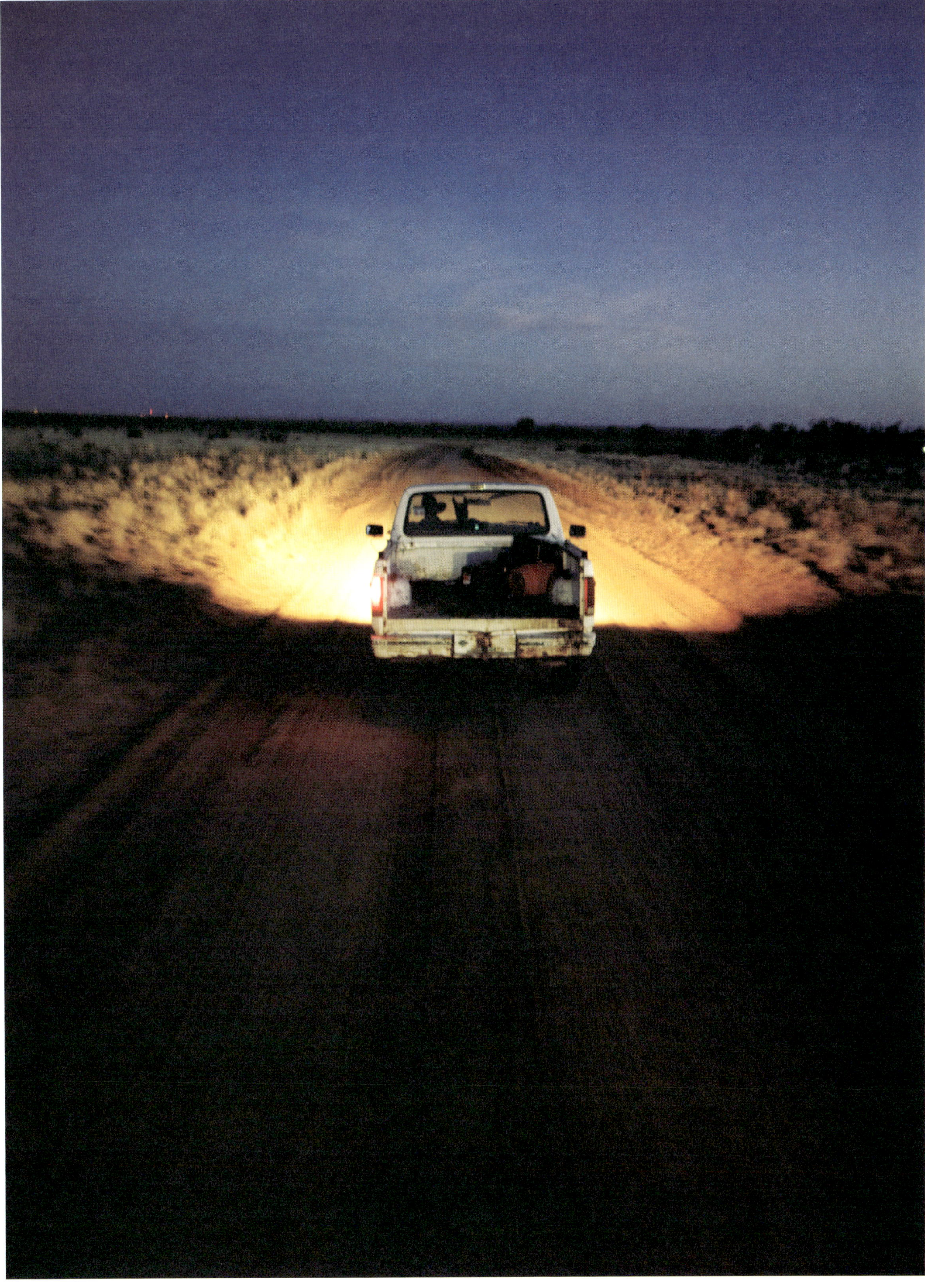

OUR CATTLE

Our cattle are the primary tool that we use to enhance and protect our rangelands. They are grazed in a migratory fashion through our pastures, emulating the symbiotic relationship between the great ungulate herds of the world and their native grasslands.

The cows must be intelligent, gentle, resilient, able to be worked and moved, and willing to thrive on their own in adverse range conditions, such as heat waves, blizzards, floods, and droughts. These special cattle come from a breed founded in the United States called Beefmaster, and they were first brought to the Chico Basin Ranch 20 years ago.

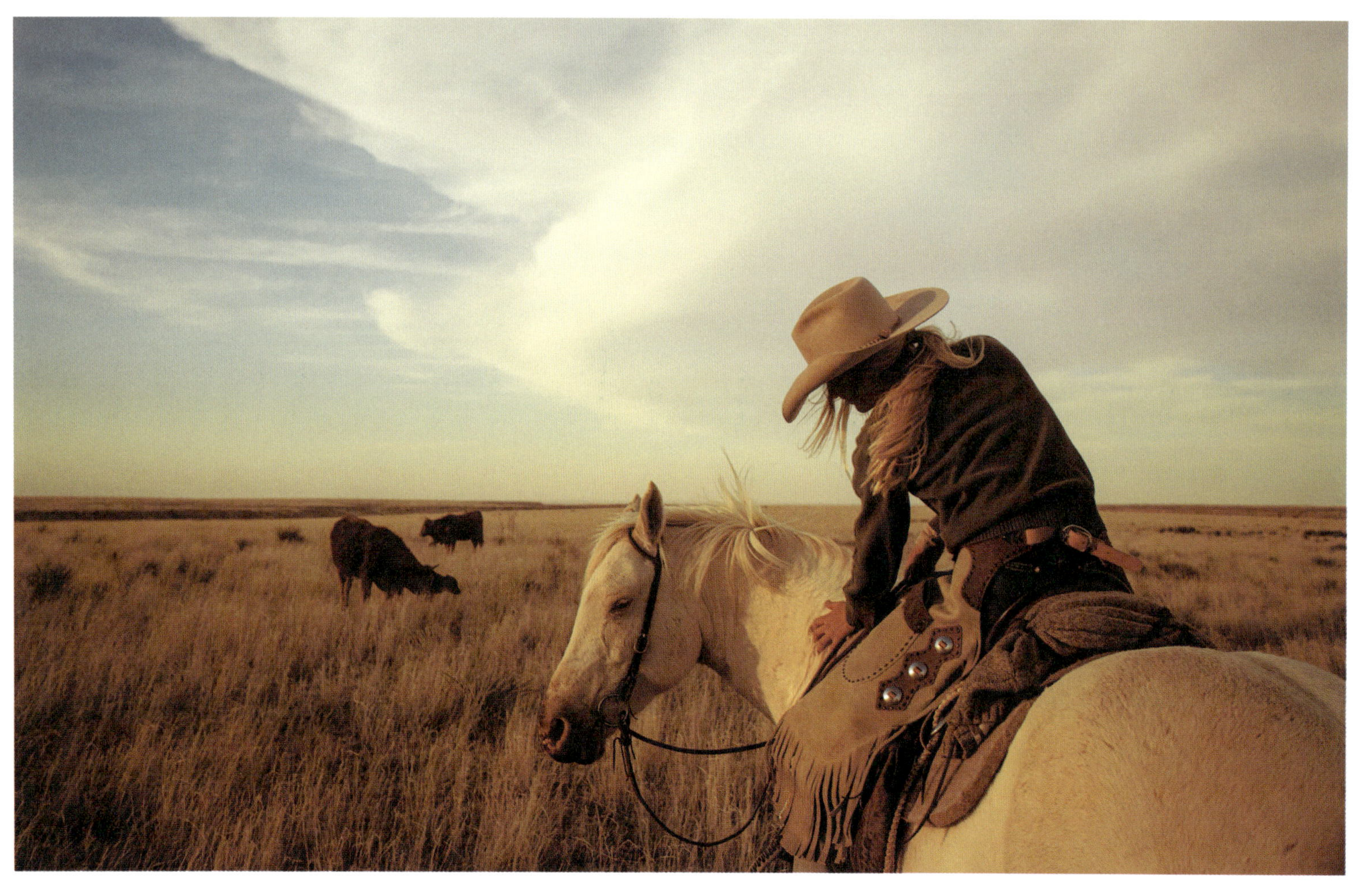

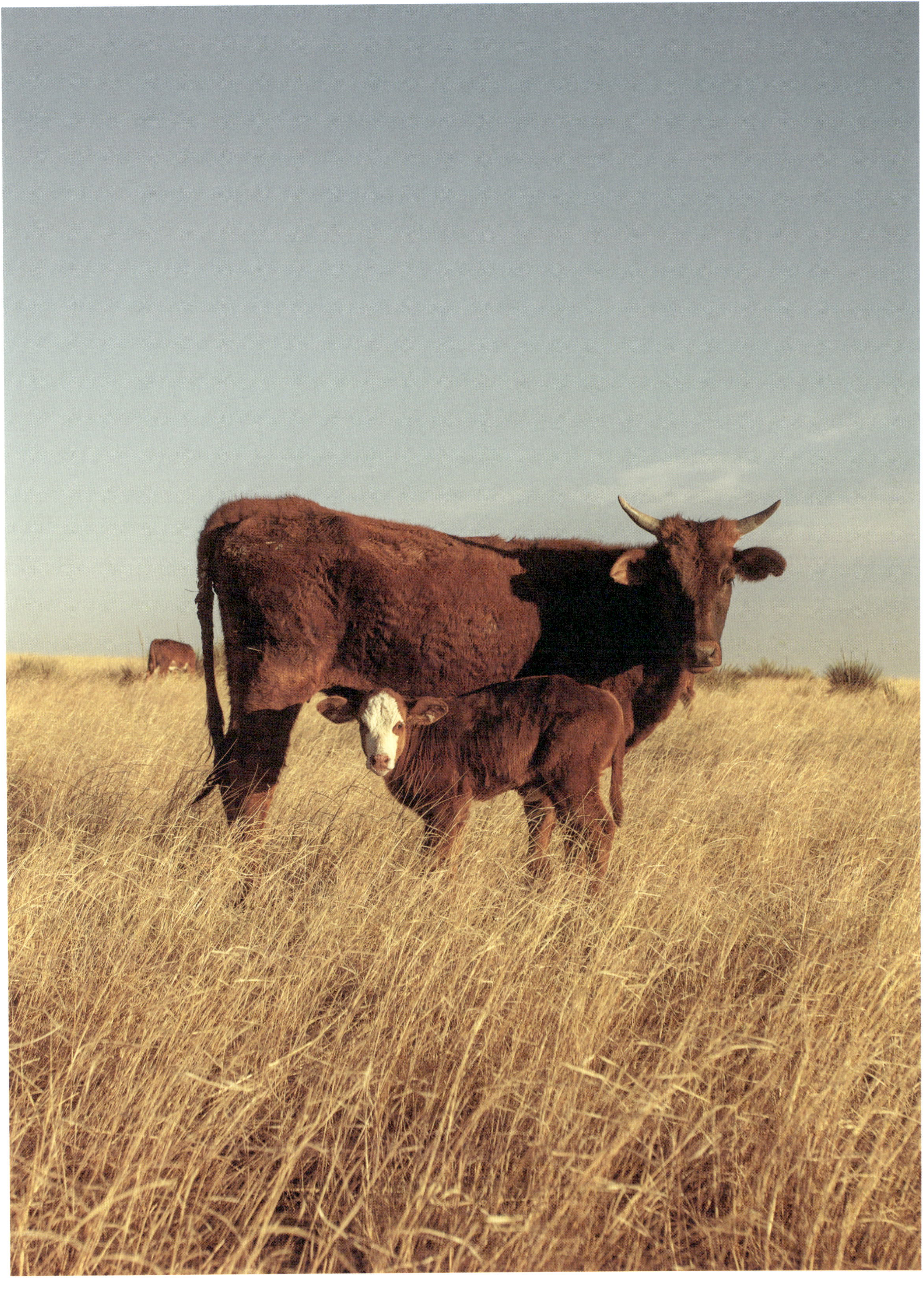

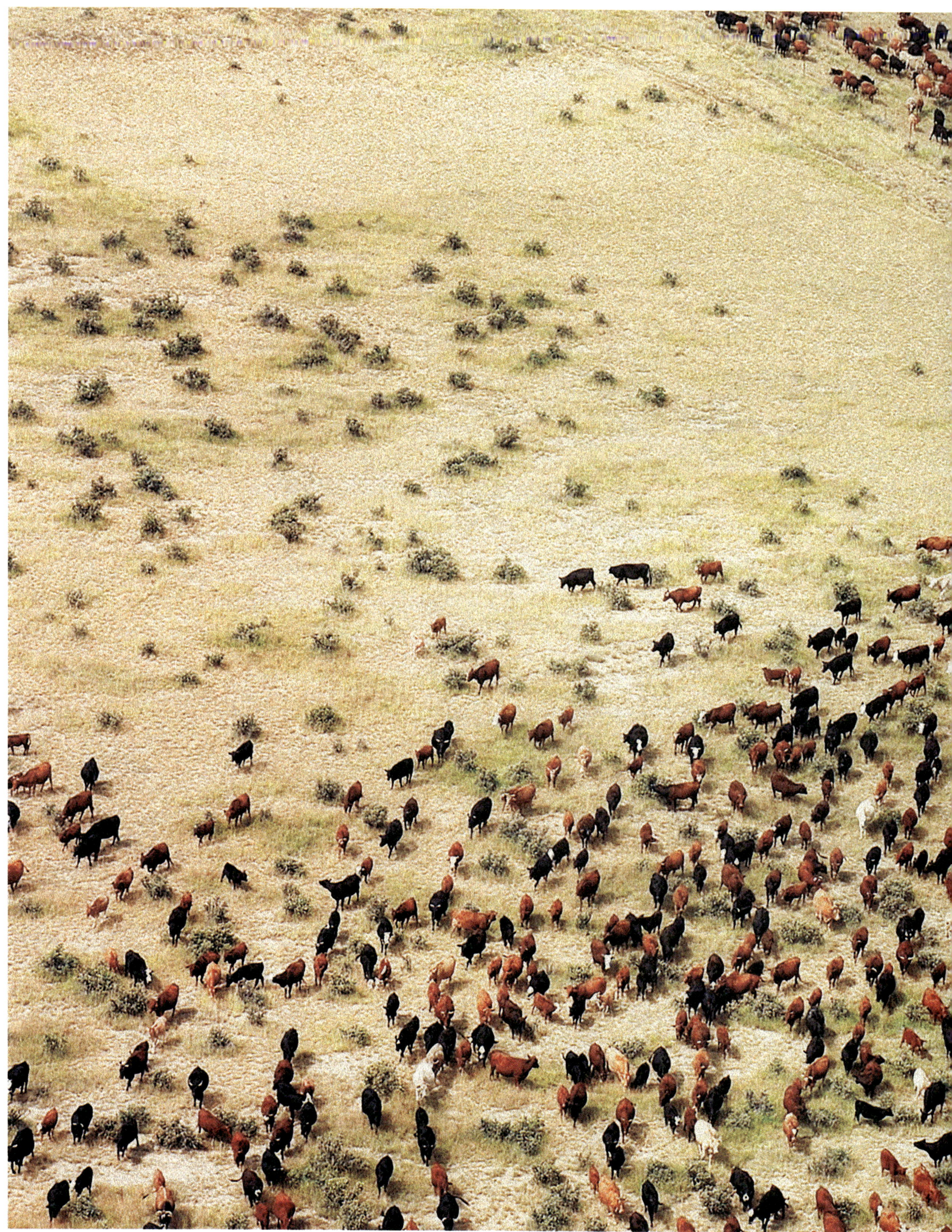

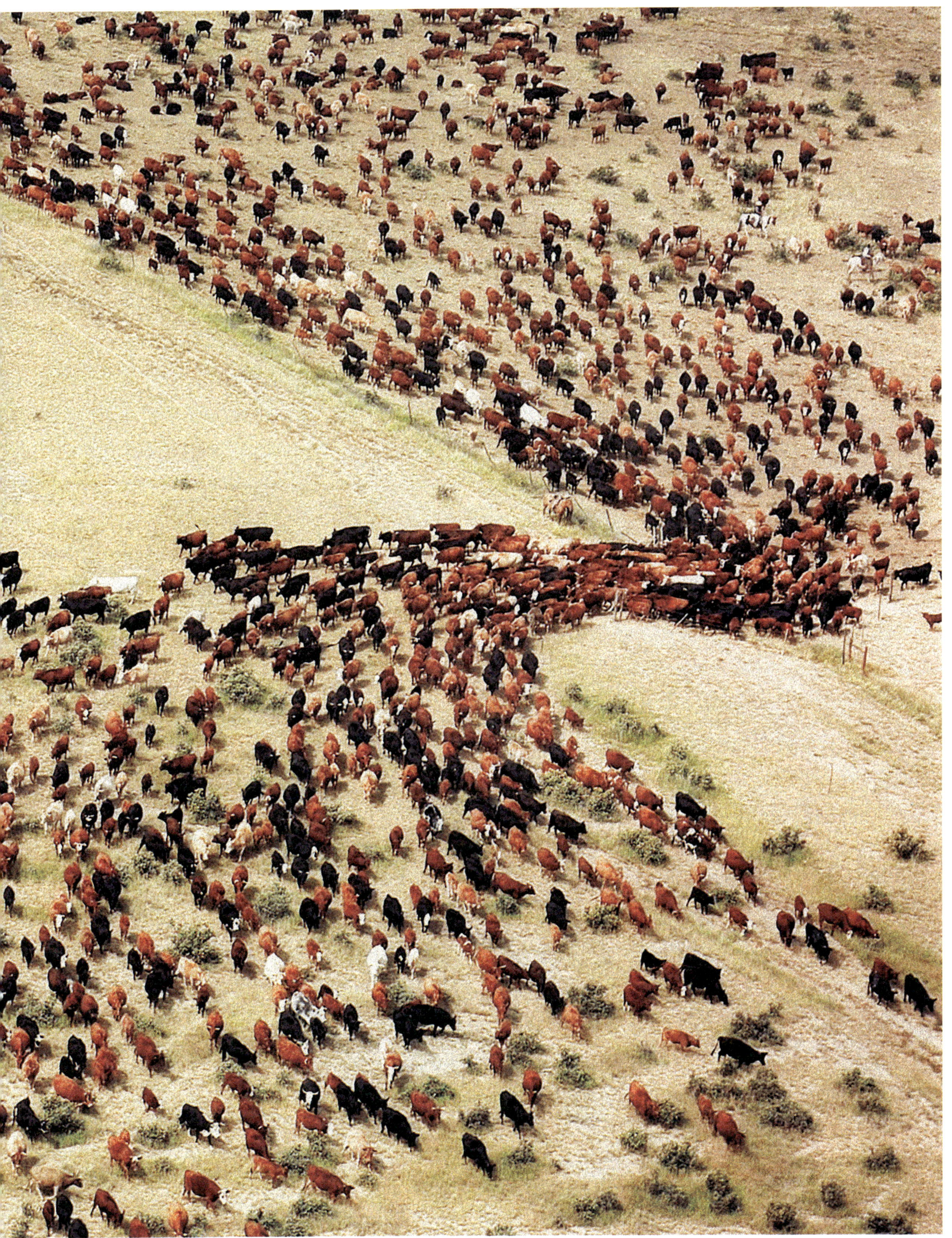

BEEFMASTER CATTLE

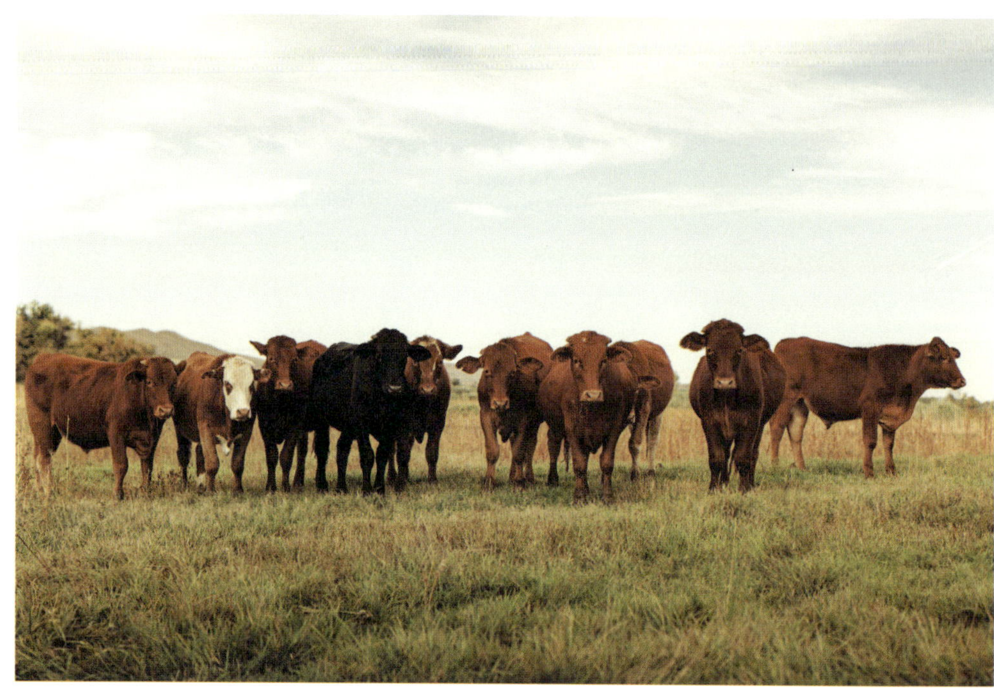

In 1989, Duke III purchased his first 15 Beefmaster cows from the Lasater Ranch, home of the original Beefmaster breed created in 1931 by Tom Lasater. The Lasater philosophy emphasizes production and functional efficiency over superficial traits like hide color or horns. What should matter to the professional stockman is not what a cow looks like but that she comes up bred year after year, can calve without assistance, has mothering abilities that include producing enough milk to wean a healthy calf every year, is gentle enough to move easily from pasture to pasture, and can thrive in the climate in which she lives. In fact, the Beefmaster remains the only registered breed of cattle in the world that has no standard hide color, as the color of the hide has no relevance to the animal's performance or to the meat on the carcass that arrives at the butcher's table.

From the outset, the Lasater family's Beefmaster breeding program was based on traits that confer fitness and adaptability in the natural environment. For example, rather than shooting or poisoning predators, the Lasaters cull any cow that does not protect her calf. Rather than providing insecticides, the Lasaters cull any cow that is unduly bothered by insects or pests. The selection and culling process reduces complicated modern breeding philosophies and techniques to a simple emphasis on mimicking natural selection; it is meant to reflect the same choices that nature might make, in the hope that each successive generation will outperform the previous one and the next generation of cattle might be a little better, so that the land itself might be a little better for the next cycle of seasons and the next generation of humans. The Beefmaster emerged from a humble awareness of the unknowable wisdom of the natural world, and that same awareness guides Ranchlands' management philosophy on the landscapes we care for. The Beefmaster cow is the product of an understanding that nature is smarter than we are, so we ought to help her work on the cattle to adapt them more effectively to their natural environment rather than imposing our own criteria. The best breeding program, then, the best artificial selection, is one that mimics natural selection.

The resulting cattle are more than just animals. The Beefmaster is a management philosophy embodied. It's a flesh-and-blood way of living with the land that represents how humans and their animals might do better to live in harmony with the indigenous patterns of the natural world rather than to try to fight them.

All of Ranchlands' female cows—be they purebred or commercial Beefmaster cross cattle—are bred exclusively by our Beefmaster bulls. These bulls, all born and raised here on the ranch, are direct descendants of the original Foundation Beefmaster herd created on the Lasater Ranch in Matheson, Colorado. In addition to allowing us to maintain a high standard of production in our cattle operations, the Beefmaster bull is part of a tradition that allows us to carry on the genetic and philosophical legacy of the Lasater herd.

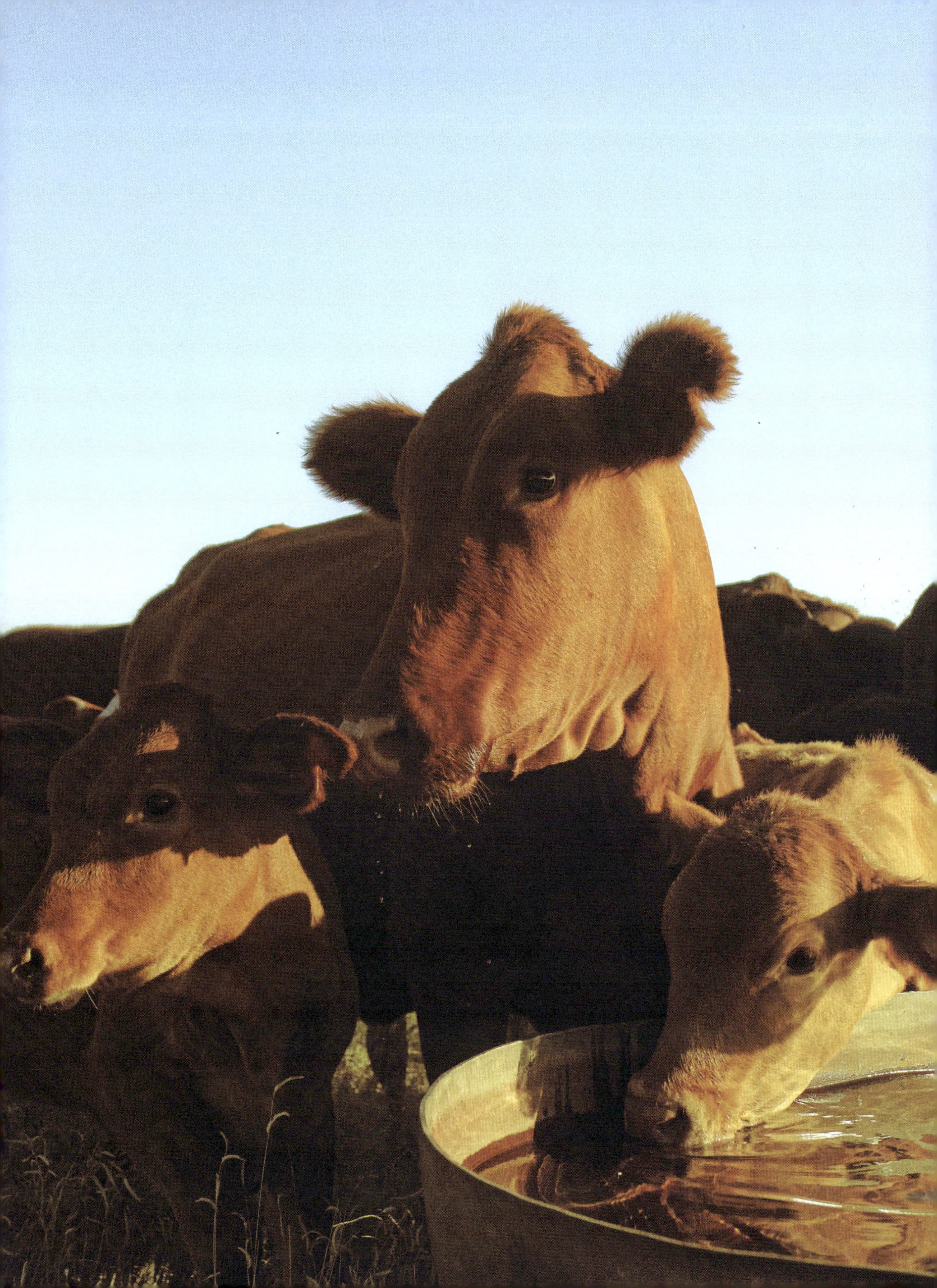

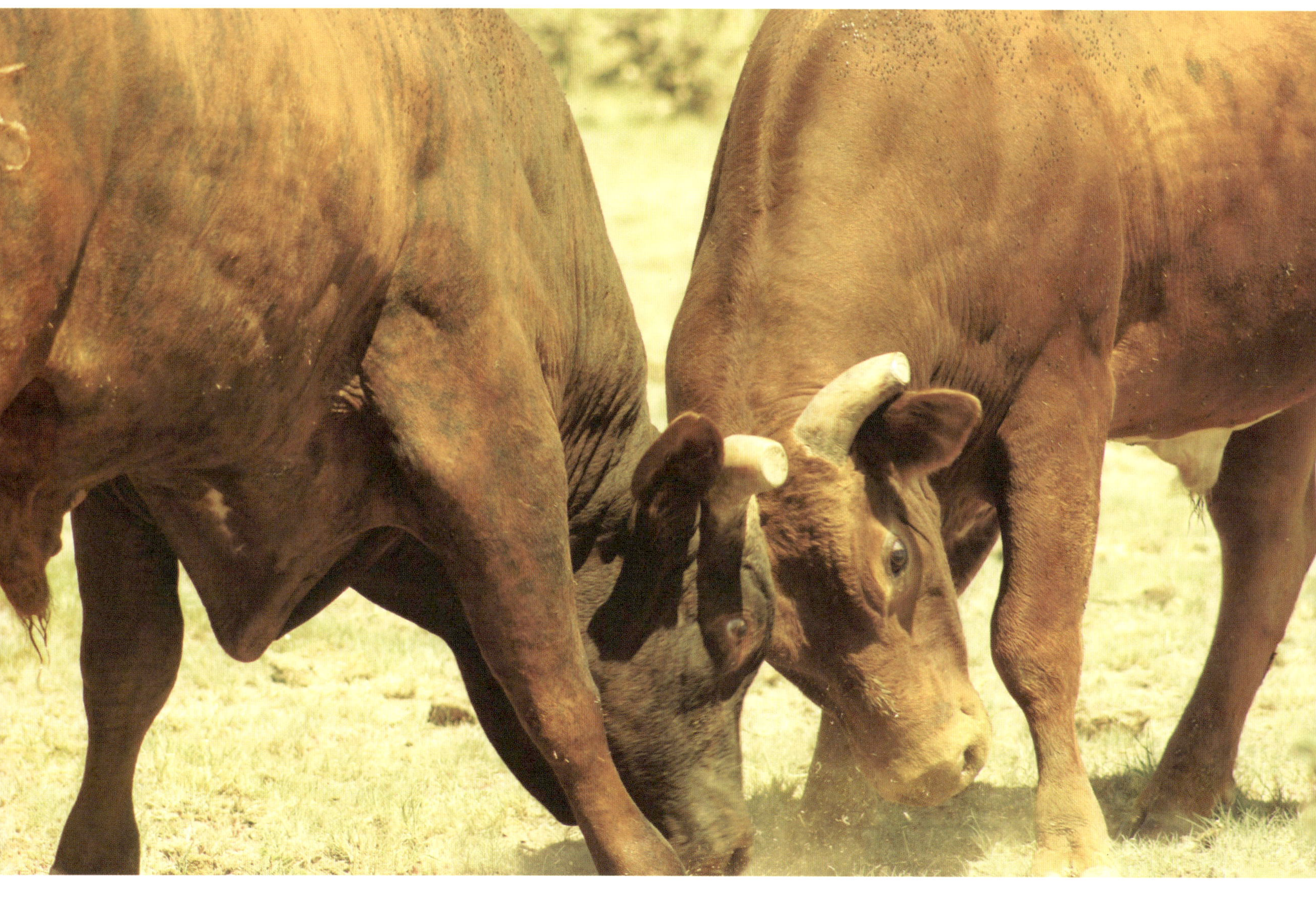

Smaller fights among yearling bulls can be broken up quite easily from horseback, but when the three-year-olds (most of which weigh well over a thousand pounds) begin to tussle, it's usually best to step aside and wait for them to work it out on their own.

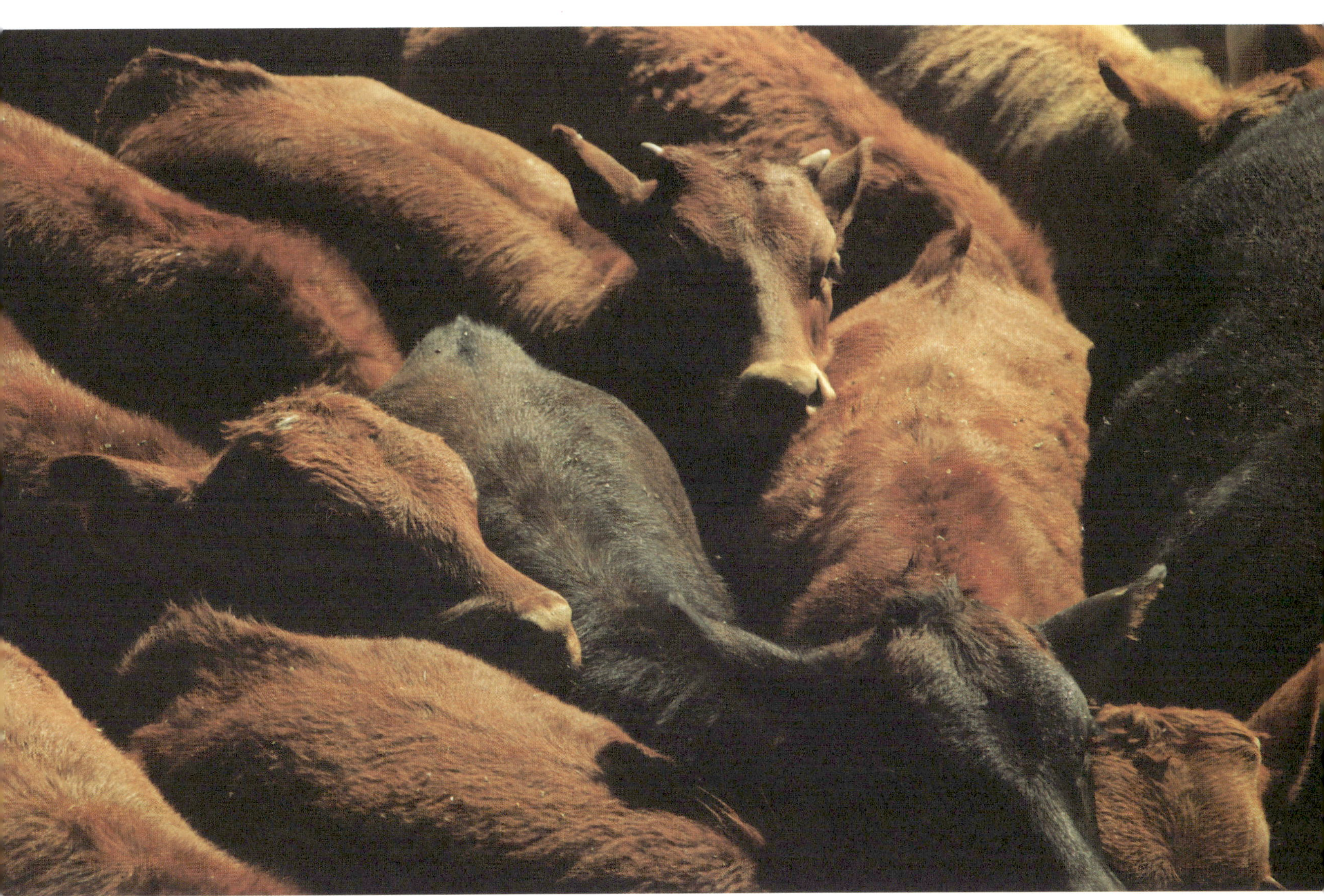

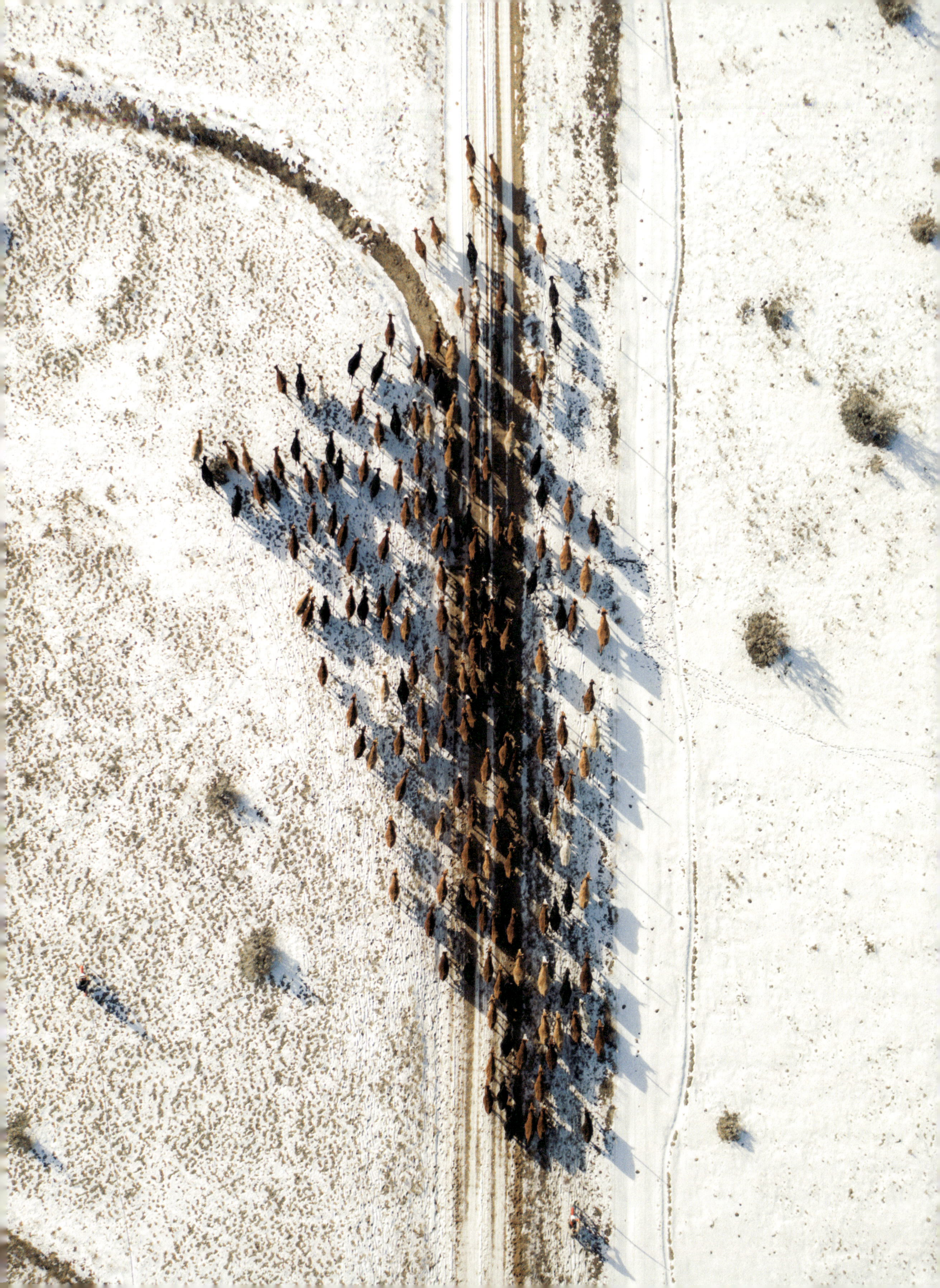

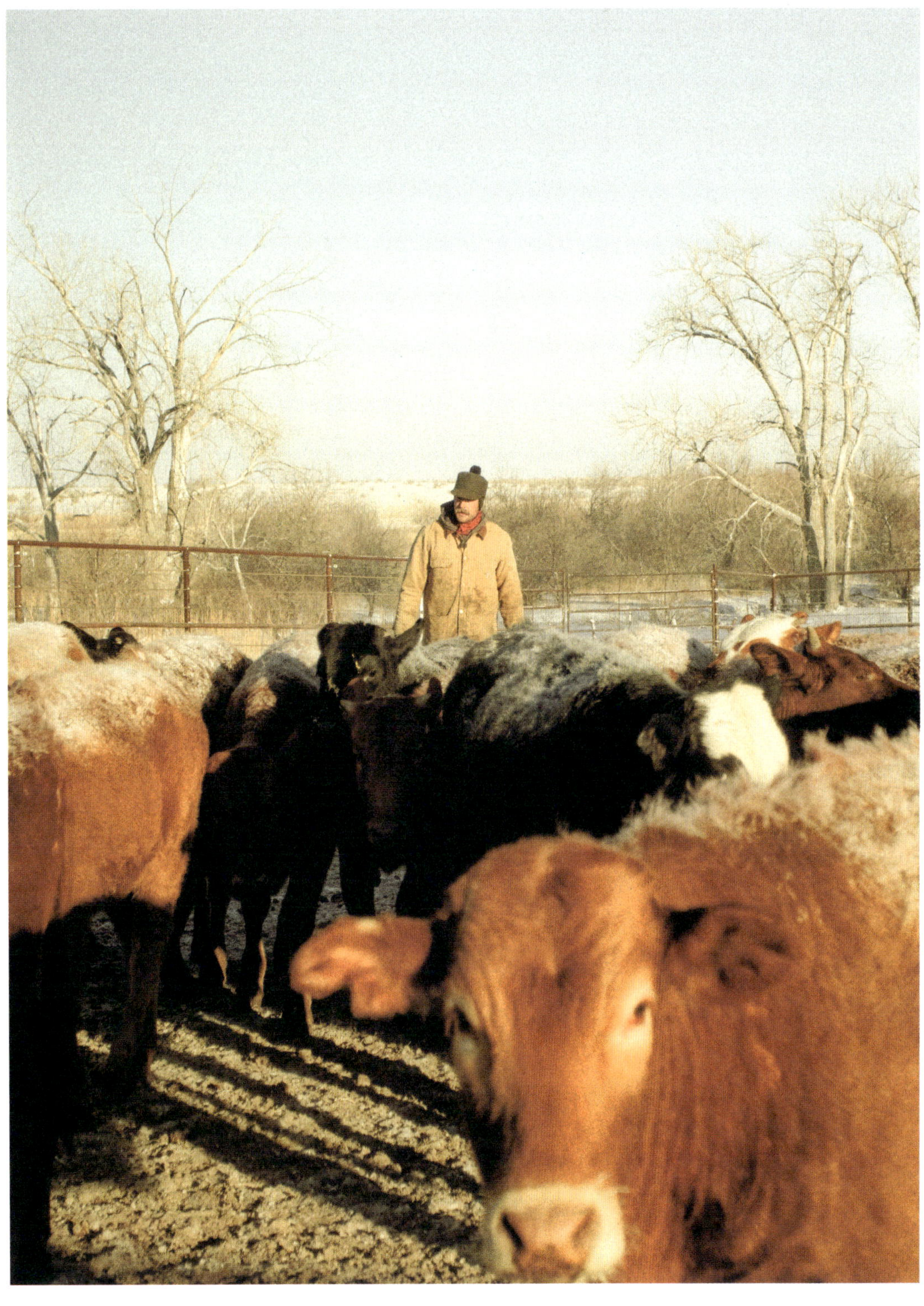

DYLAN TAYLOR

"We work hard today so that we can hold onto the hope that the future will be one of growth and continued evolution, alongside traditional ranching. The sustainability of anything in this world hinges on its ability to adapt and evolve. Whether it be land management or livestock management, there's always something to learn and new paths to be discovered."

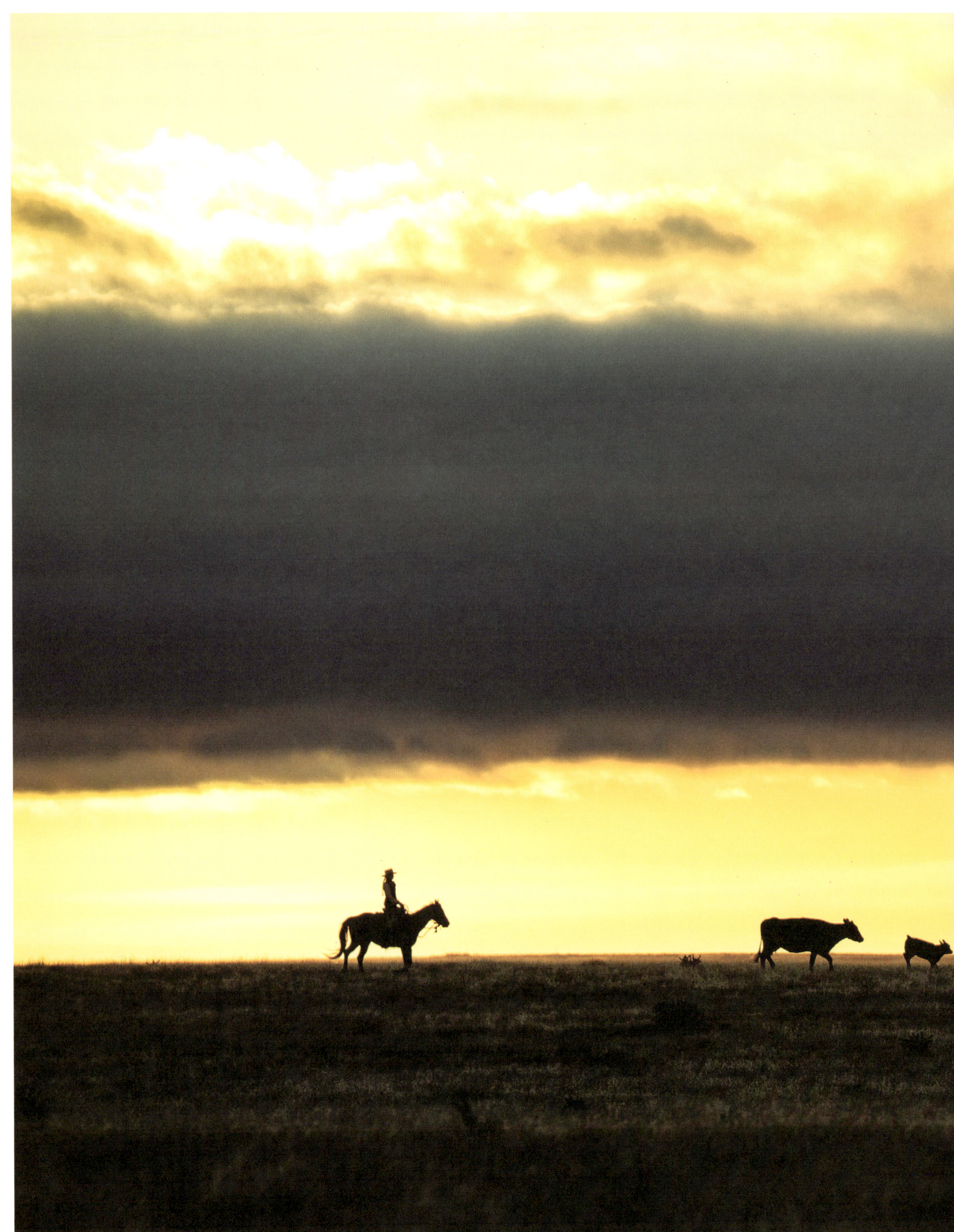

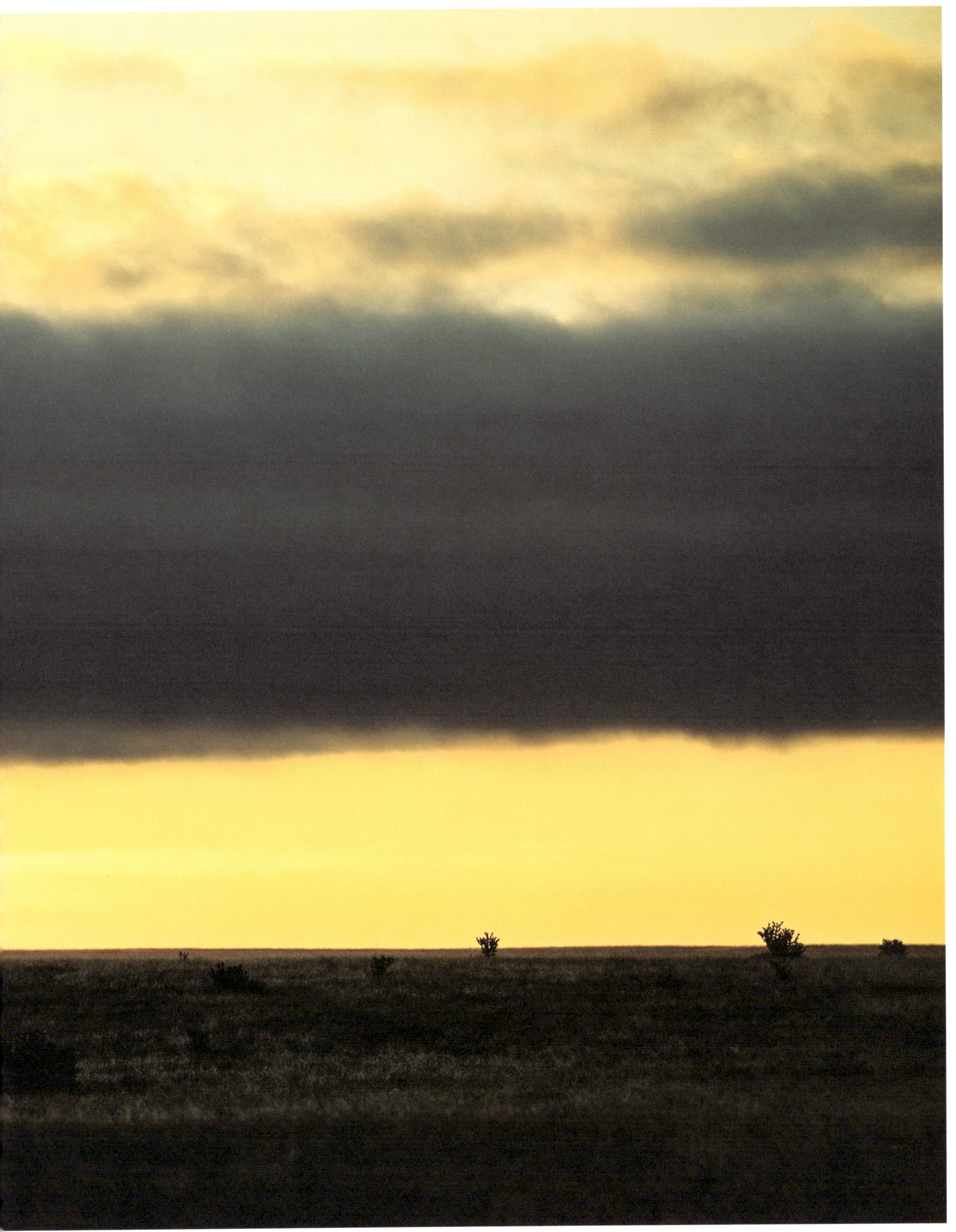

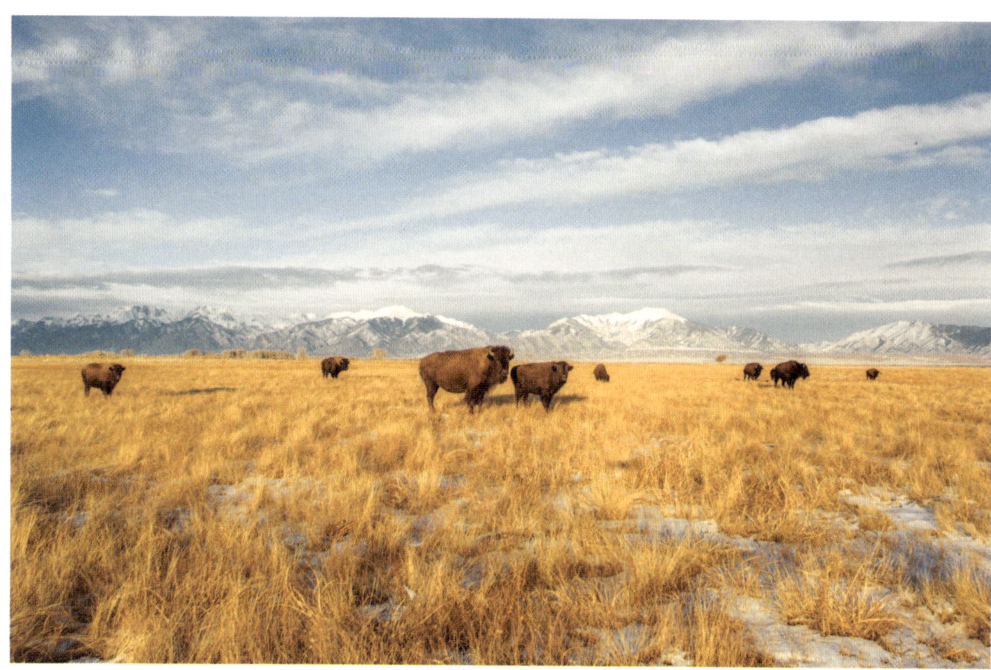

BISON WORKS

By the mid-1880s, the herds of tens of millions of bison once inhabiting North America had been reduced to a total of fewer than 2,000 animals, driven nearly to extinction by unrestricted commercial hunting and indiscriminate slaughter. Today, thanks to collaborative conservation efforts that began all but too late, several hundred thousand exist in North America, mostly in private herds that are treated as livestock. Only a small handful of American bison—roughly 20,000—live today as truly wild free-ranging animals.

Our management of The Nature Conservancy's bison herd is as hands-off as possible. The bison are left to roam freely, as they might in the wild, across 45,000 acres of unfenced rangeland. Once a year in the fall, the herd is brought into the corrals, and we cull a select number of animals, trimming the herd to fit the range.

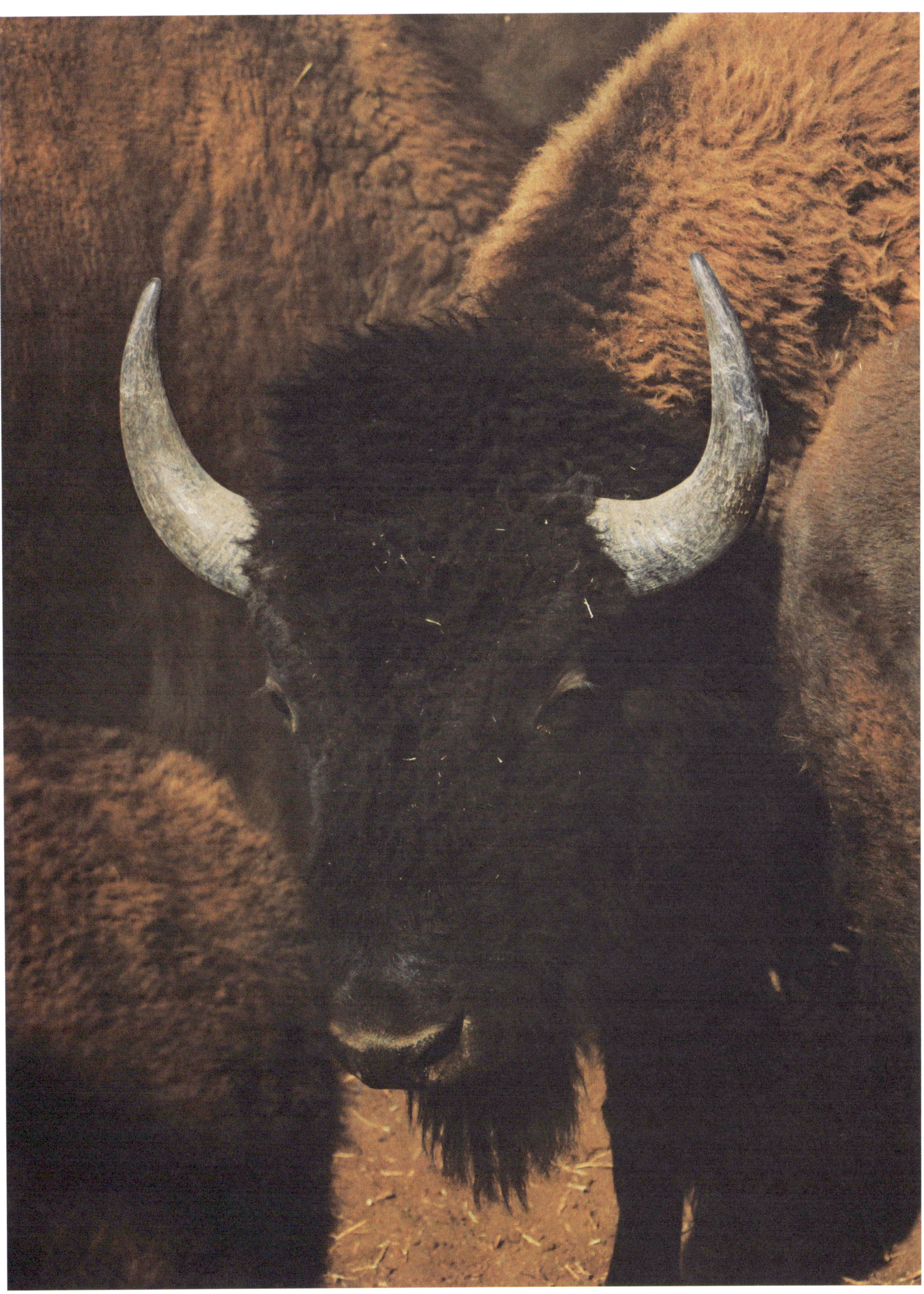

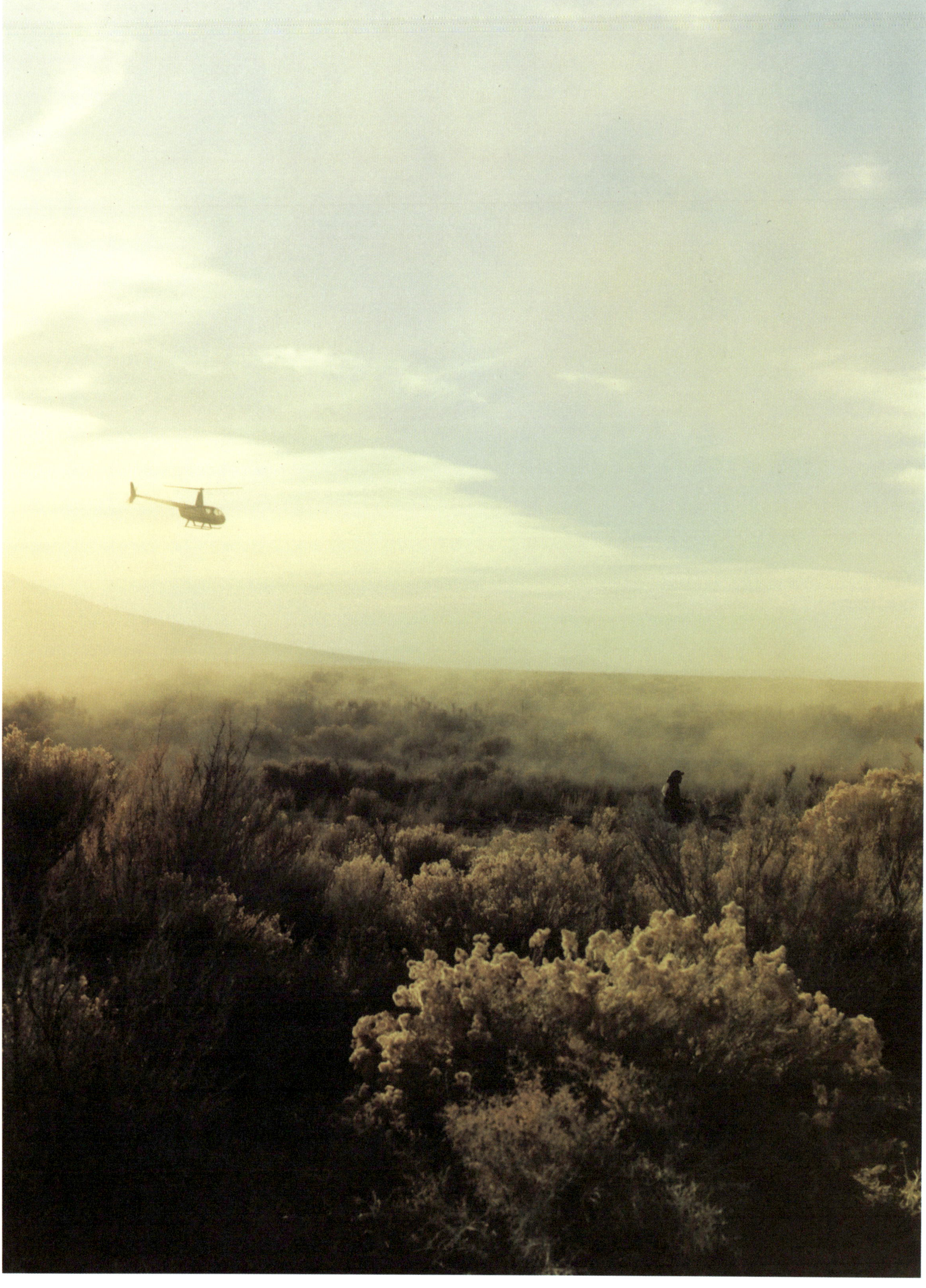

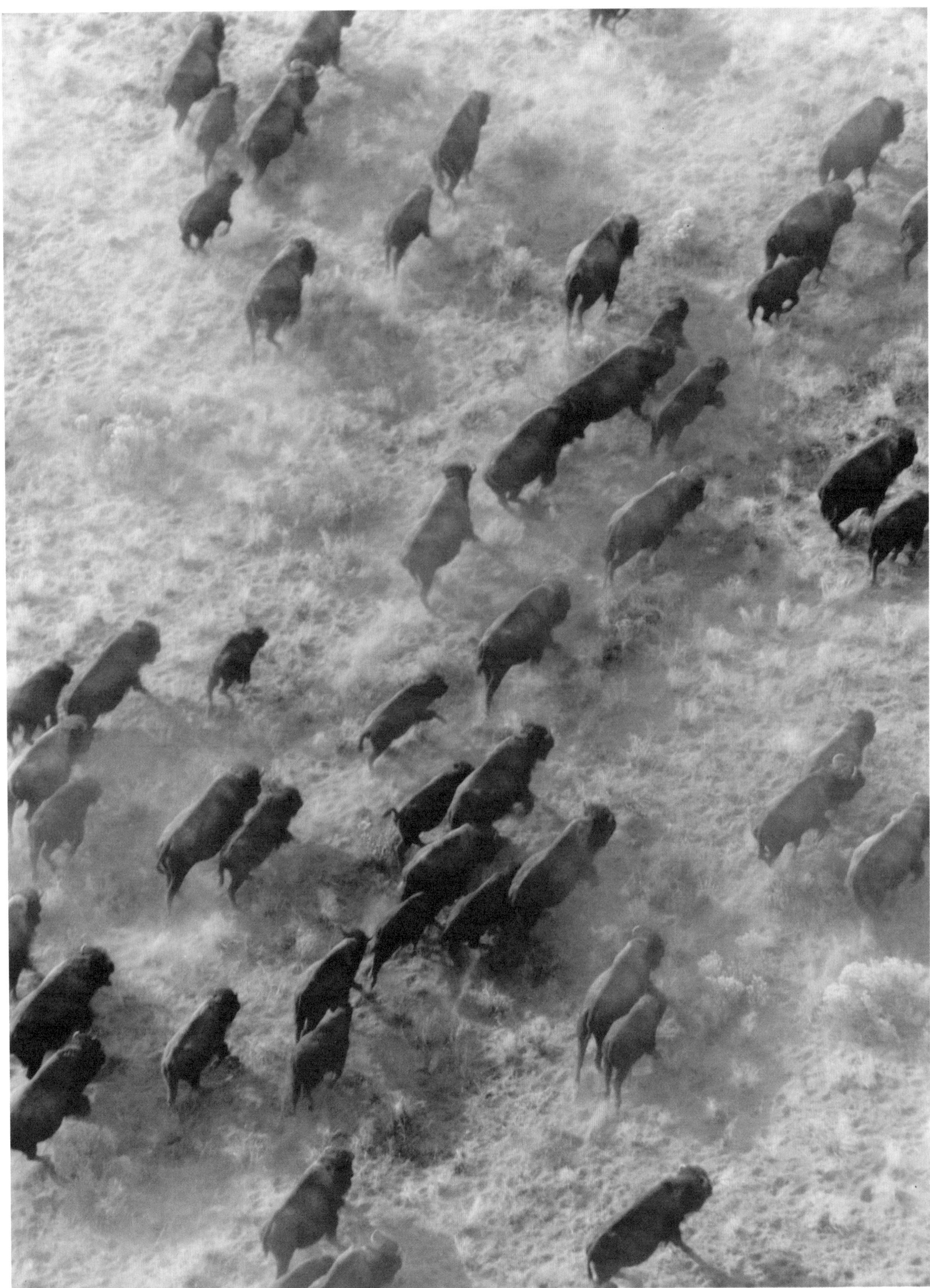

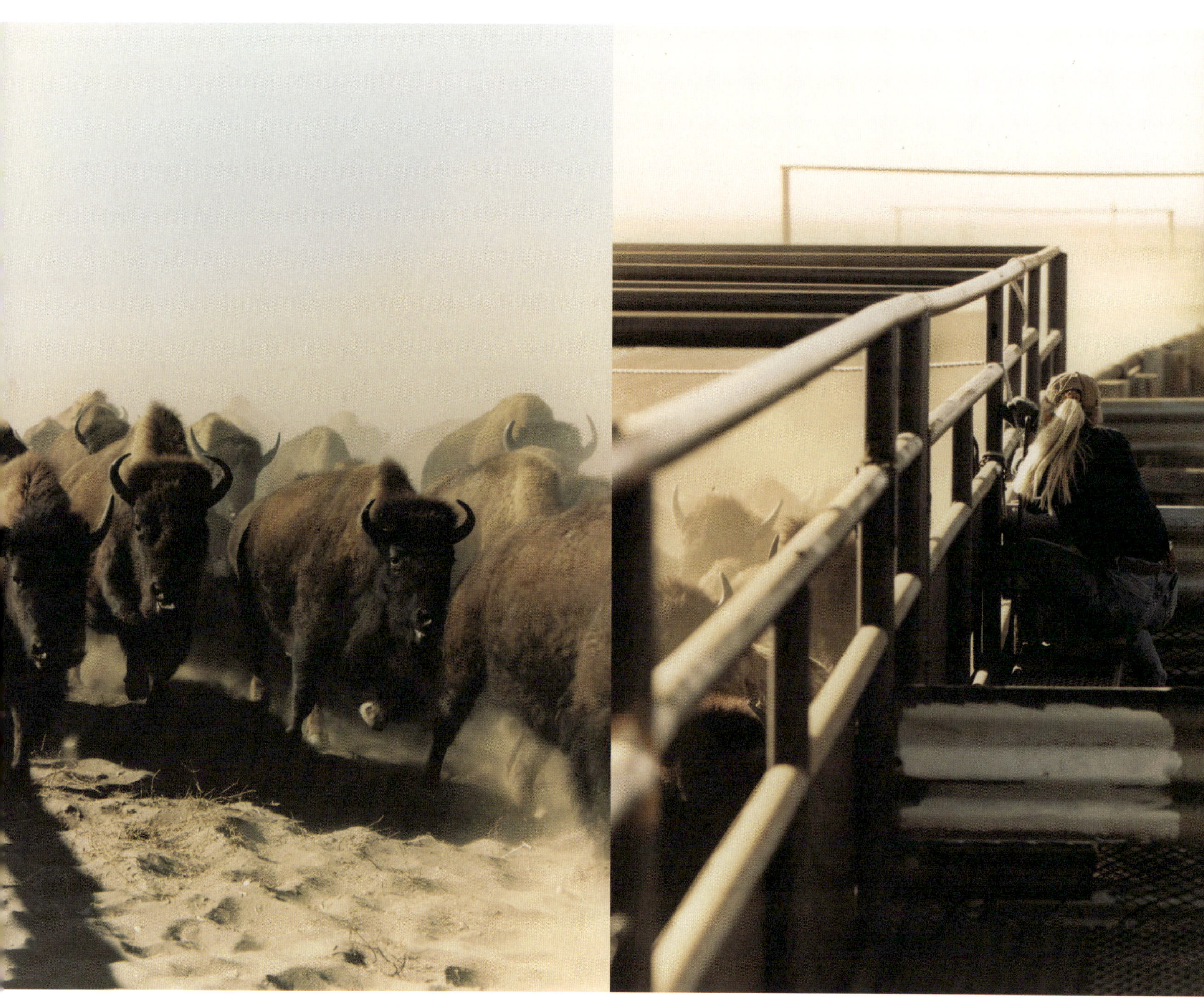

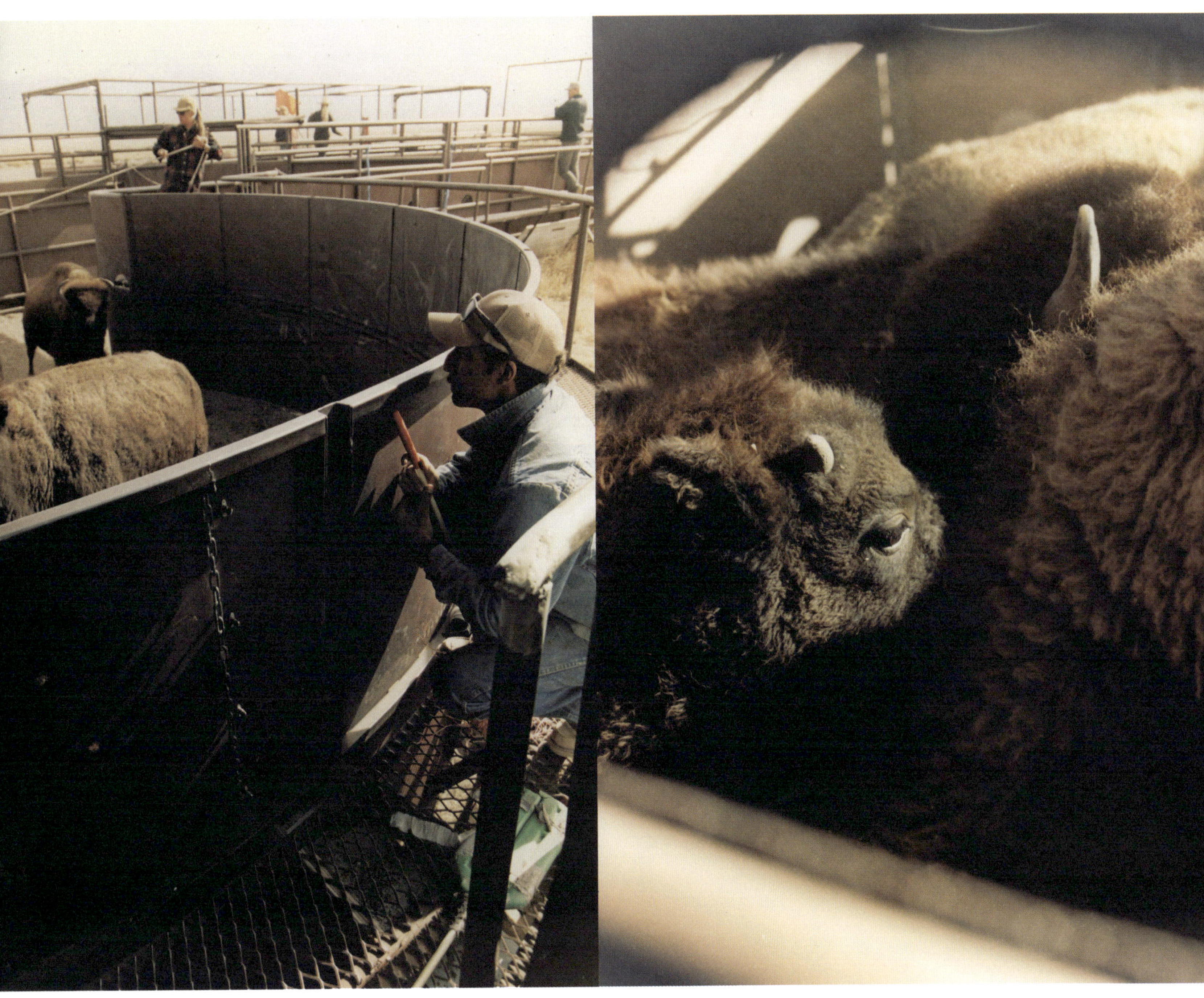

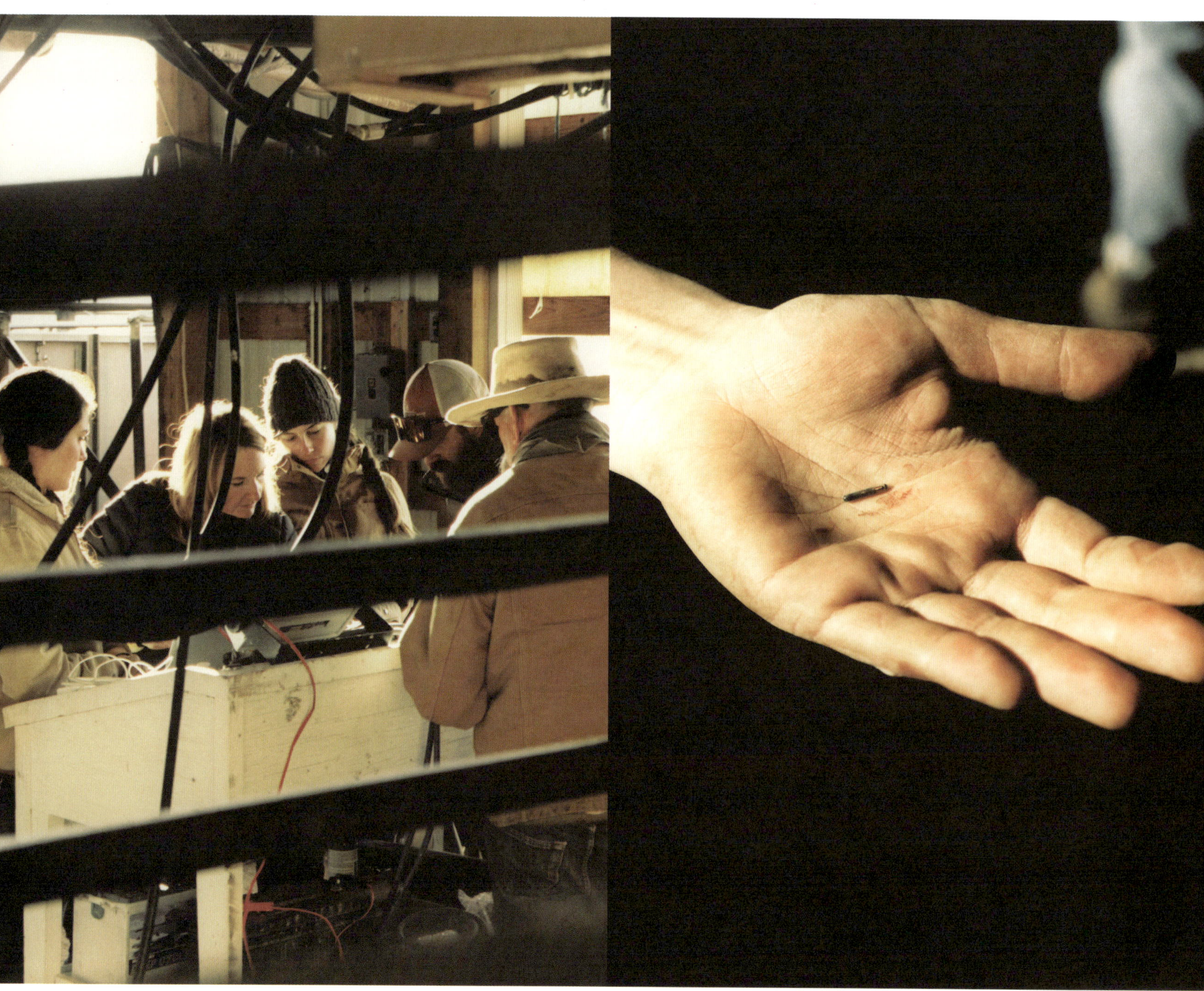

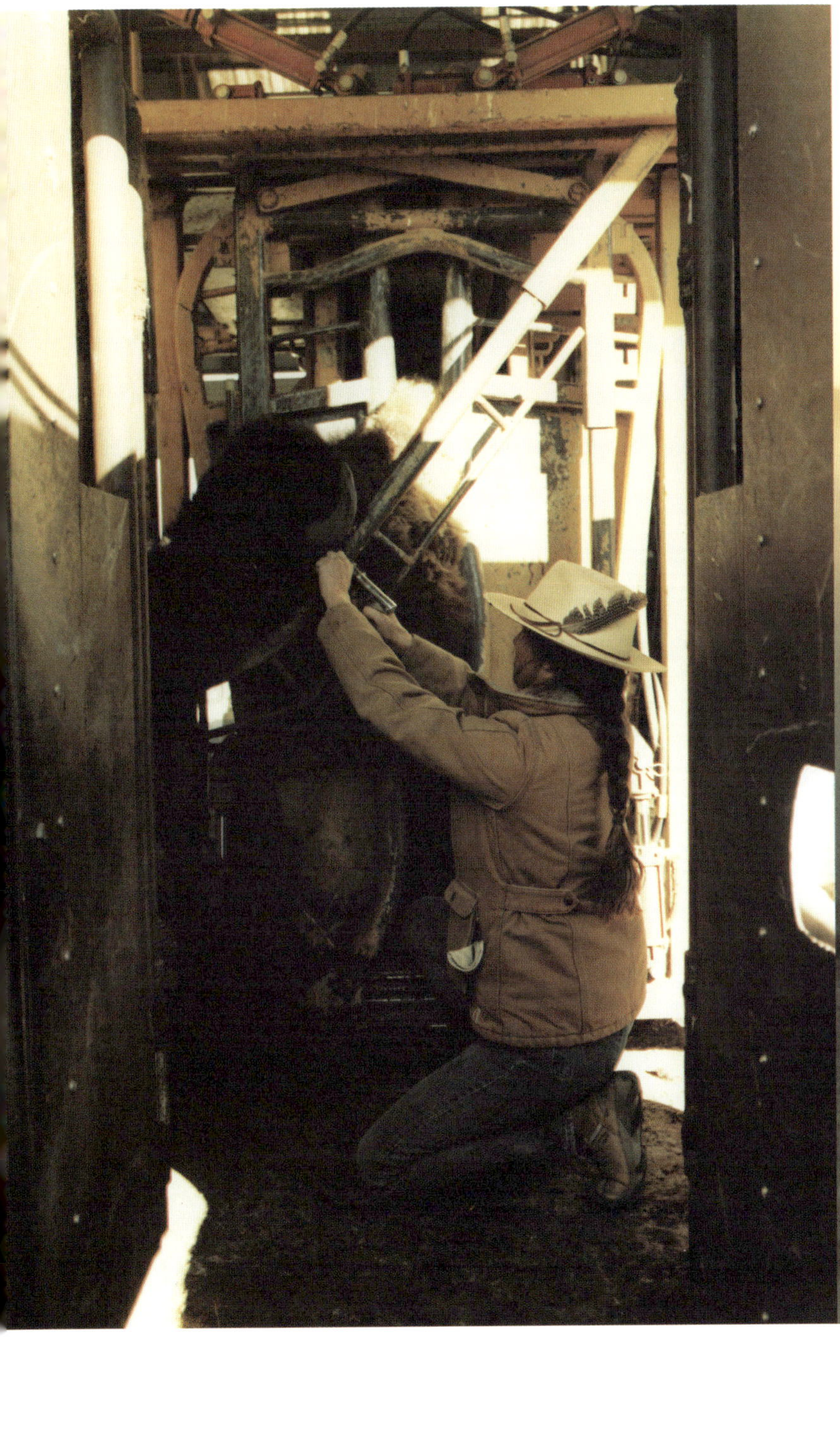

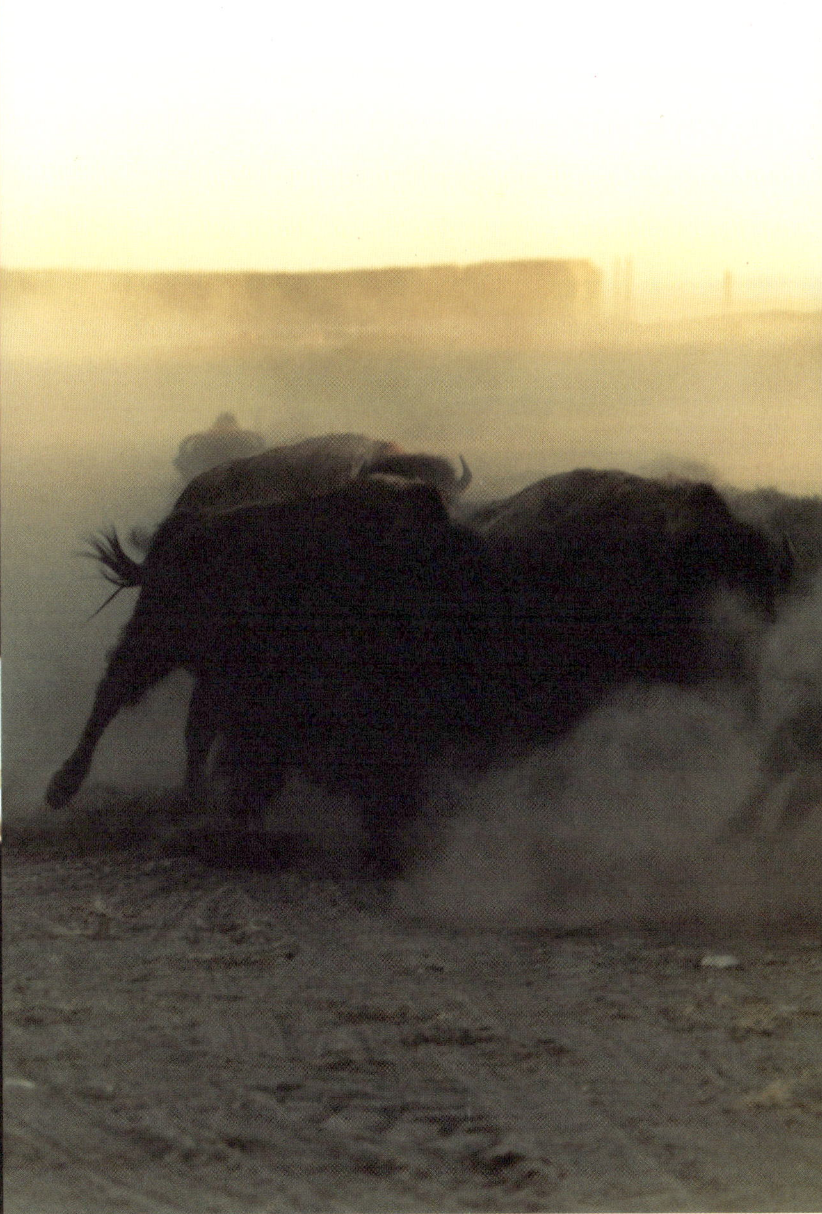

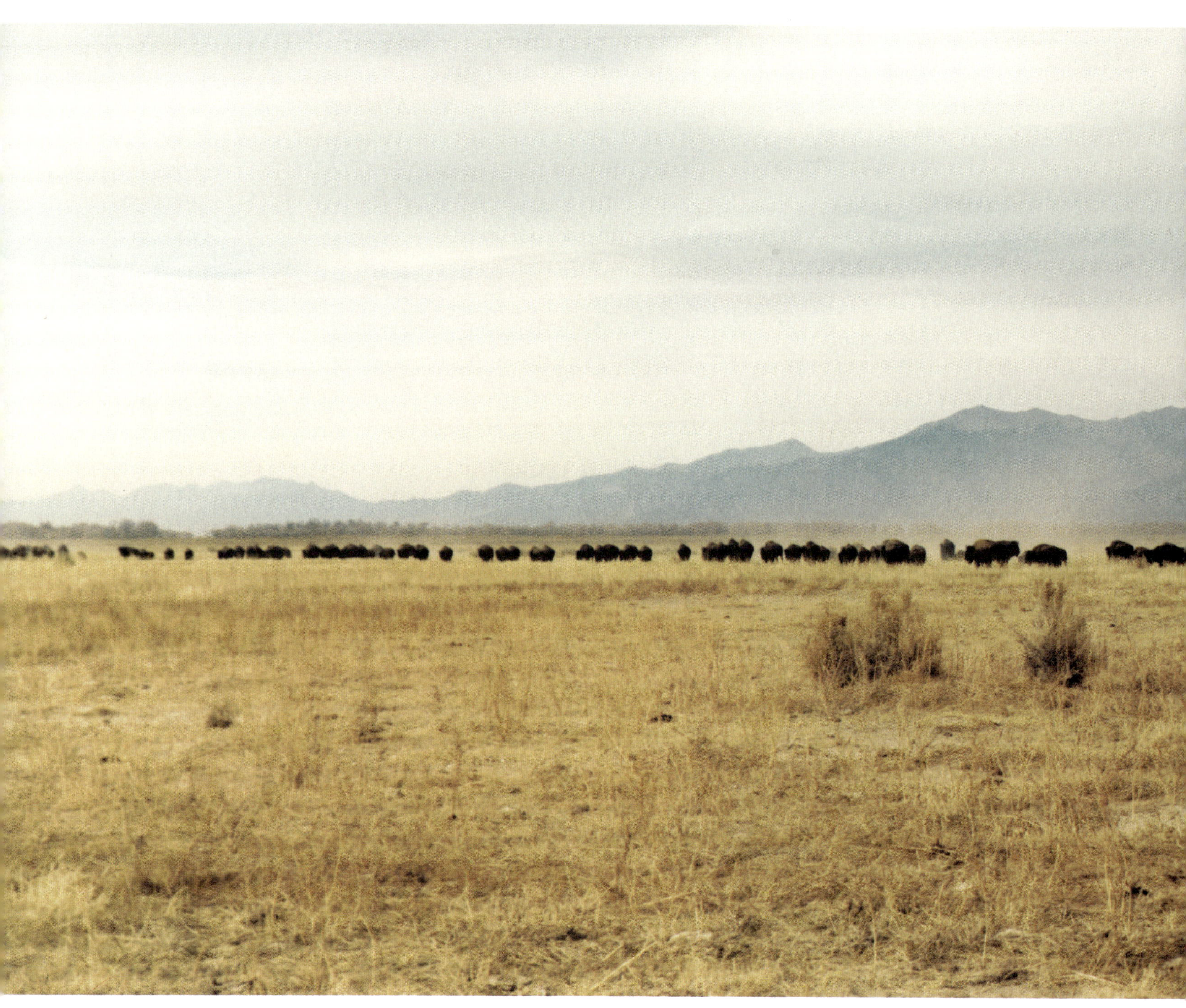

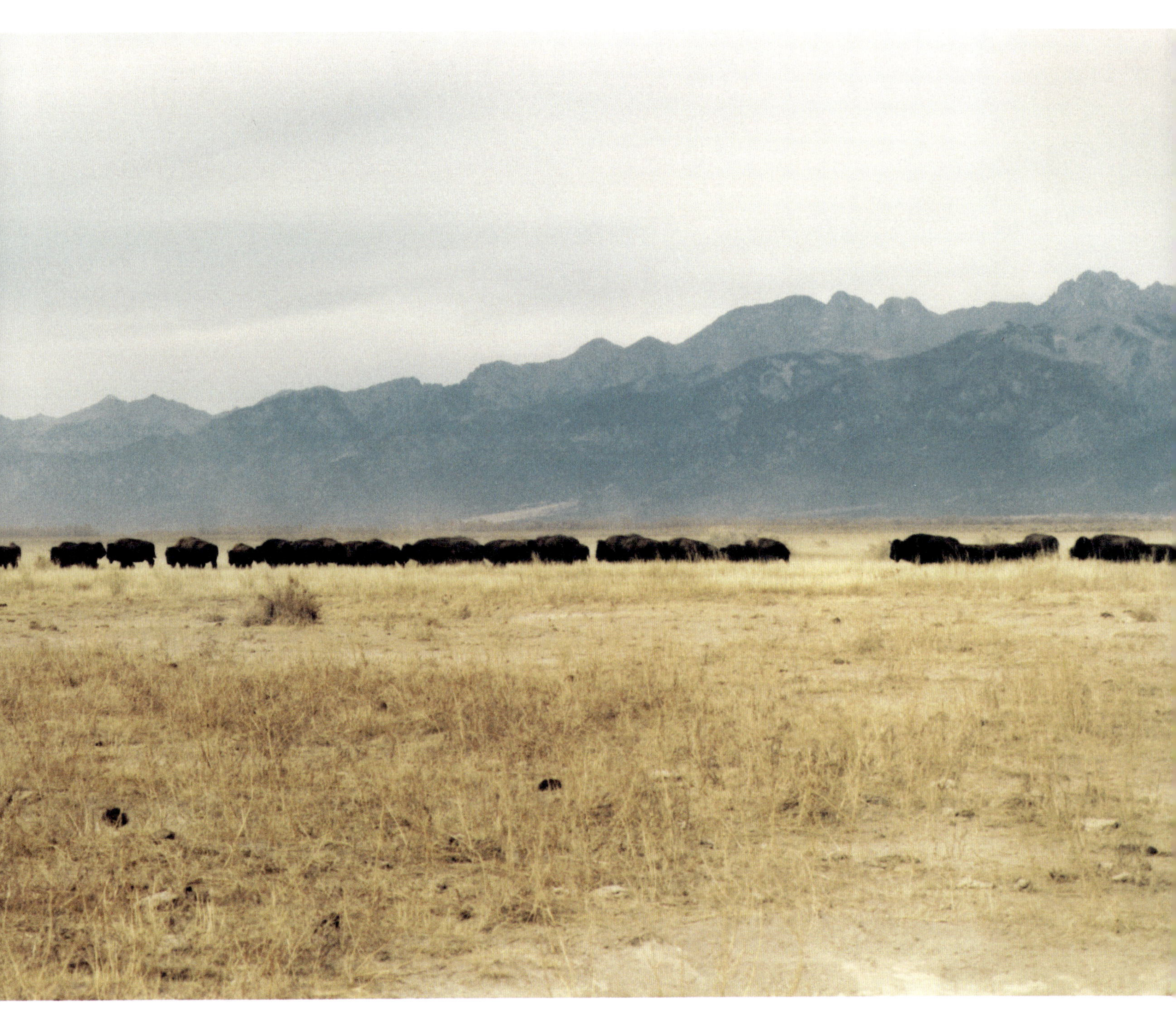

To many people, the prairie isn't particularly inviting. It looks largely the same as you drive by, humming along at highway speeds: flat, sparse, and empty. But slow down and spend some time walking or riding across the open prairie, and you'll encounter the wildlife that makes the grasslands a rich community of biodiversity.

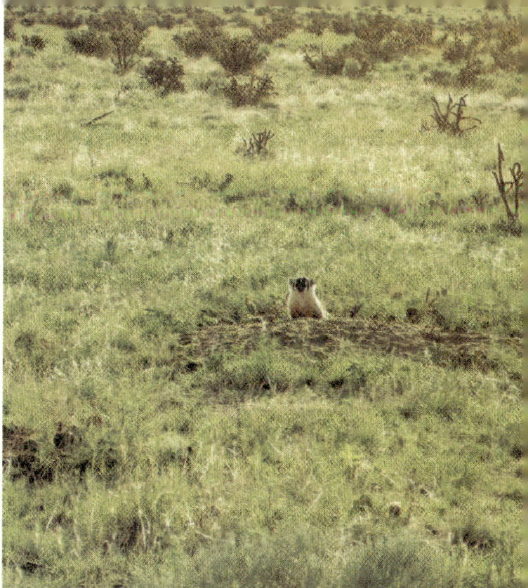

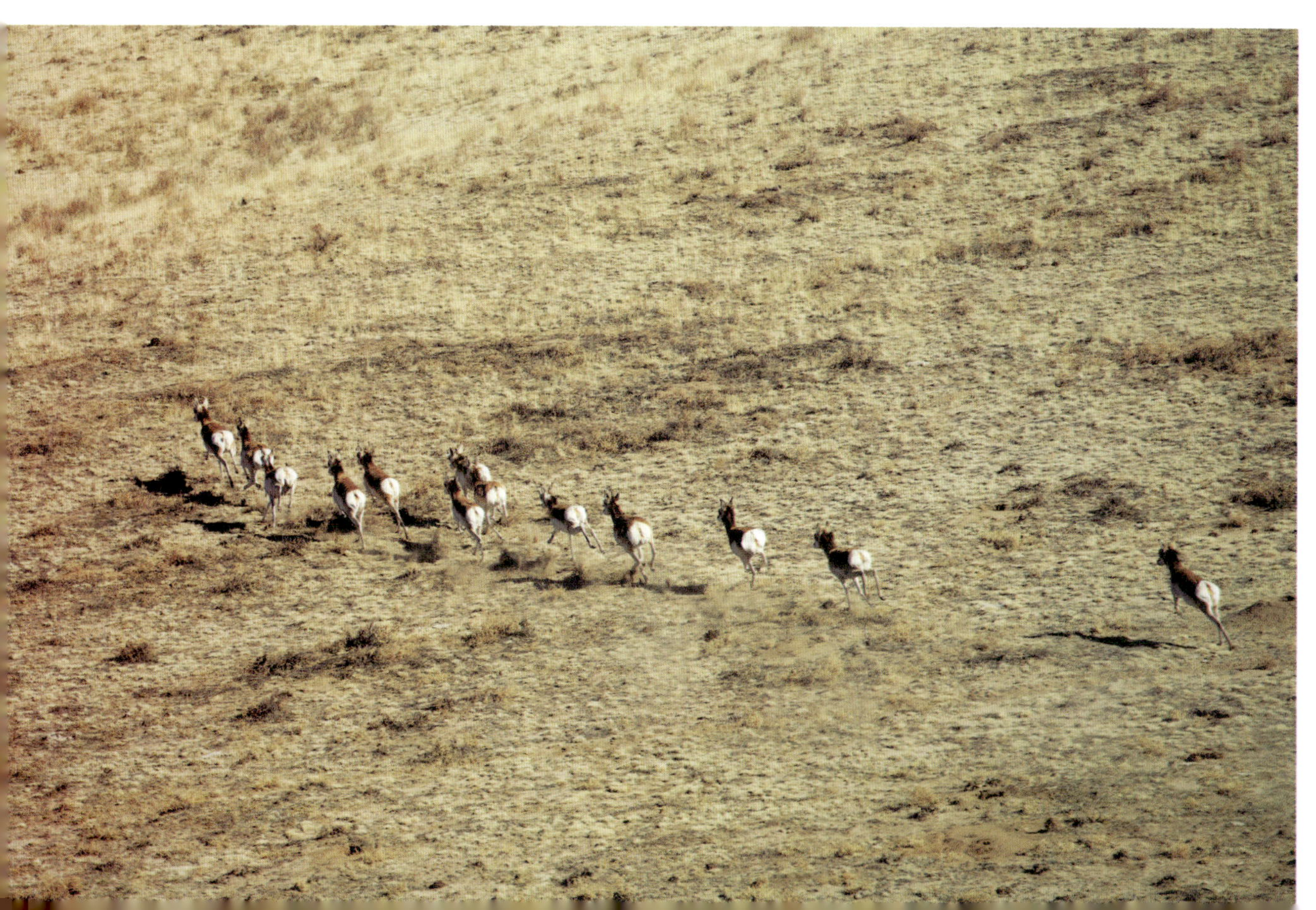

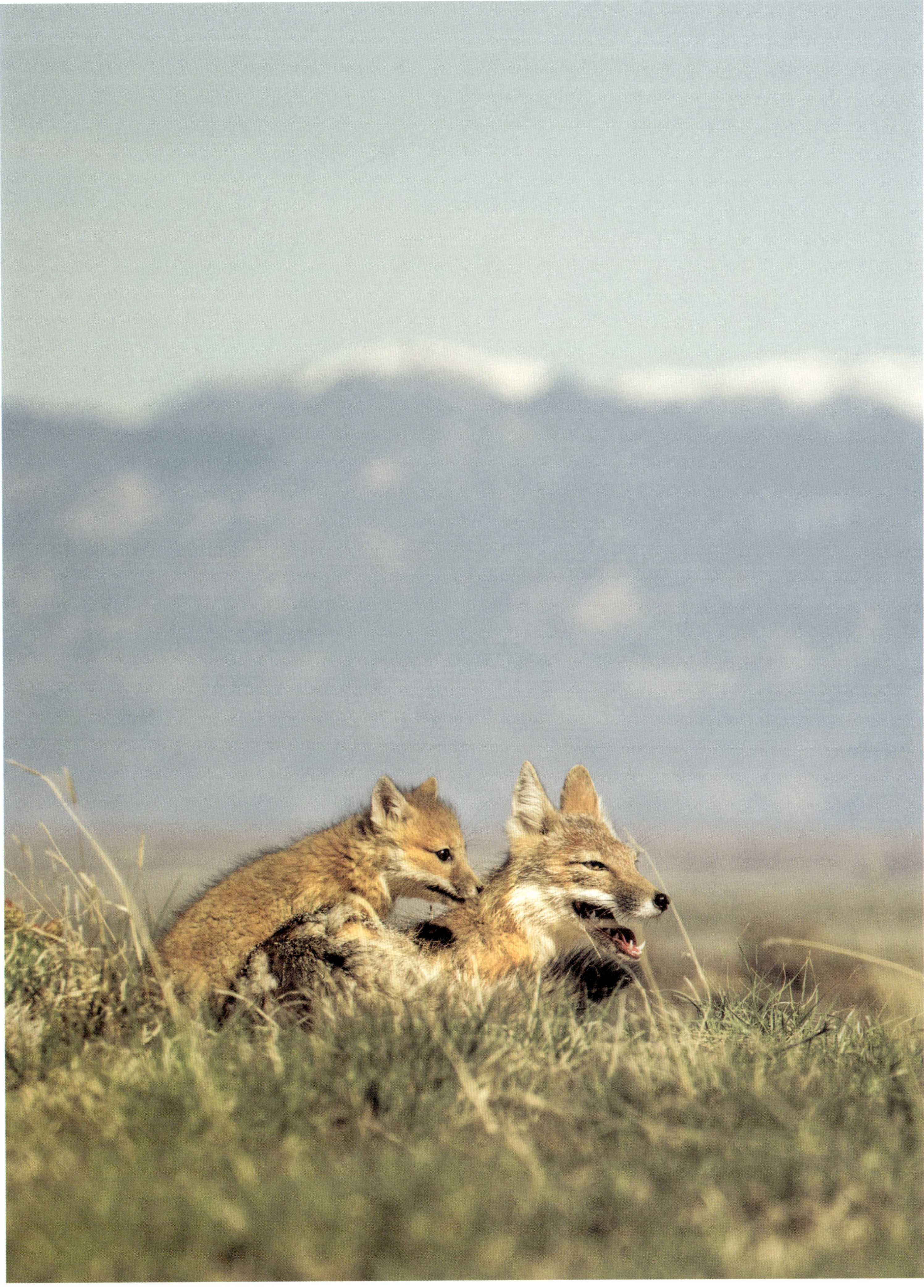

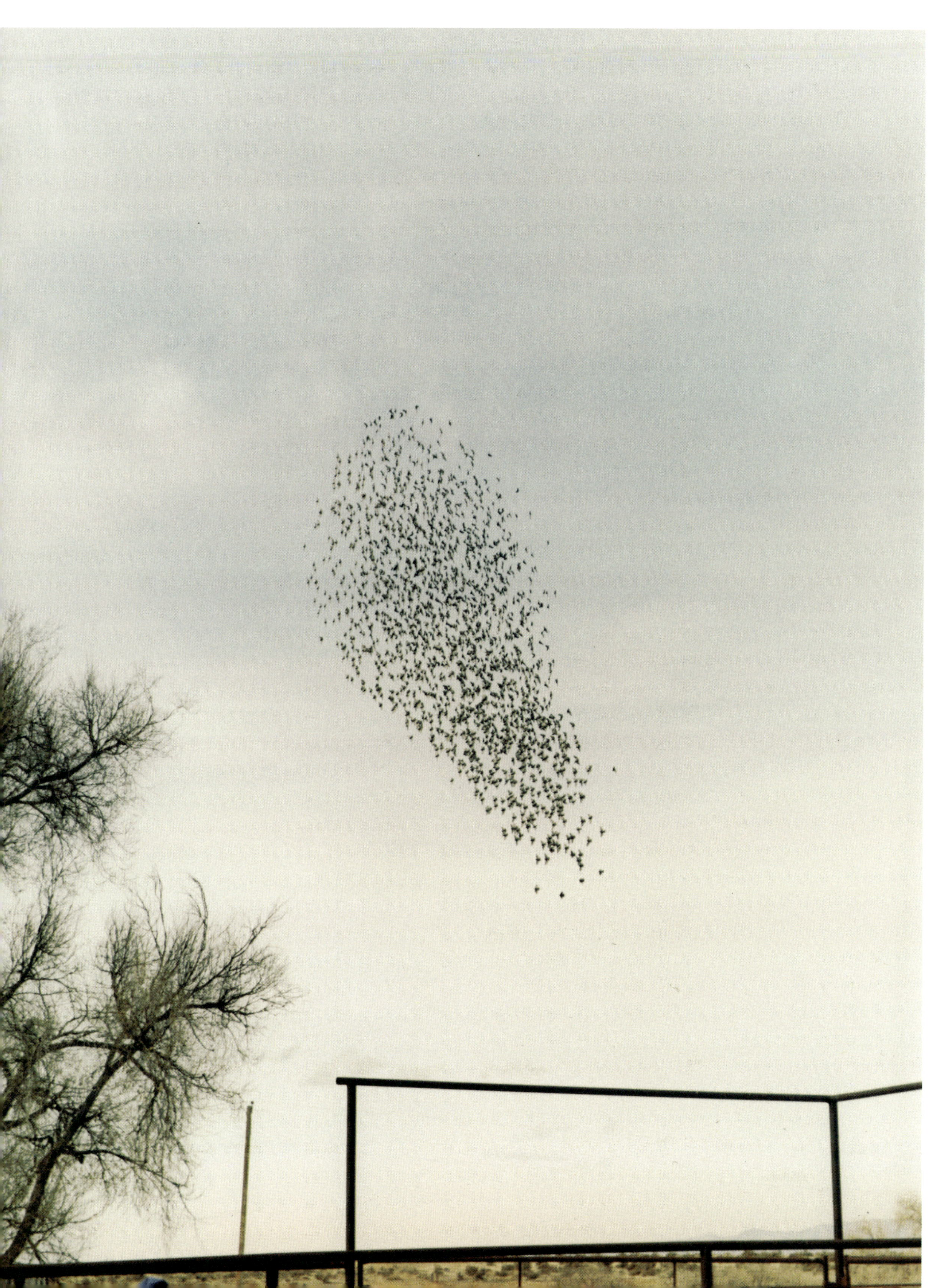

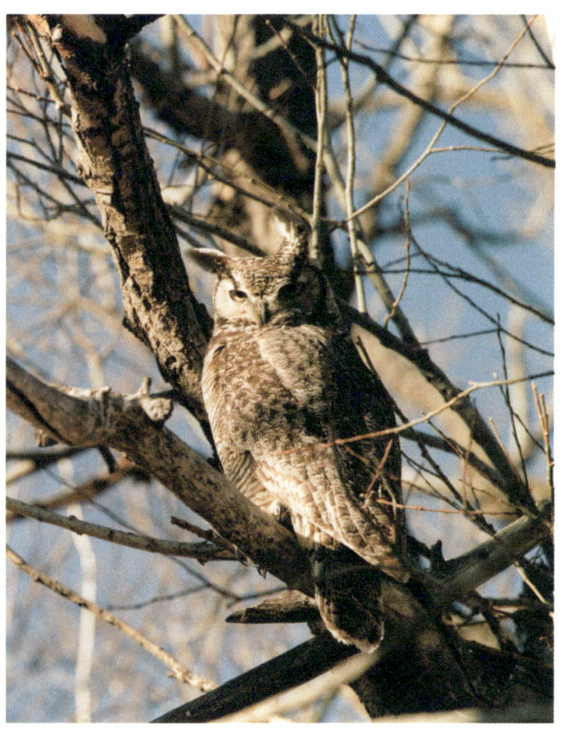

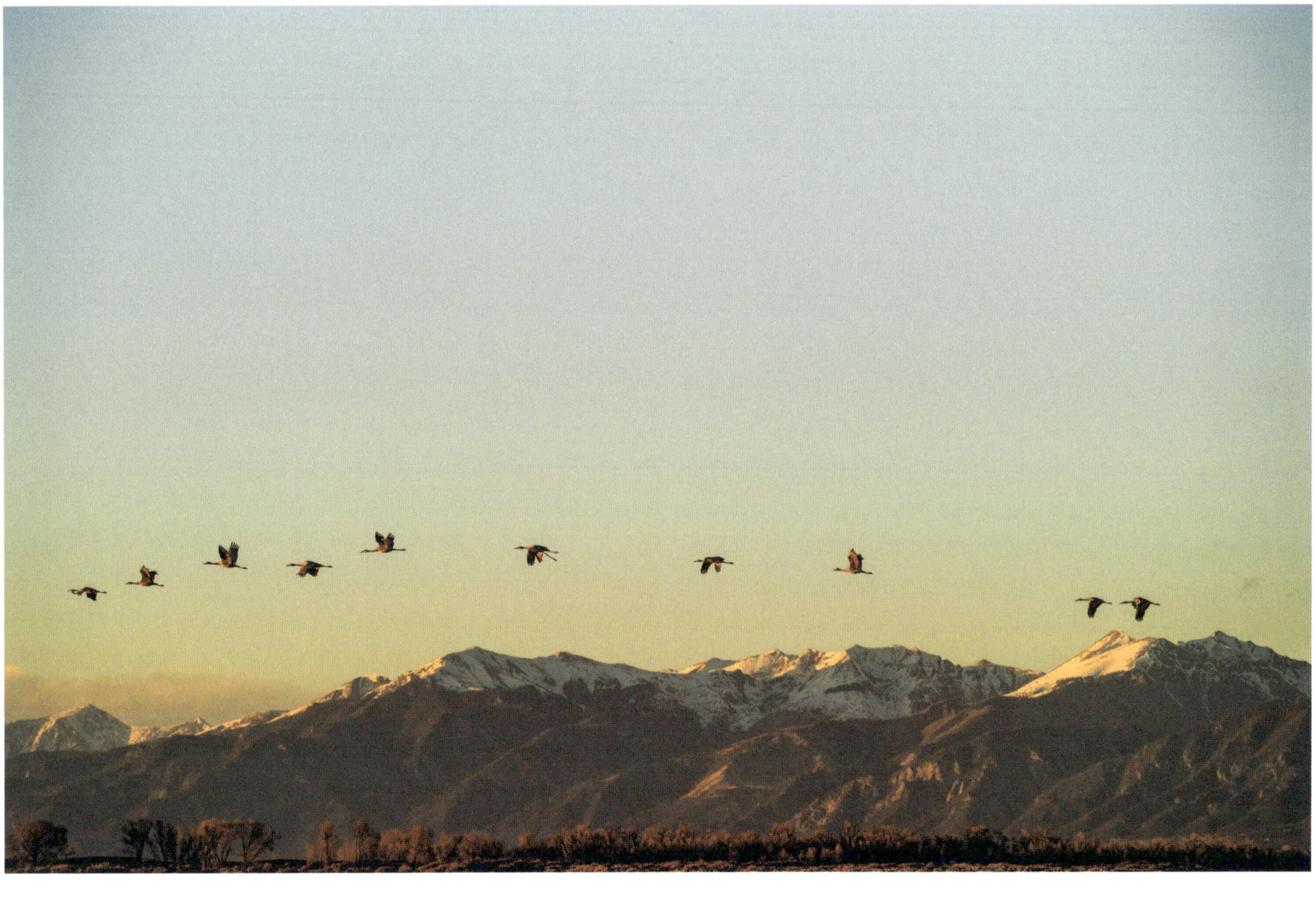

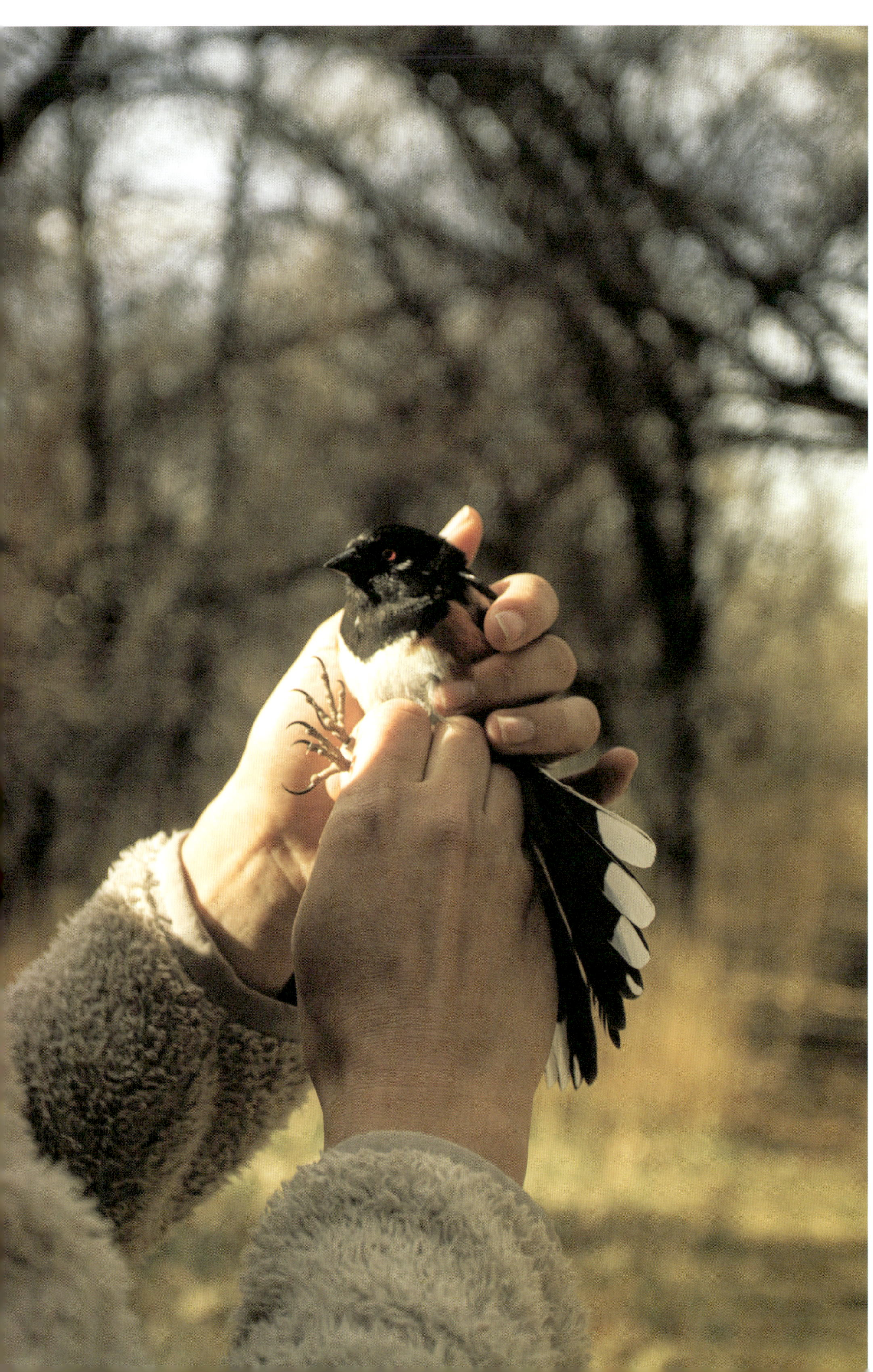

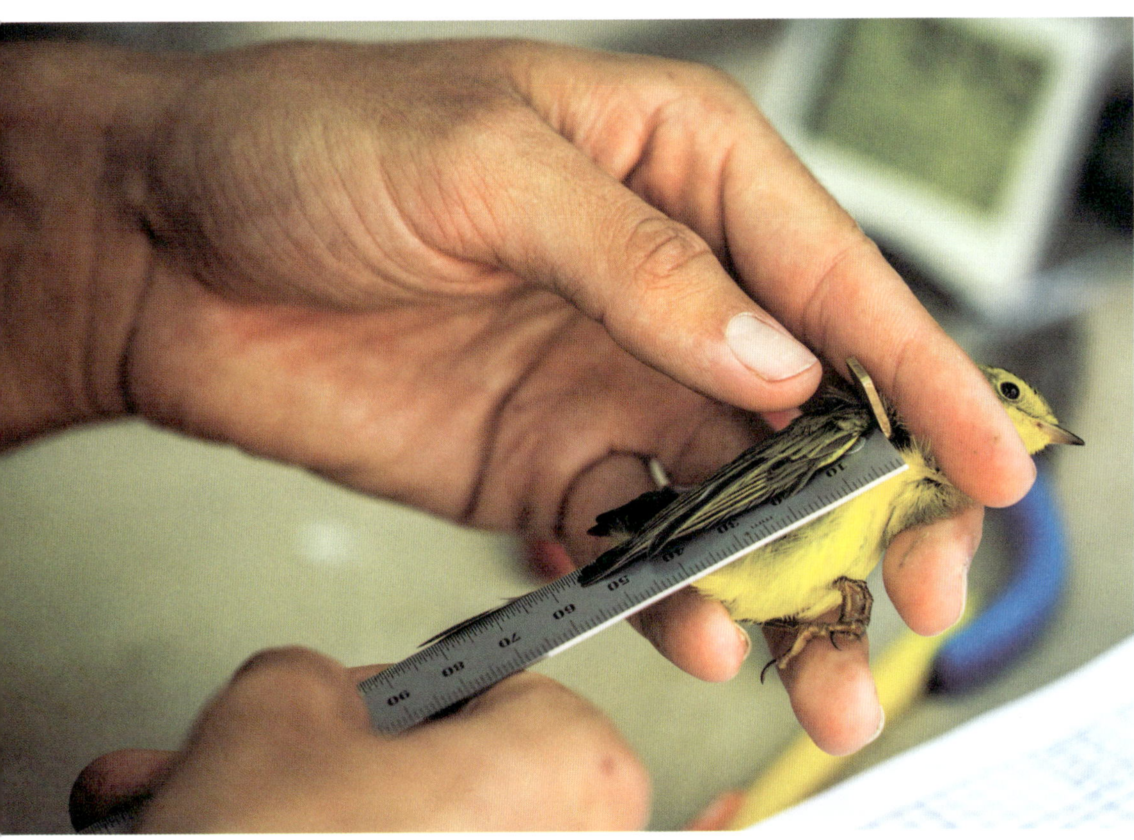

The Chico Basin Ranch's abundant springs, creeks, lakes, and other habitats provide refuge for hundreds of species of migratory birds. Each spring and fall, we maintain a bird-banding station in partnership with Bird Conservancy of the Rockies, where visiting biologists net and band hundreds of these migratory birds.

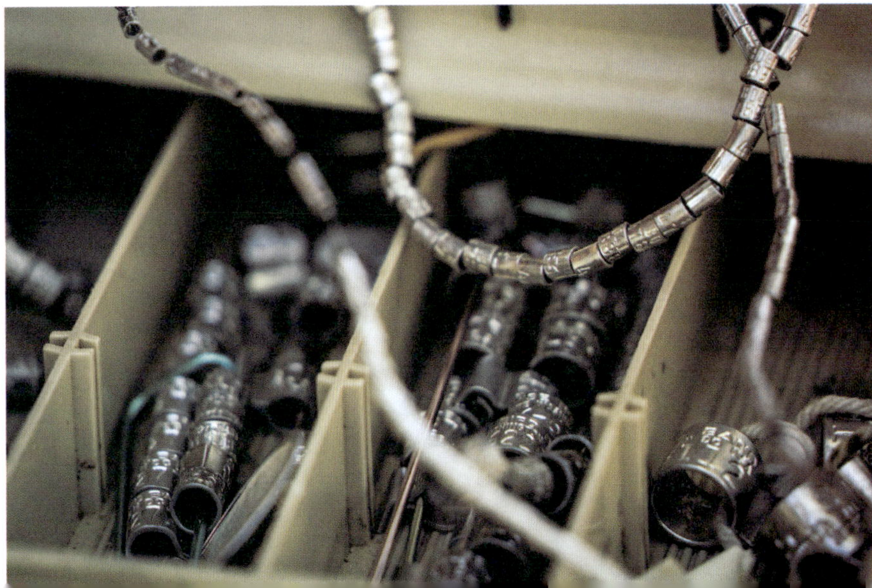

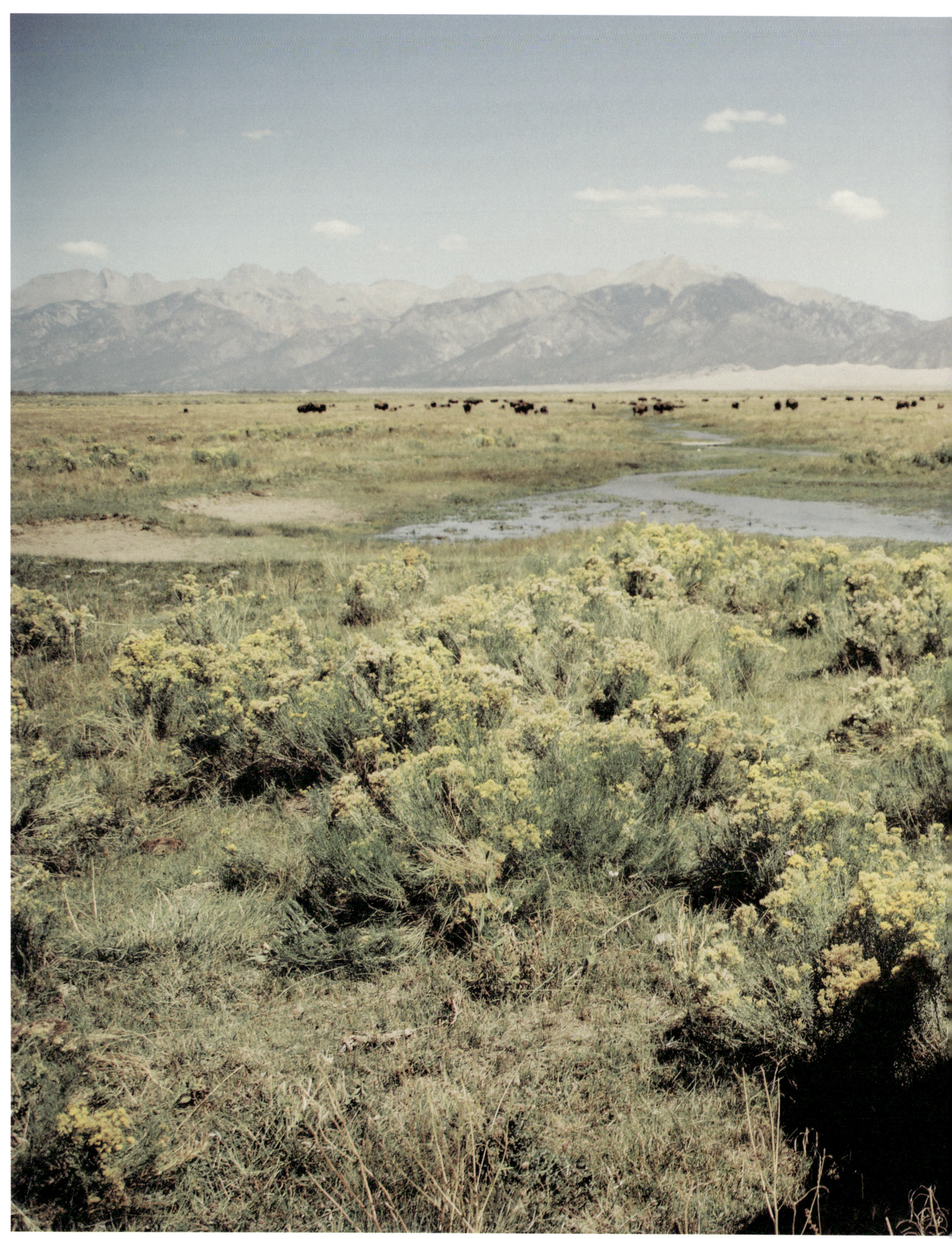

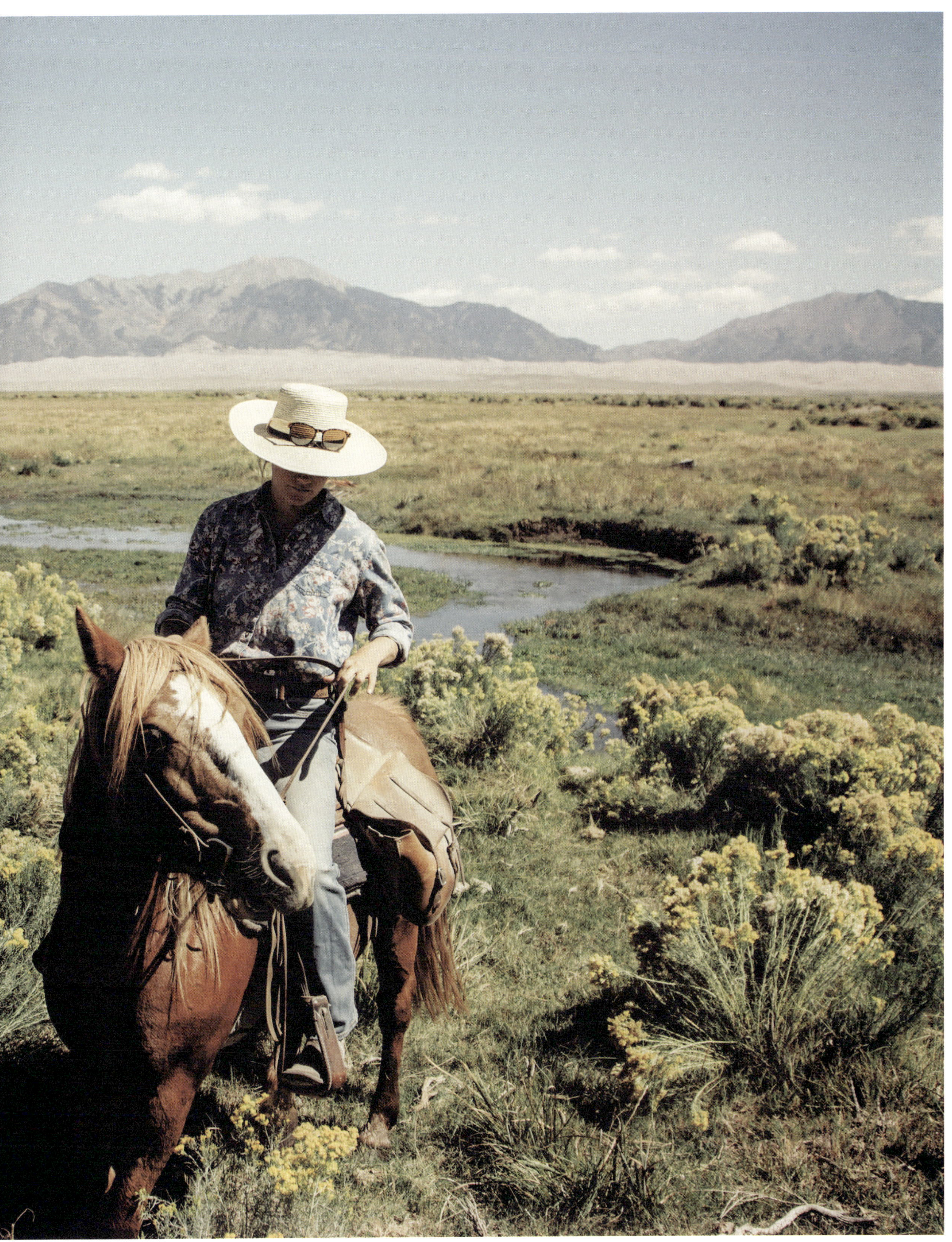

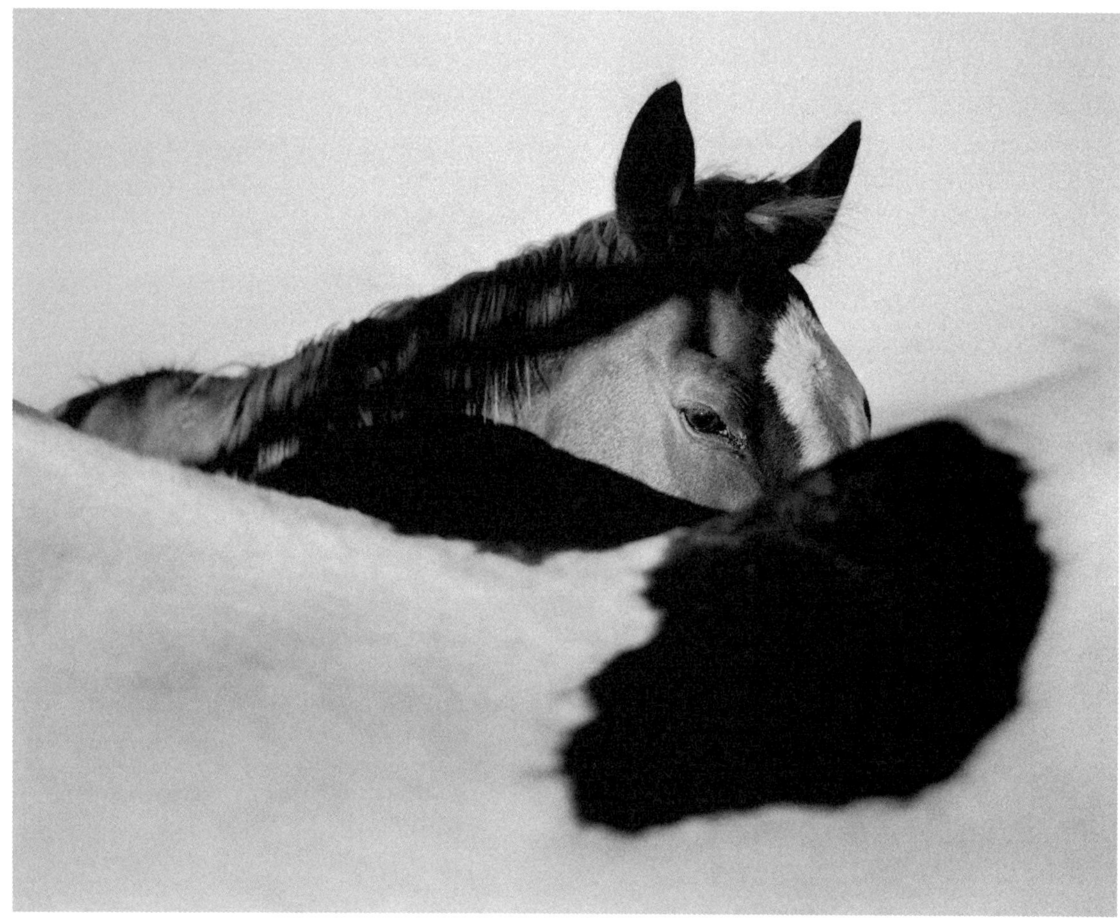

We ride washed by water and wind, hoofprints pressed into wet soil. This is the life we have come to breathe.

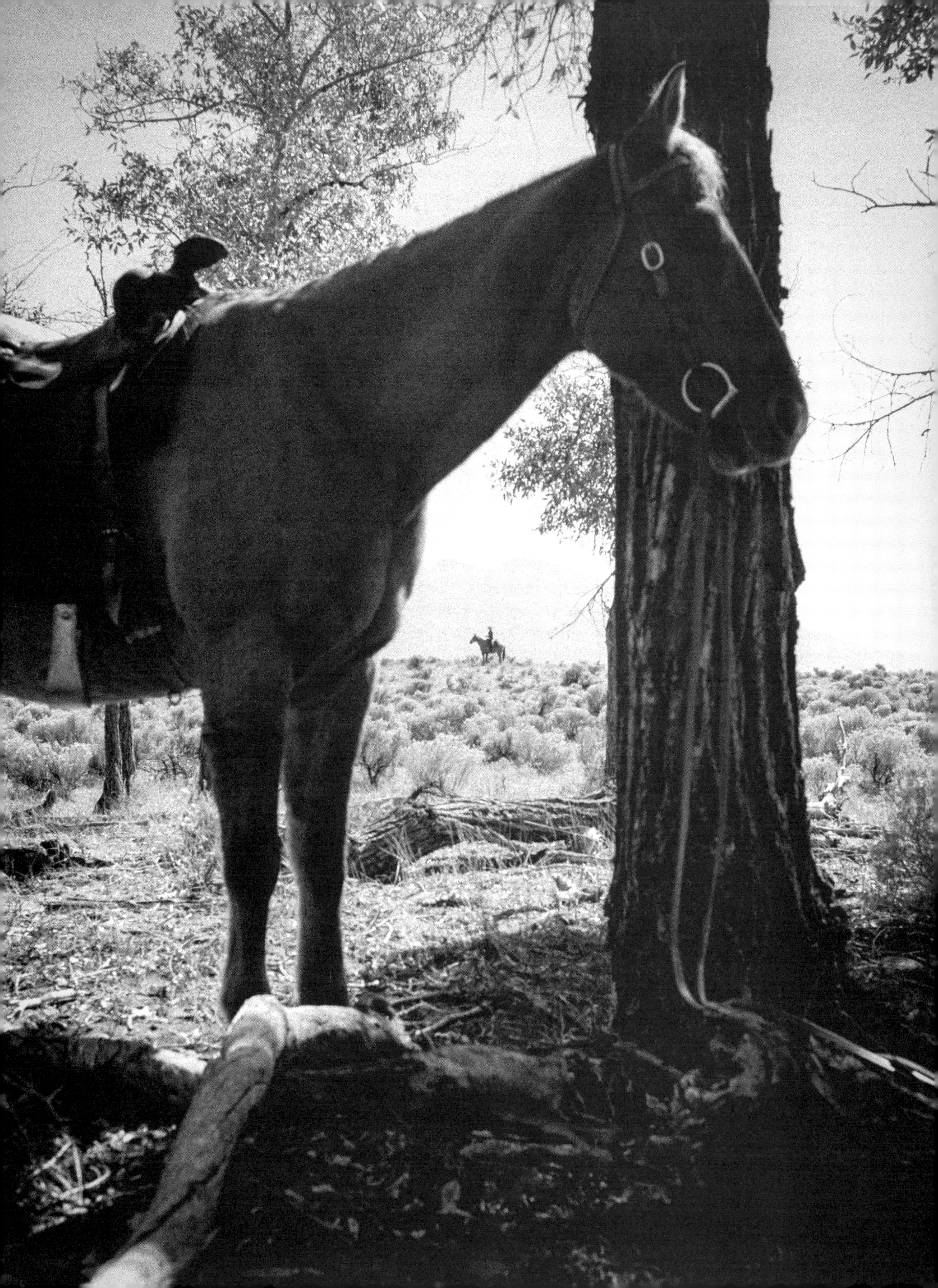

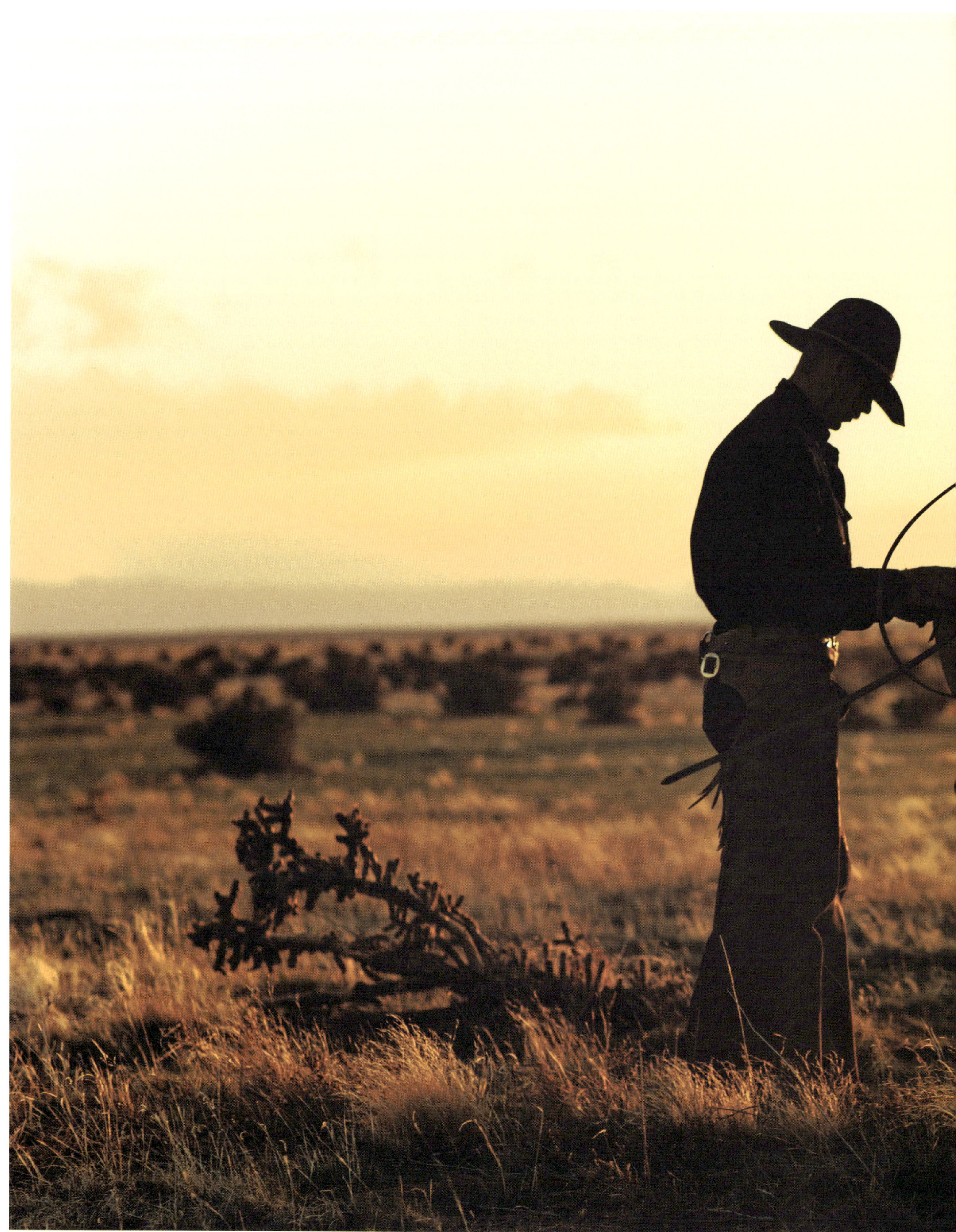

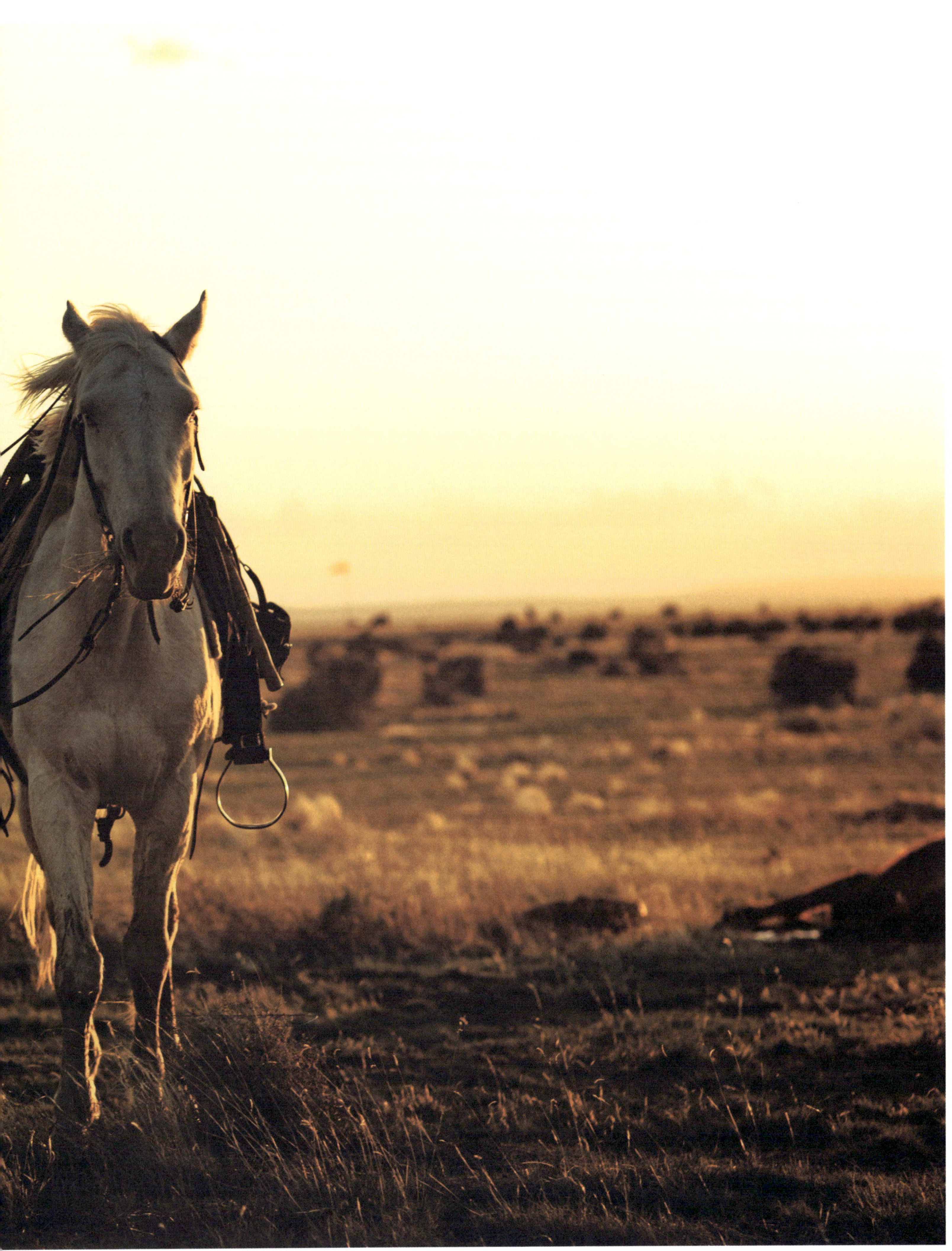

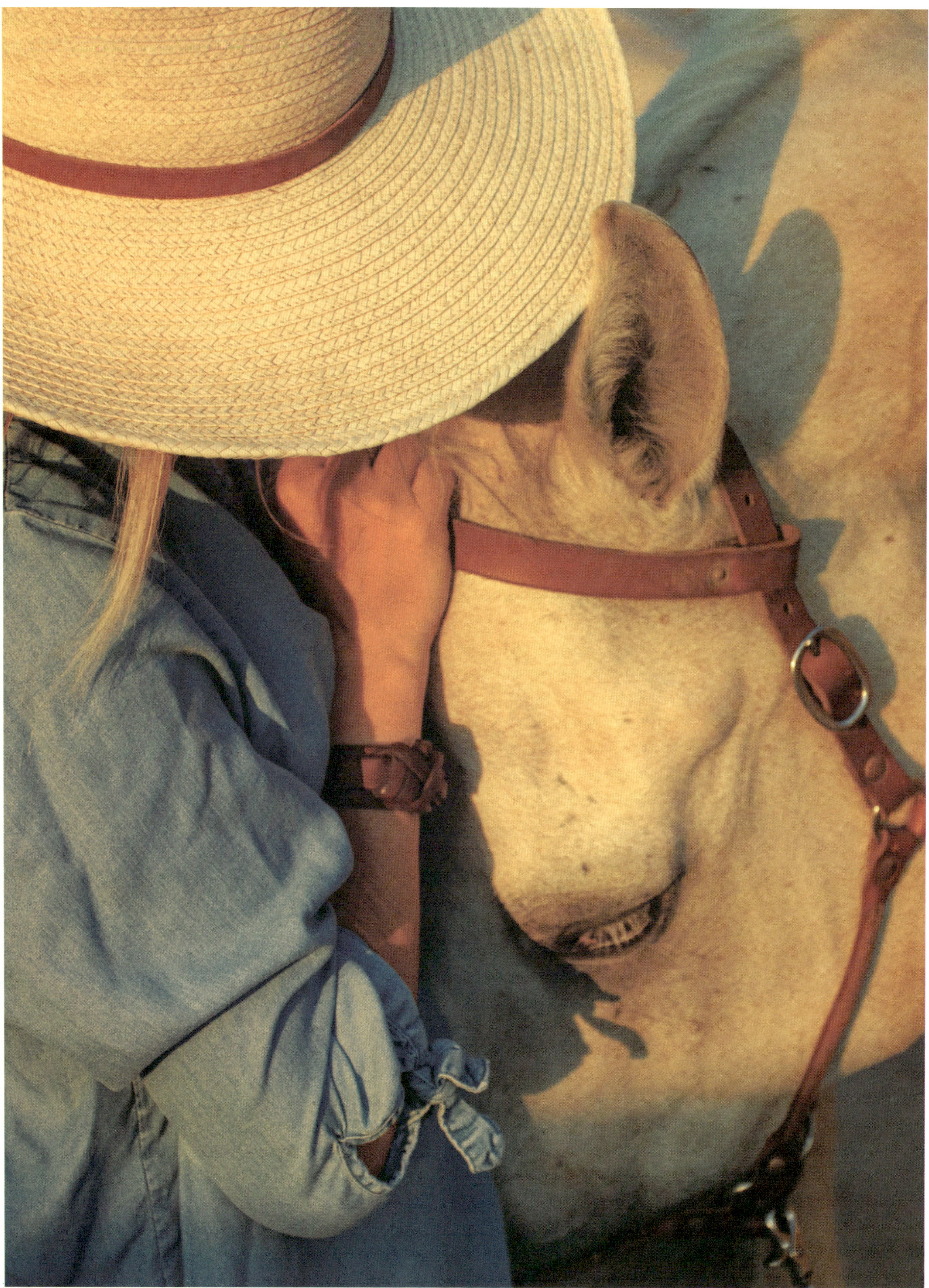

BONNEY MACDONALD
Neighboring Rancher

"Moments of glide are what we seek on horseback—moments of concord, harmony, and balance—when two different creatures locate a shared intention, and where each is, for that time at least, both follower and leader."

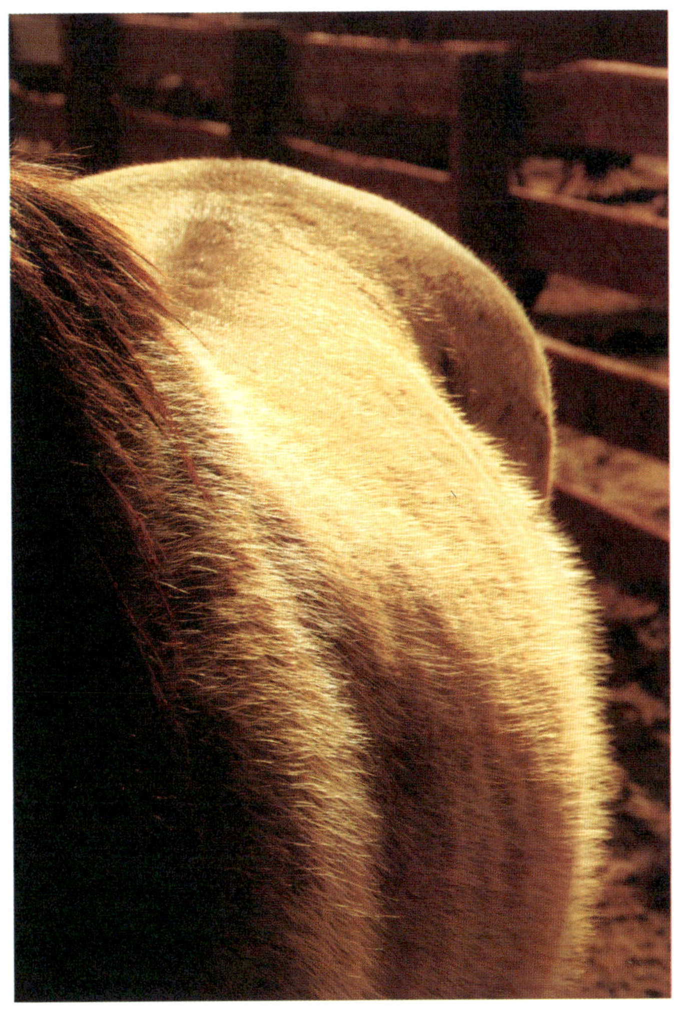

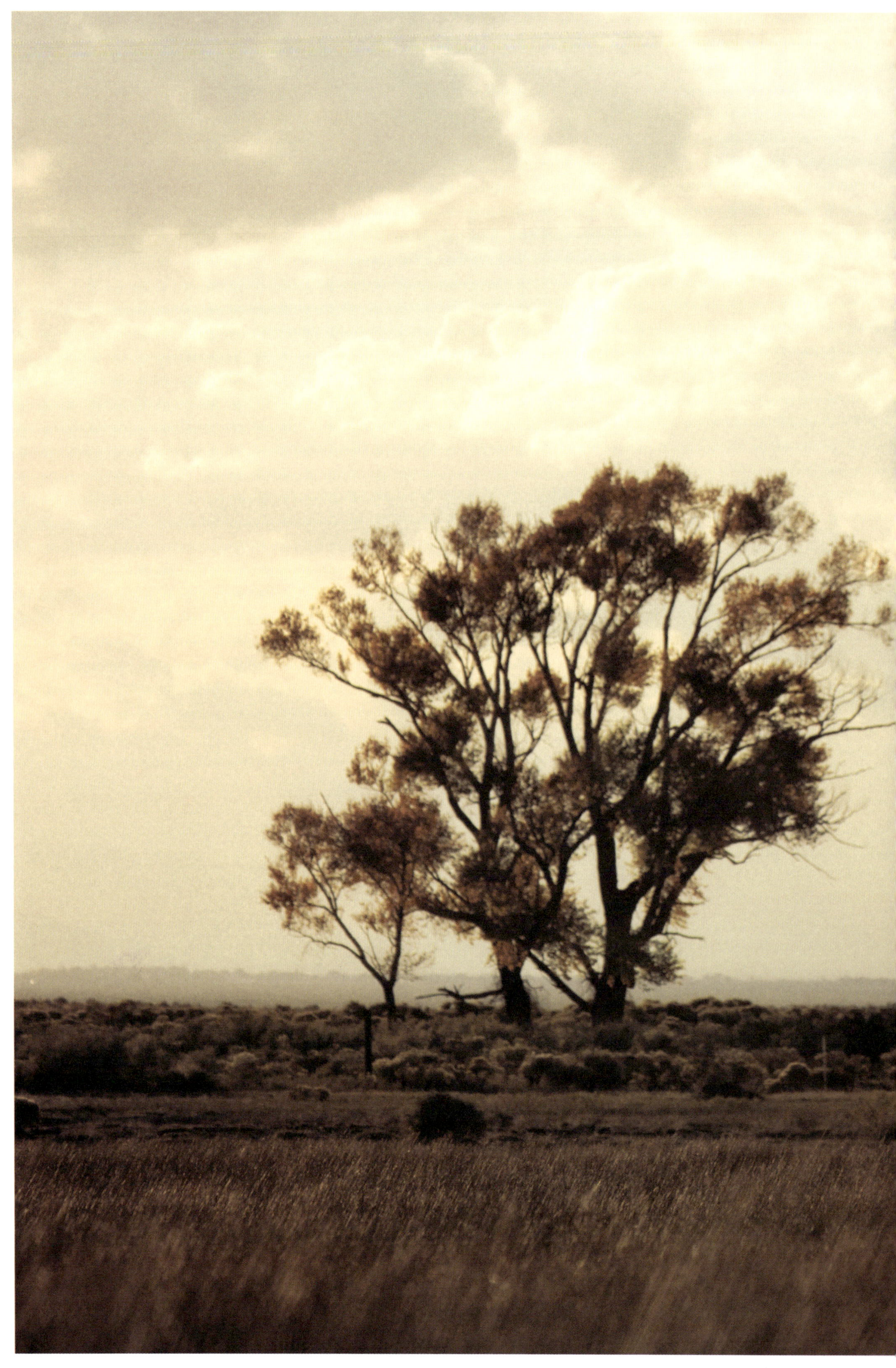

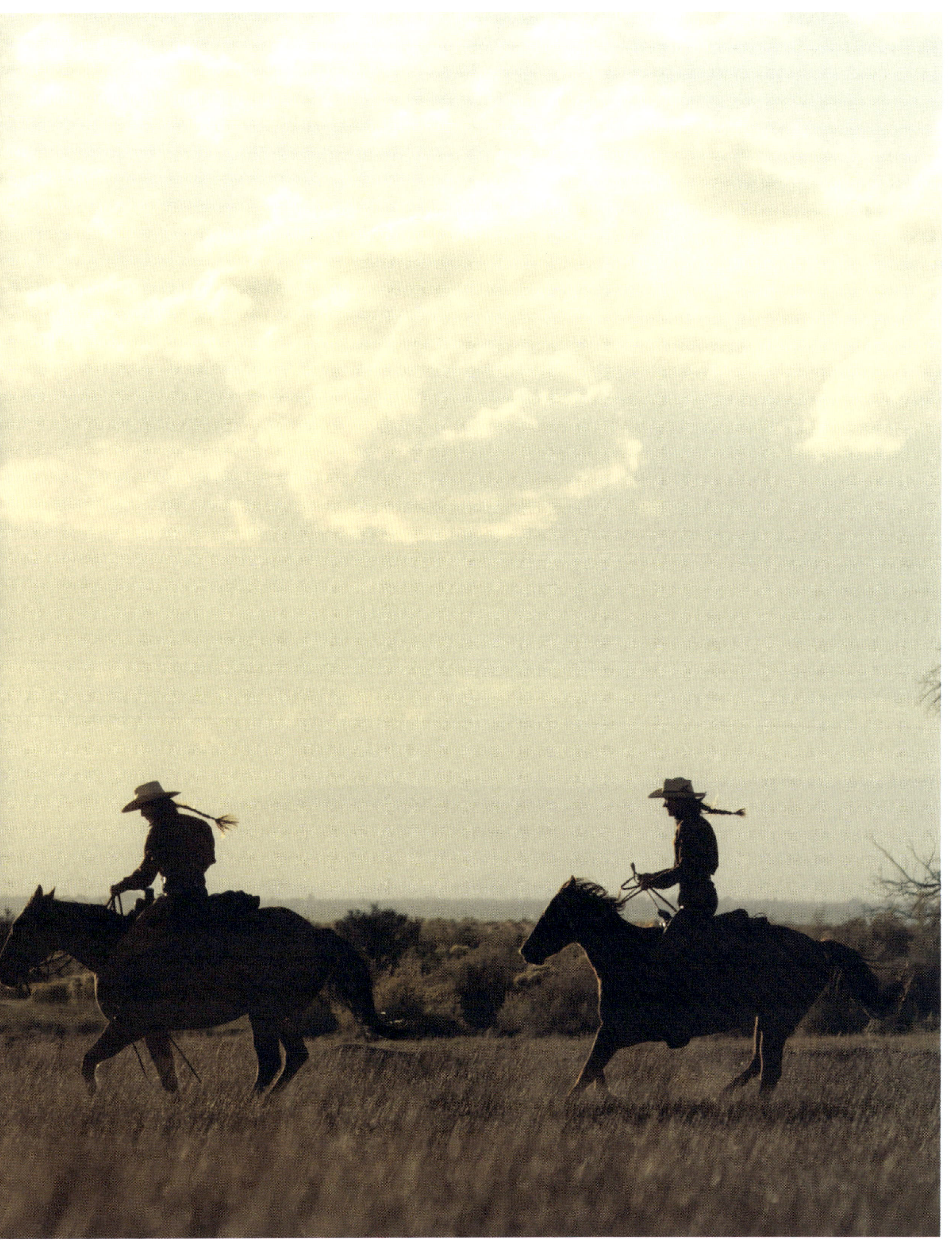

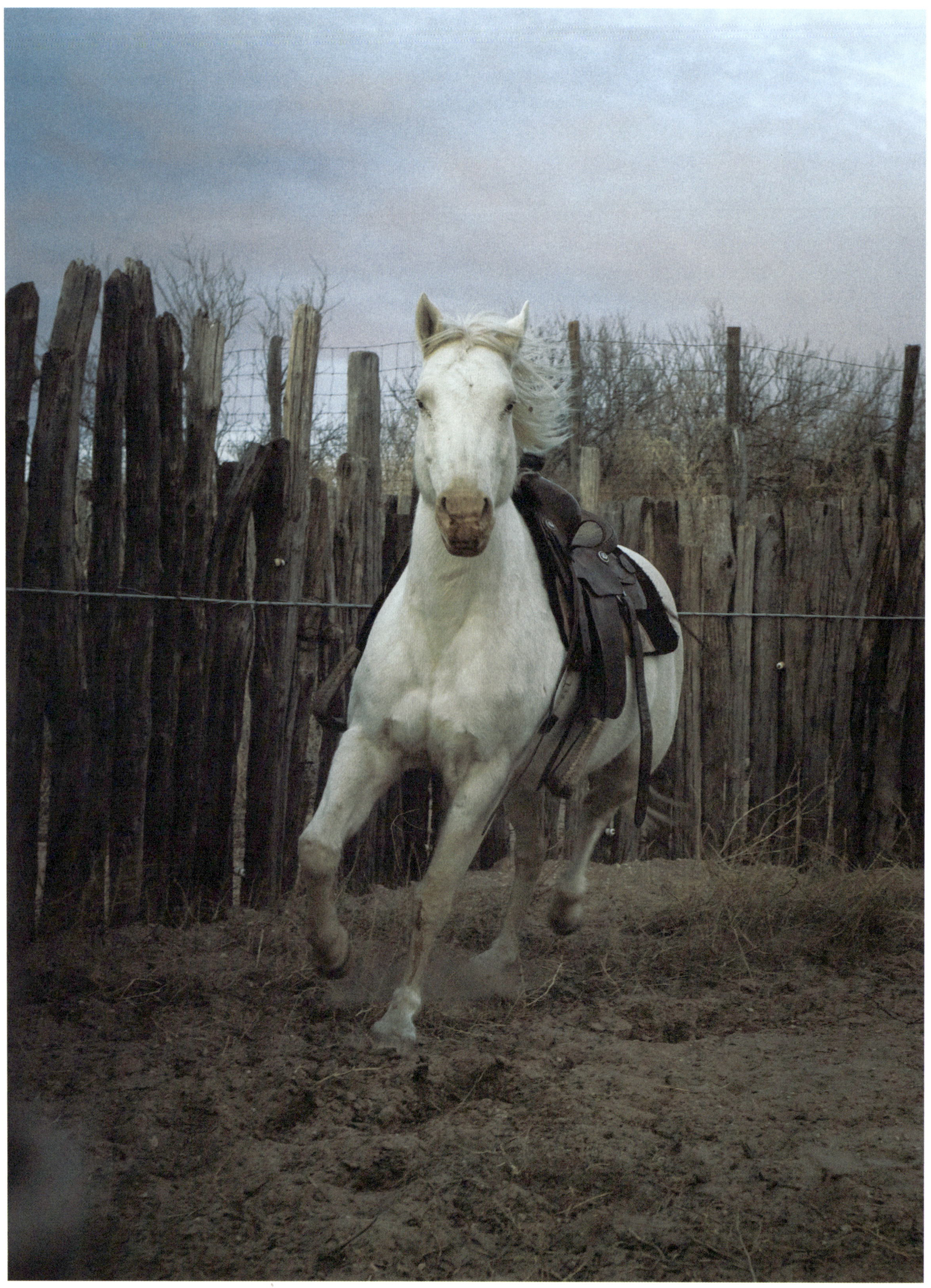

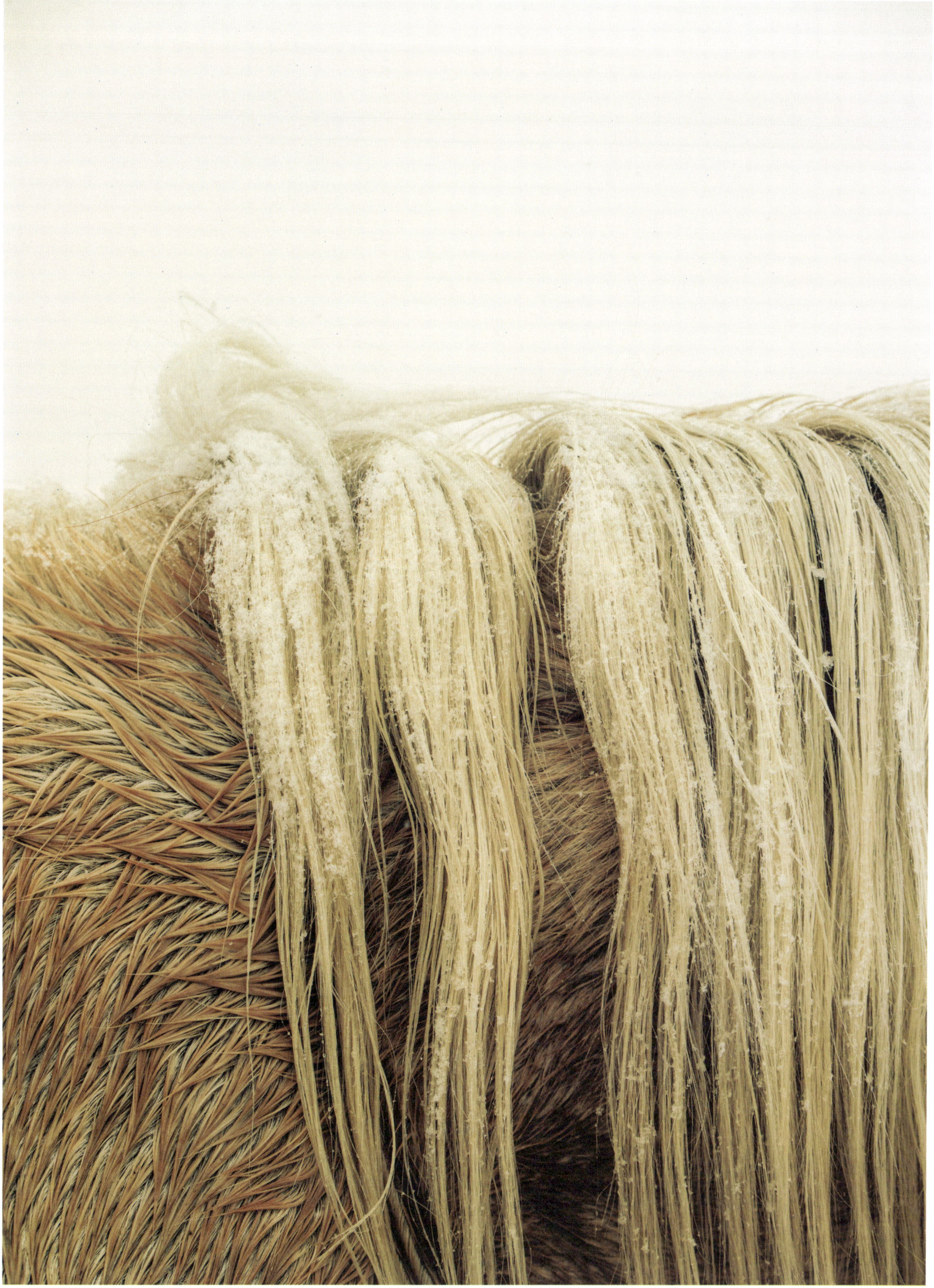

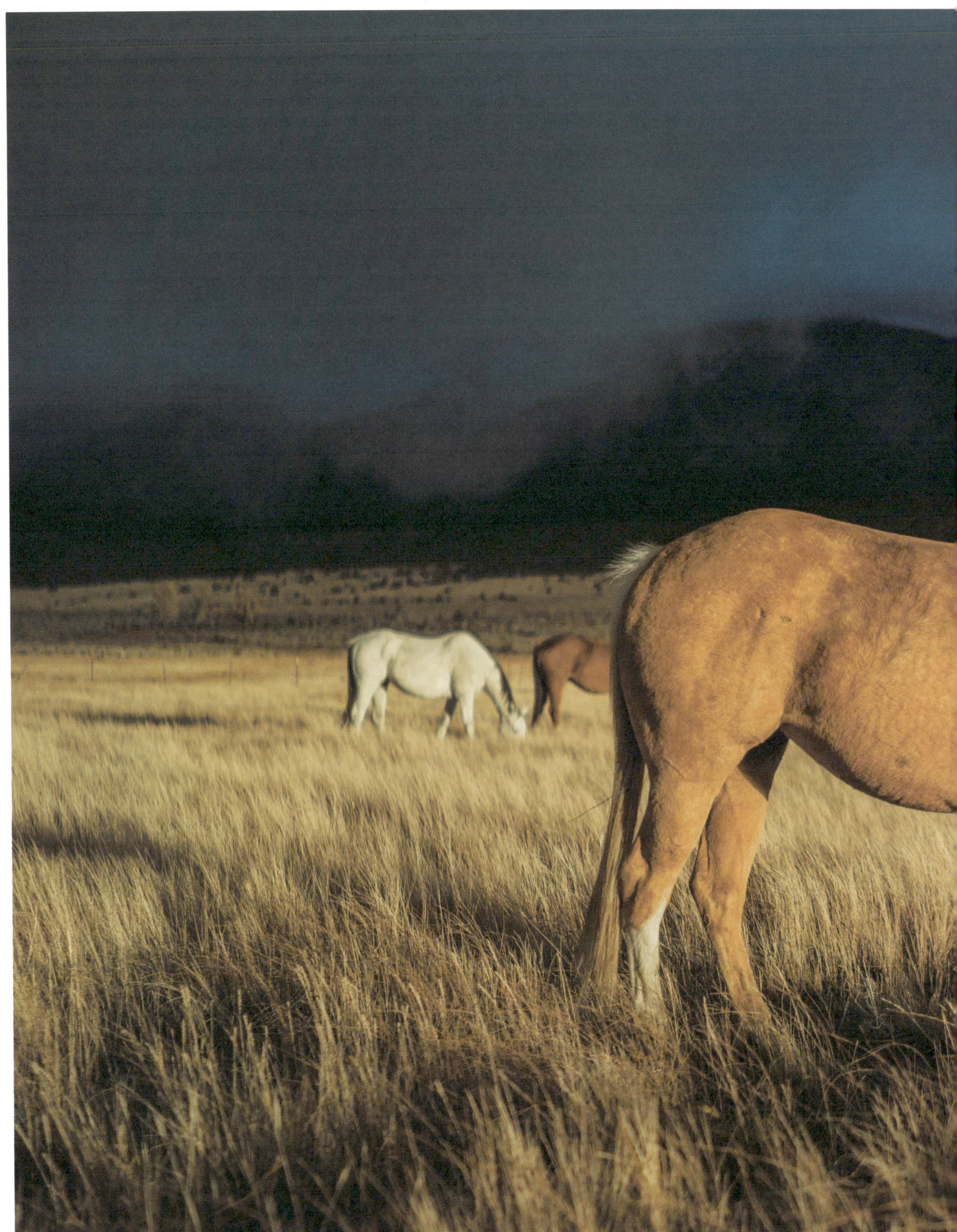

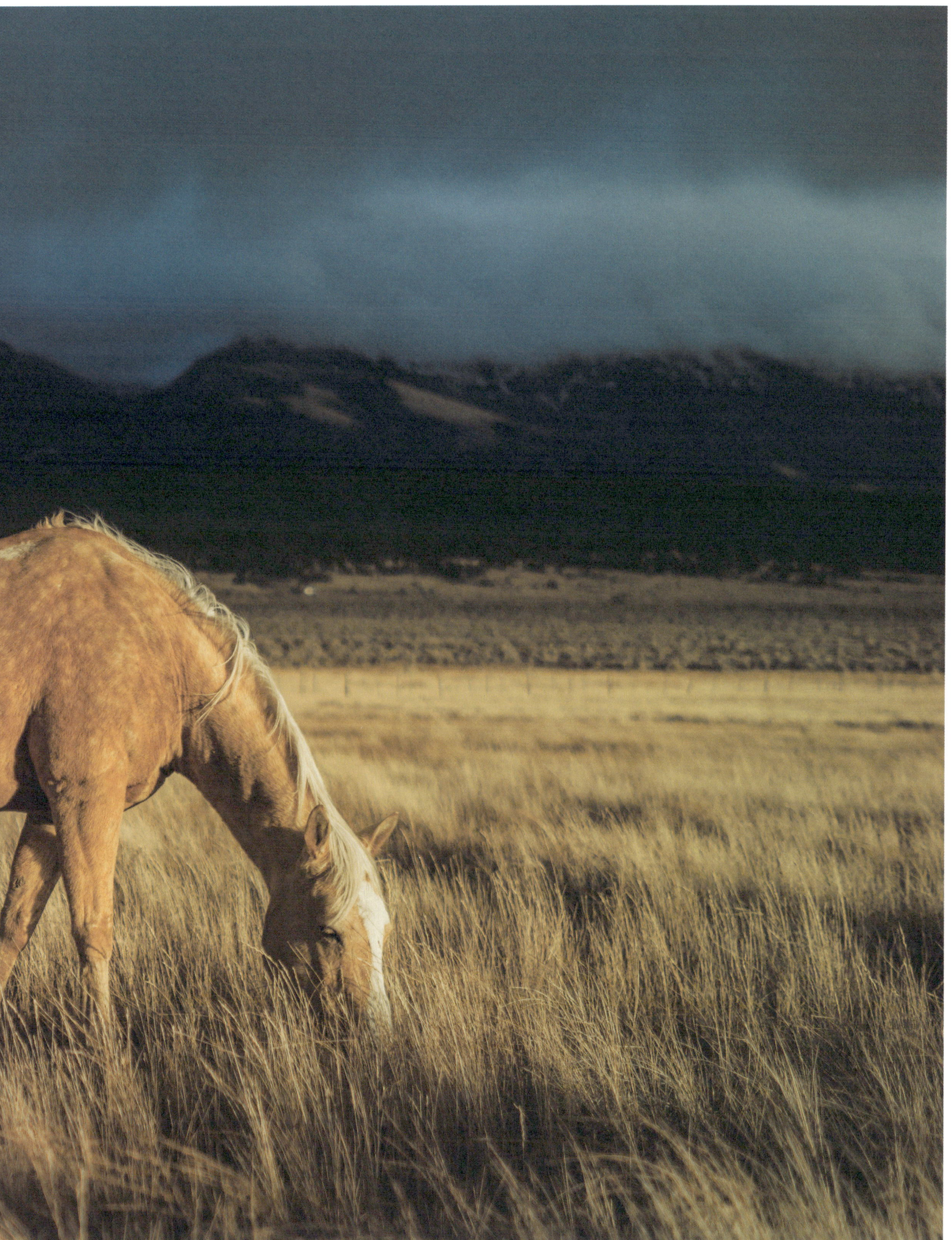

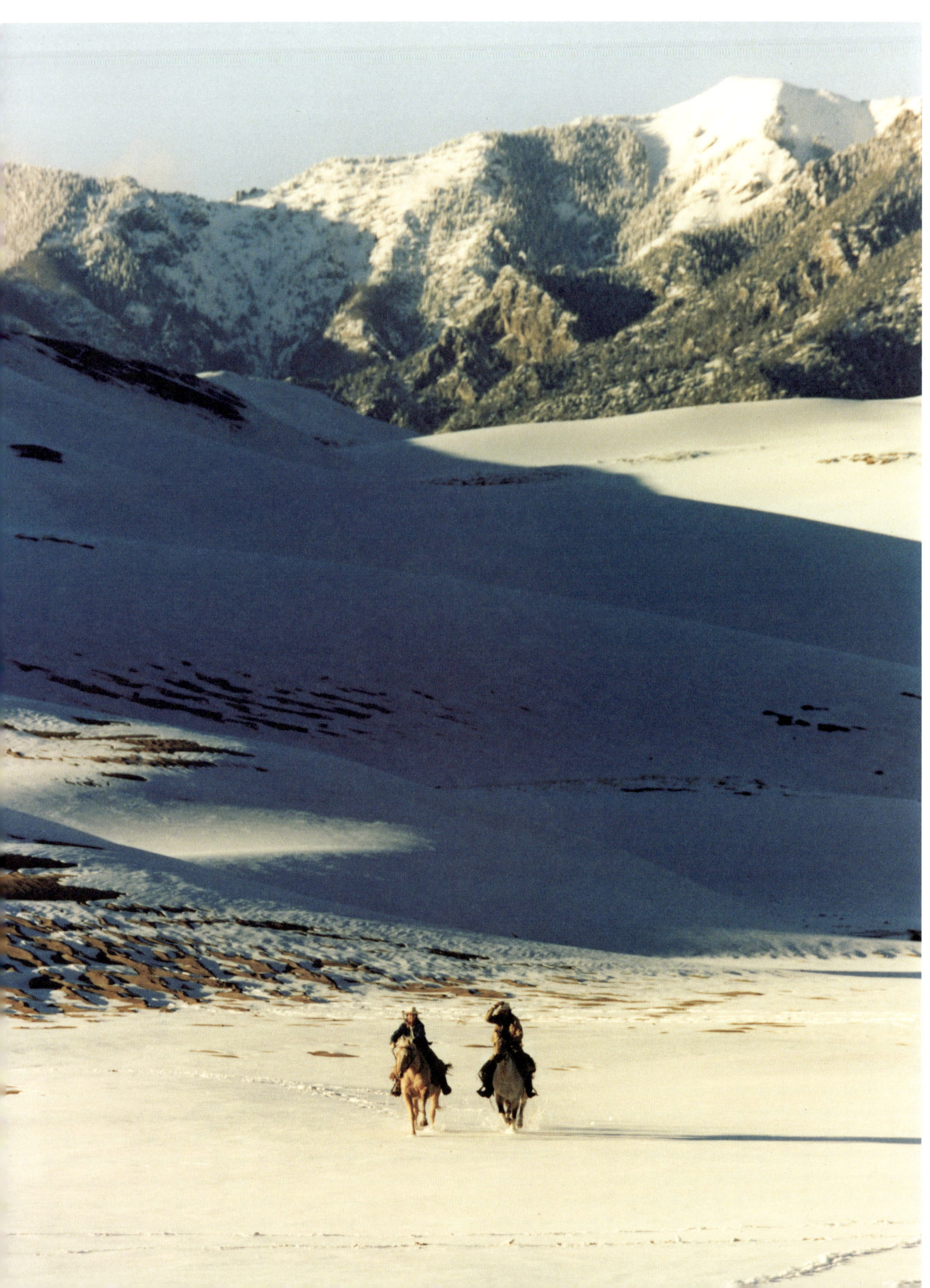

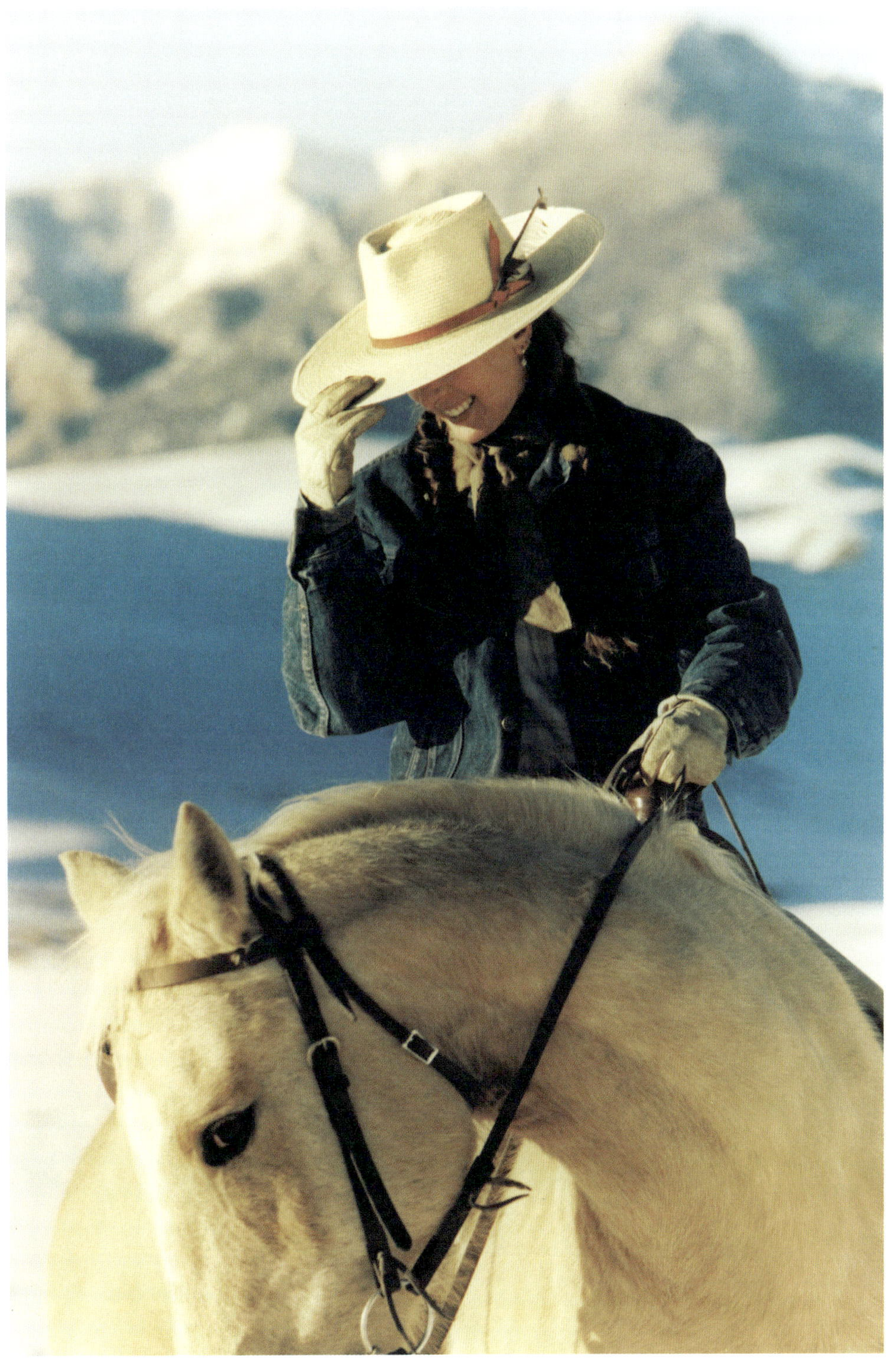

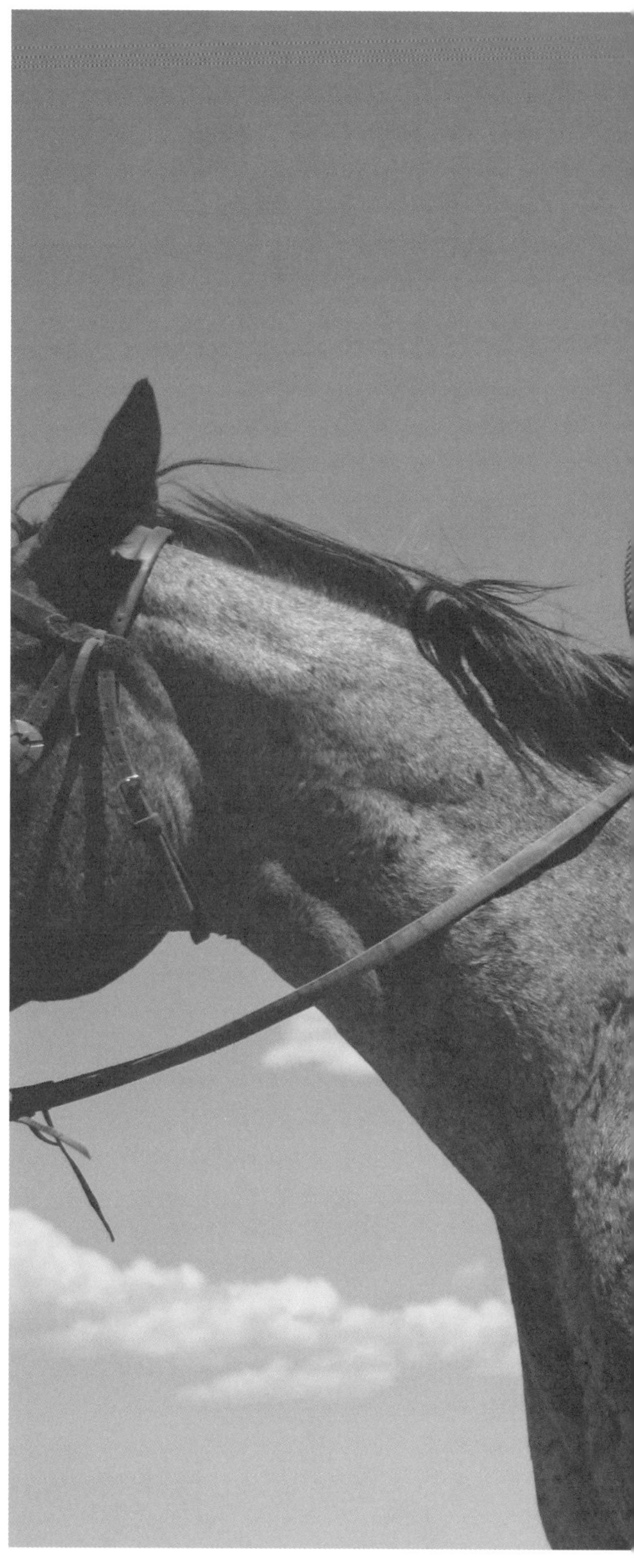

As a young boy living on a ranch, I grew a fondness for traveling everywhere on horseback,

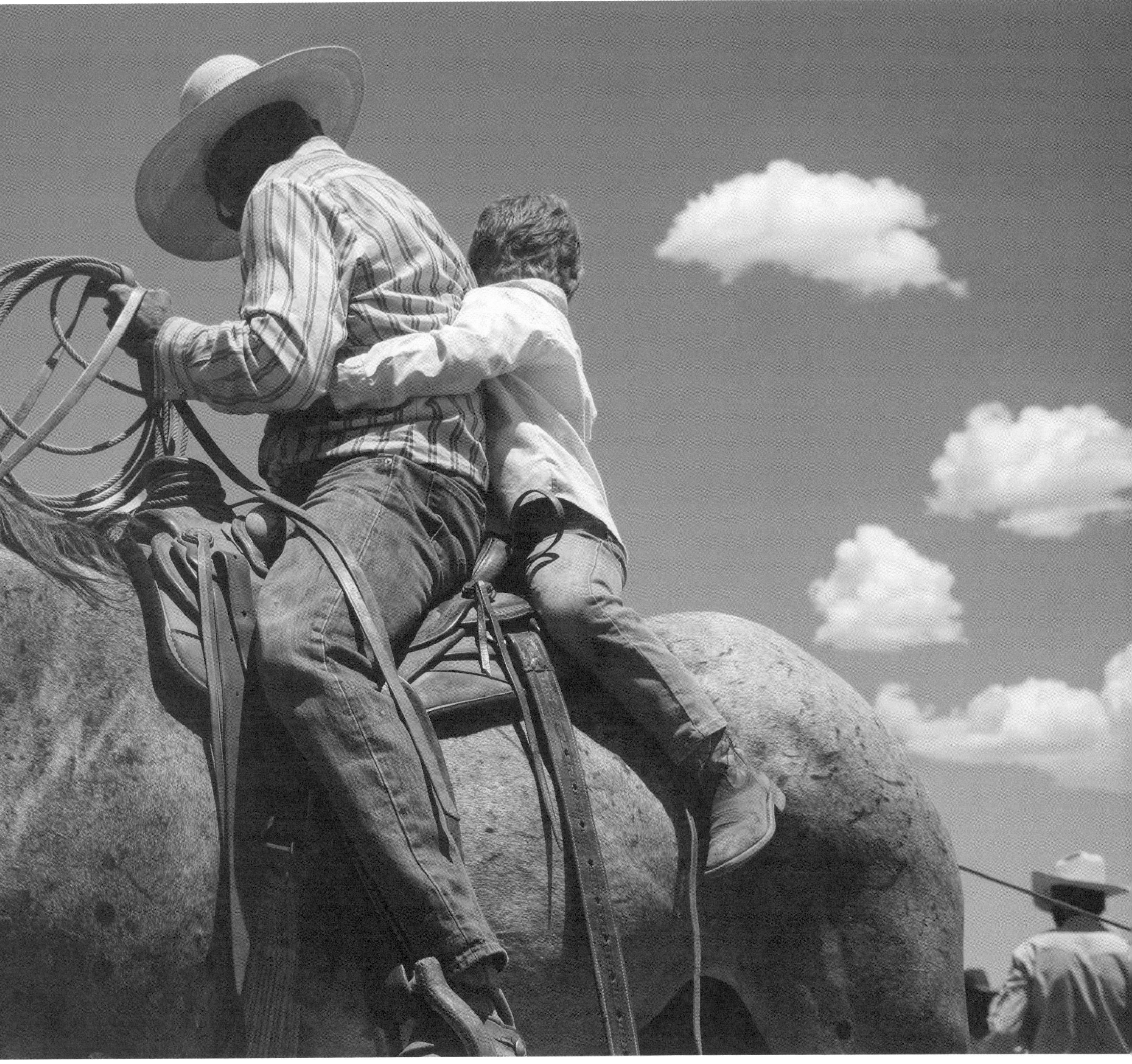

for wild lands where few people had been, and for caring for these places and animals.

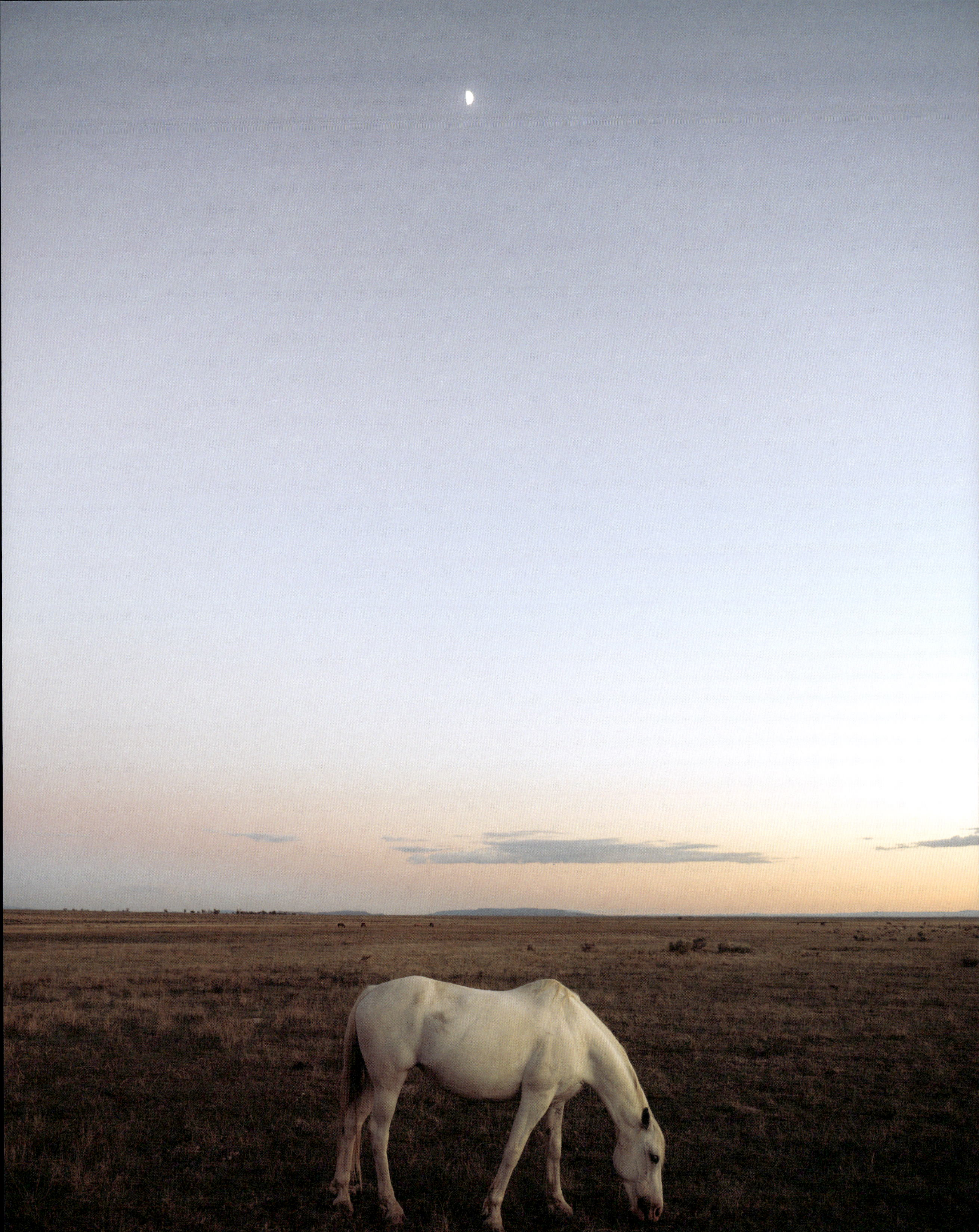

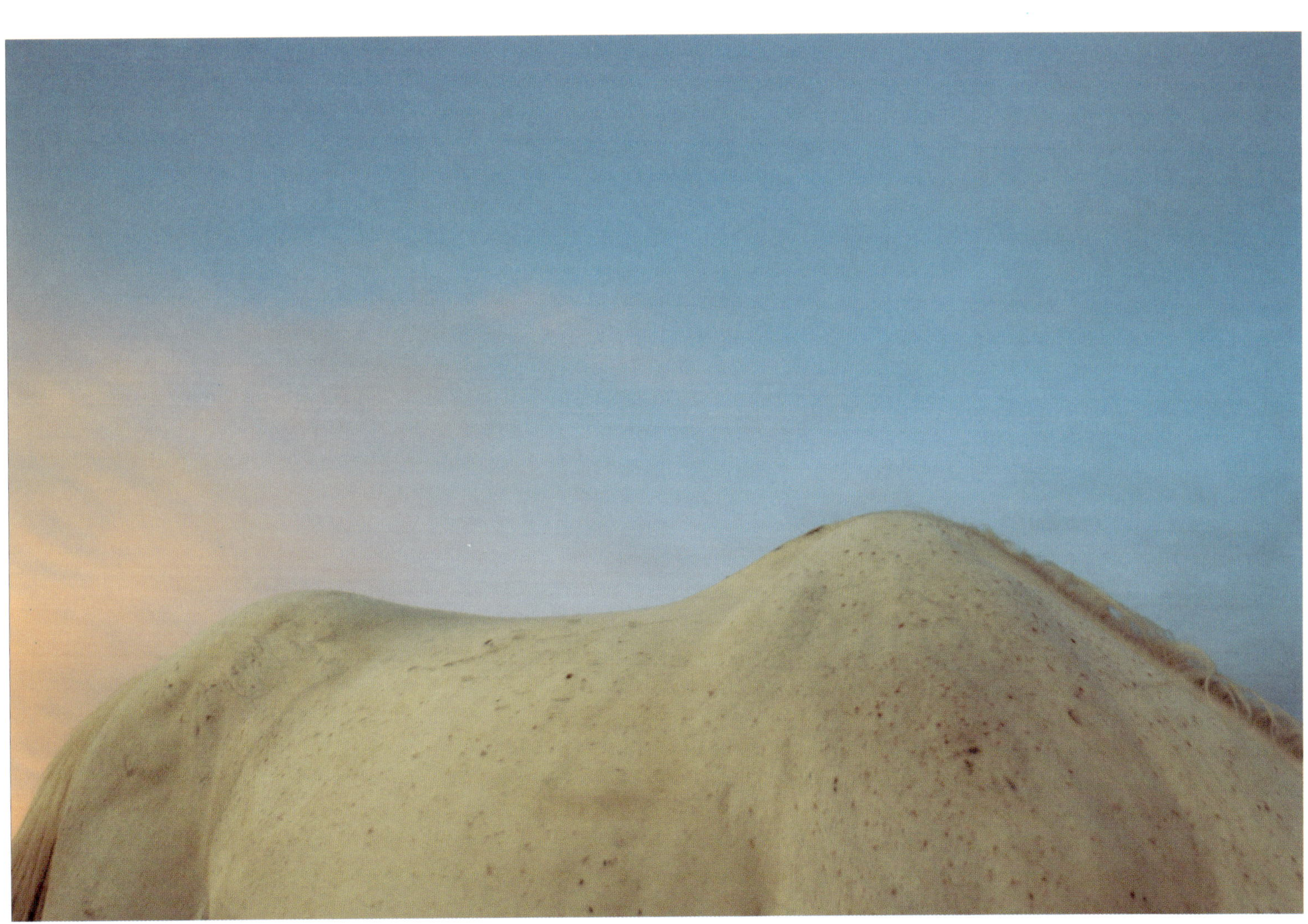

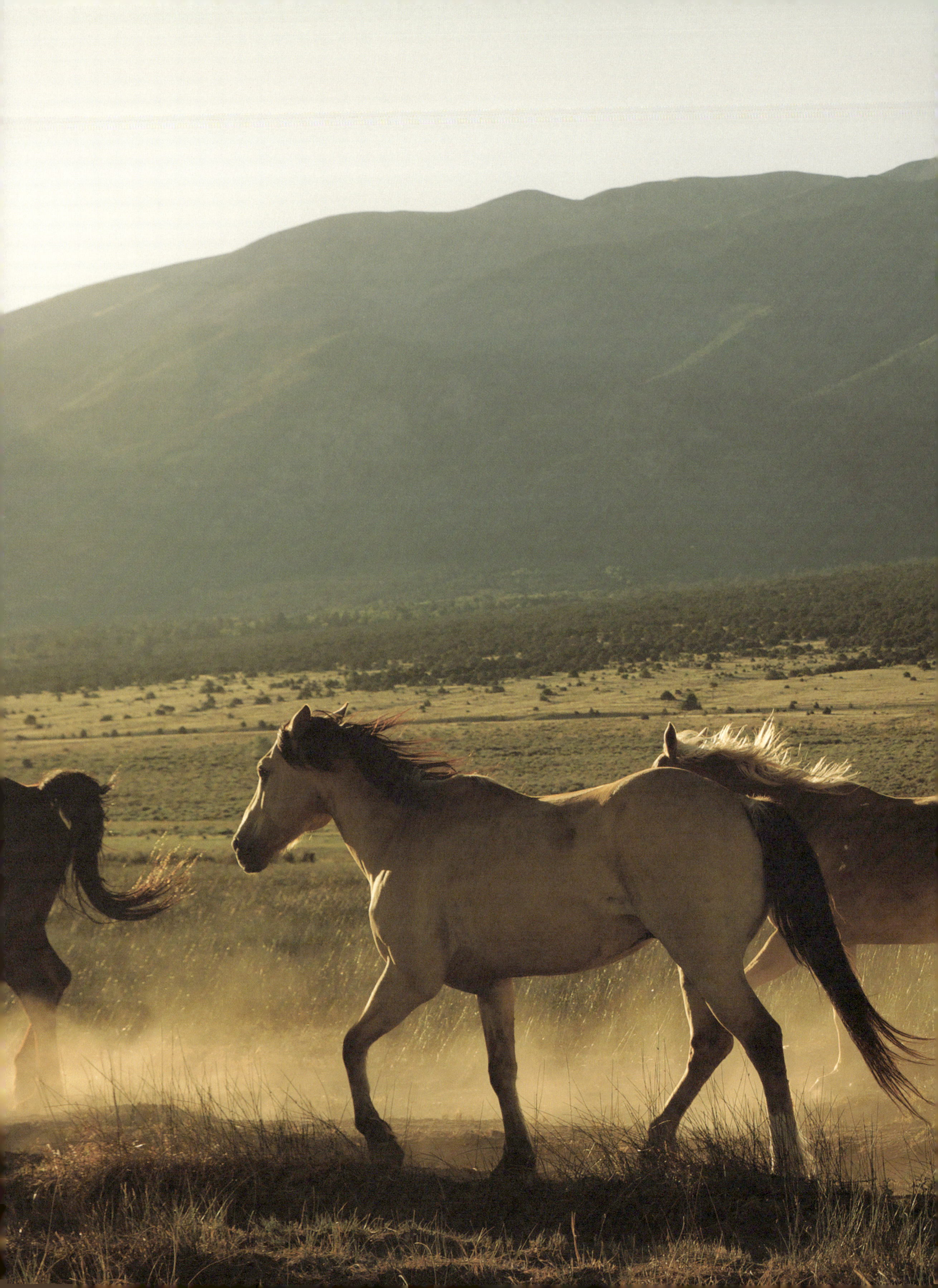

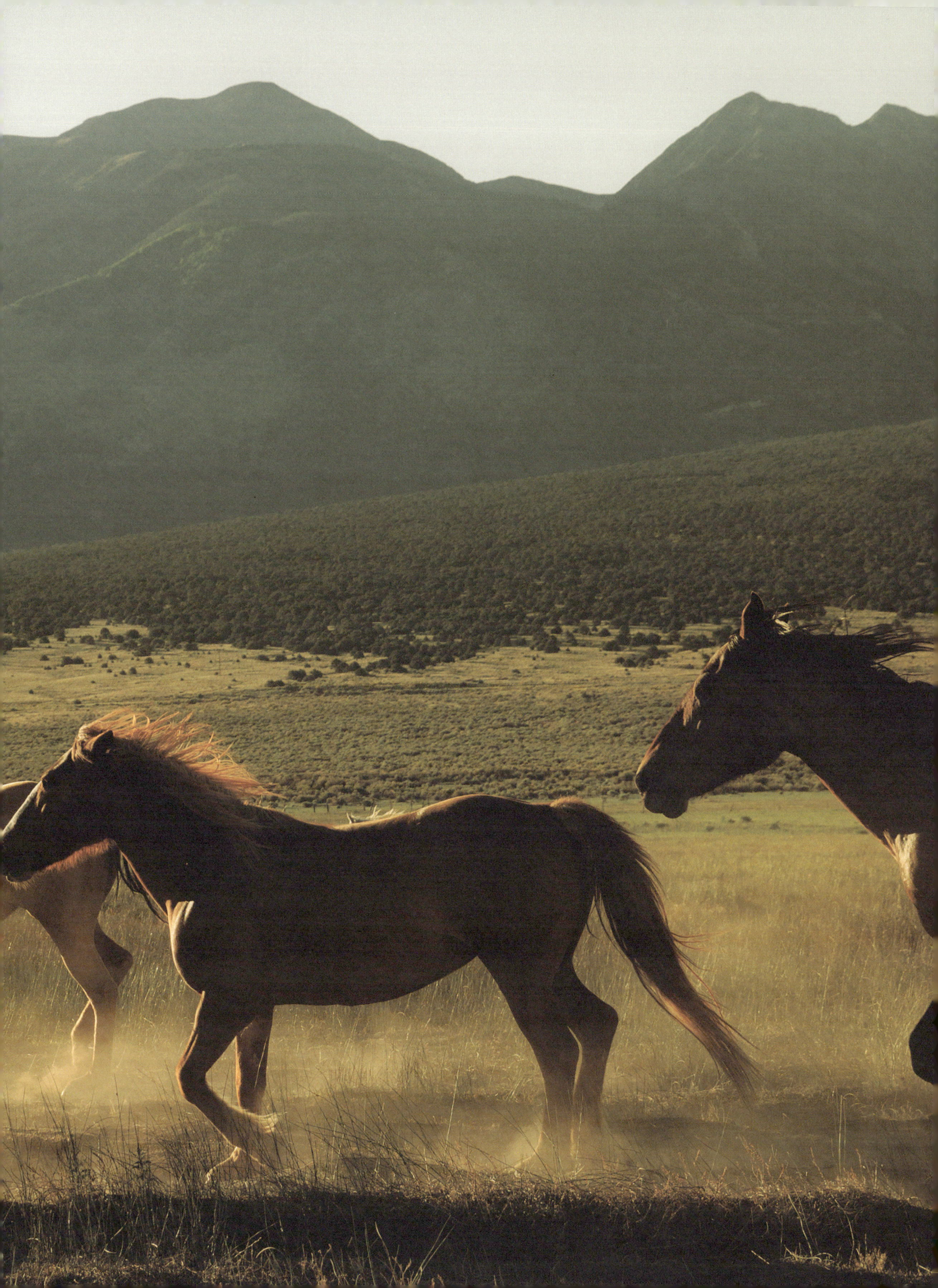

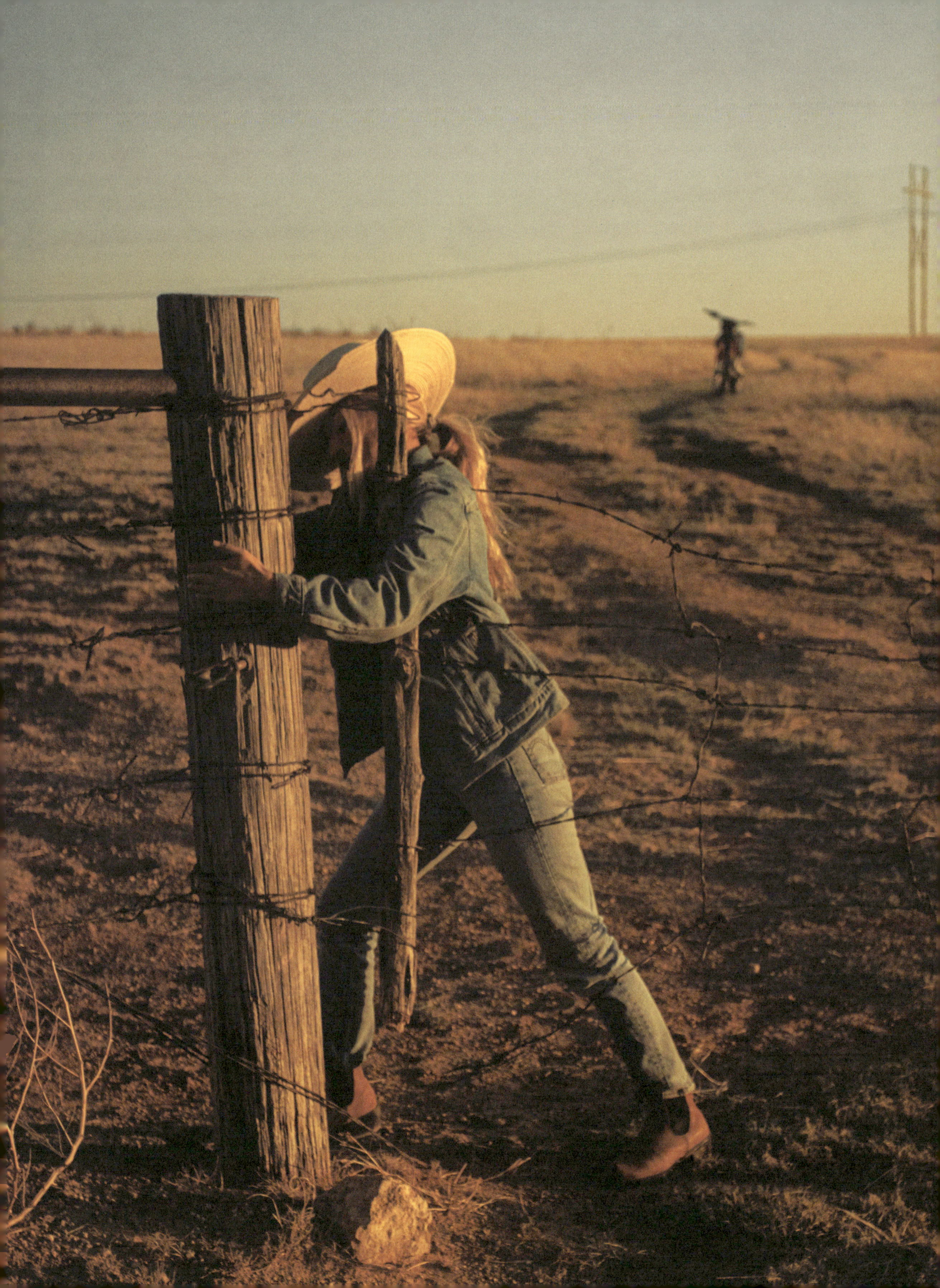

THE

WORK

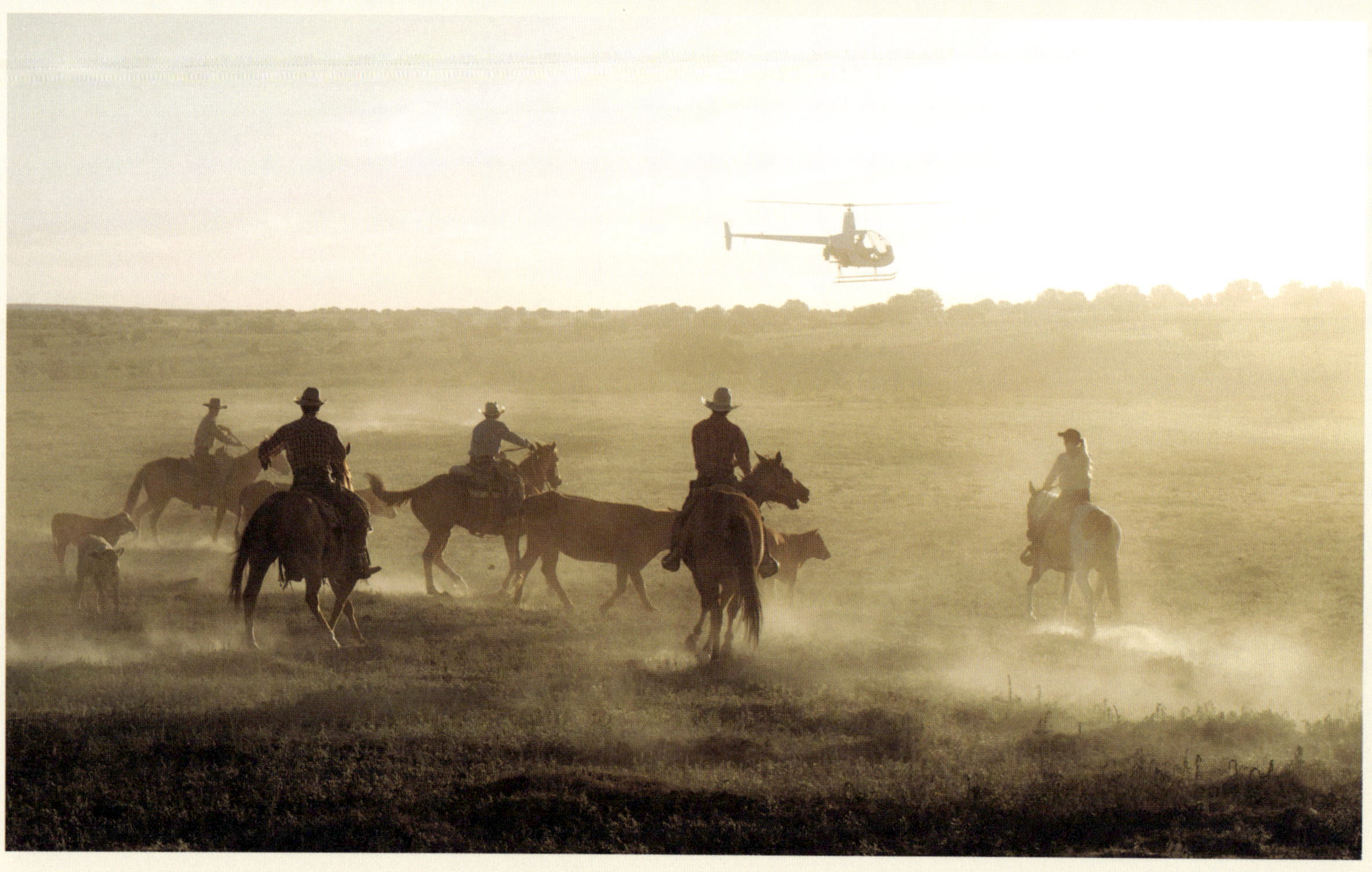

In the dawn light, on one of Ranchlands' properties on the Colorado high prairie, a pilot prepares his aircraft for flight. Diligence proves paramount—the rotor blades, fuel lines, fluids, and controls must function flawlessly in order to maintain safety. He knows his helicopter well after thousands of hours seated at the controls. He double checks the components, and then again for good measure. The engine starts and the blades come to life, increasing in speed and slicing through the chilly morning air. The noise increases in rhythm and volume until a constant drone of rotor wash drowns the quiet songs of prairie birds. He seals the cockpit, fastens the five-point harness, and lifts off, pointing the nose east toward the rising sun, stirring dust in his wake. A massive expanse fills the windscreen, and long shadows fall across the prairie far below his skids. He can see for miles.

Back on the ground, a group of riders on horseback has covered many miles already. They trot in unison, with drumming hooves and ringing spurs keeping the tempo of their movement. The sun, barely peeking over the horizon, illuminates the lead rider. He reaches for a leather bandolier slung across his chest, raises a two-way radio to his face, and makes contact. The pilot answers, his voice punctuated by radio static and rotor wash. They have arrived in a conjoined effort to move the herd from the Arrowhead Pasture to the Dune Pasture, by land and by air.

The riders spread out along the back of the pasture like a long net and begin to travel in an even line toward the watering trough at the other end. The cattle begin to move, as the airship gently nudges them to start the long trip across the pasture. Soon, long strings of cattle can be seen walking calmly down the beaten trails toward the water trough. The helicopter pilot, Duke IV, guides riders from above in order to keep the slowpokes moving in the rear, and then lands on a hill to watch the cattle drive unfold. The sun has barely broken the sharp edge of the horizon.

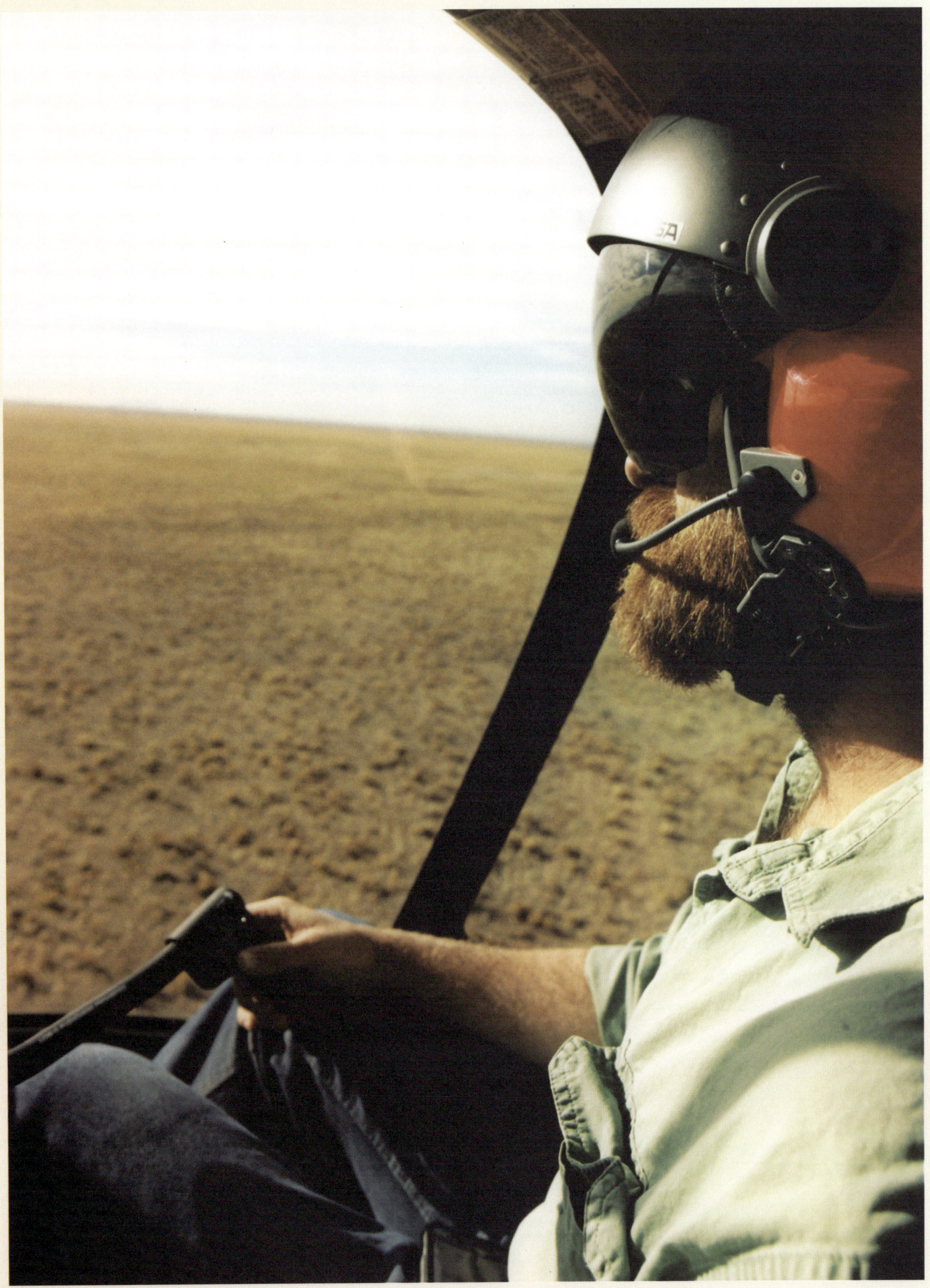

Ranchlands is a conglomeration of working ranches in the truest sense. As in all ranches, we are surrounded by working animals: working cows, working horses, working cattle dogs, working machines, and working humans, all set within a working landscape. More often than not, our coworkers are animals, and we quite like it that way.

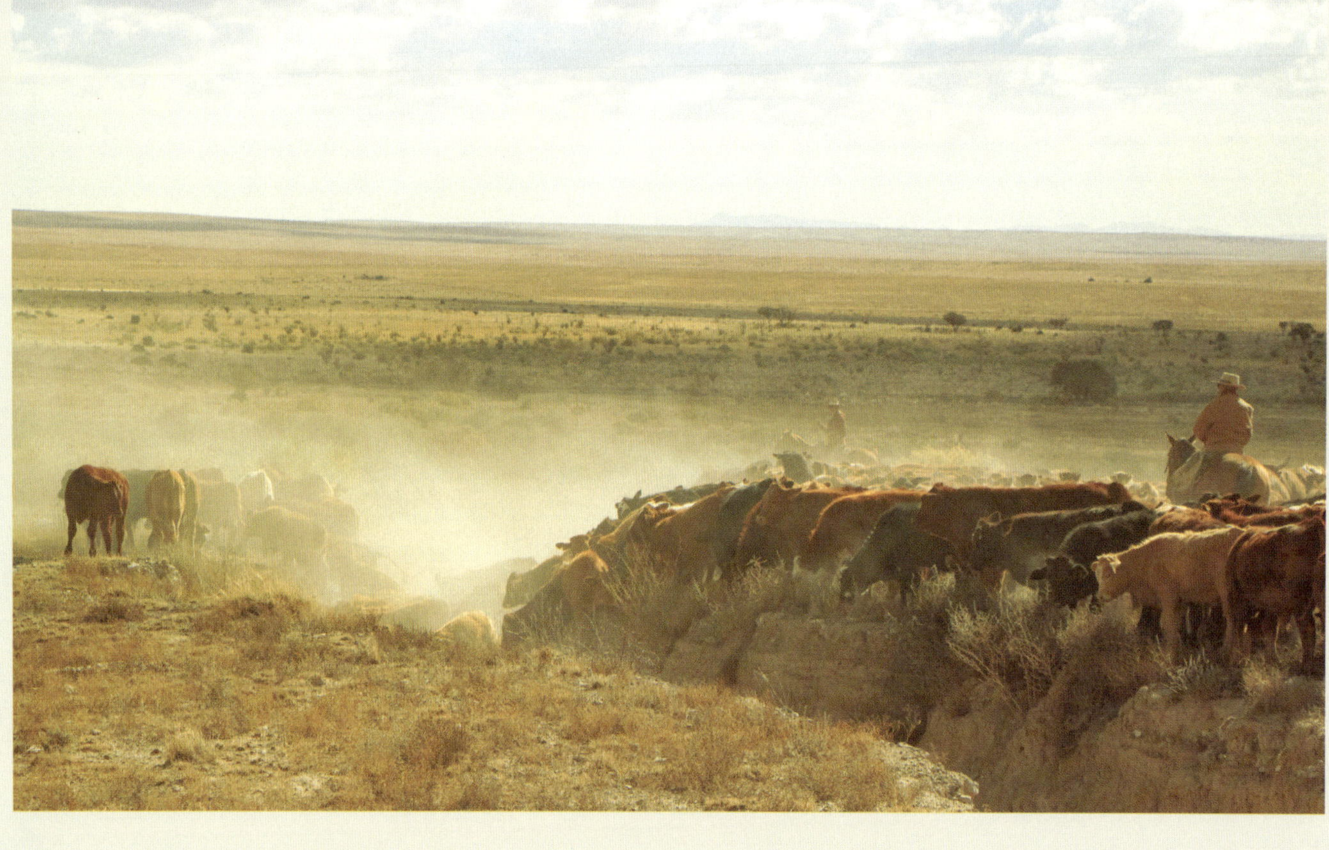

THE CATTLE

The cows, too, are our coworkers. We rely on their life cycle to make our living, and their eating habits heal the land with the help of the management we provide. Our cows are expected to work to produce, in the same way that all of the humans here are expected to work every day. They have to graze land that cannot be farmed in any other way, breed, bear healthy calves, and bring them into the weaning pen every year. They play a key role in helping shape the restless ecological processes of nature: cycles of water, cycles of nutrients, and cycles of growth and decay.

We love working cattle more than anything, especially the animals that we have raised. They have not been spoiled, have no bad habits like running away from us, and rarely get sick or injured. We get along well because we respect one another and know one another's tendencies. All we have to do is think ahead of them and then get out of their way. Handling cattle takes a long time to learn. It's a feeling that is gained by experience and practice over an extended time, like learning a language.

Handling cattle is an art that we take great pride in. Imagine artists making deliberate moves in and out of a herd, back and forth, guiding cattle that don't realize they are being guided because of the slightest of adjustments being made while walking and following the lead animals. That is the key to maintaining the entire herd moving: keeping the lead moving. If the animals in front are moving, the entire herd will follow with gentle prodding here and there. Our job is to use the momentum within the herd to establish movement. We see the herd moving like a current in a river—dynamic, with eddies, fast water, slow water, deep pools, tributaries, and waterfalls. It is highly challenging because it is always different depending on the weather, the cattle's mood, whether you're traveling uphill or downhill, with yearling cattle or cows with calves. Of course, it also depends on the crew you have riding with you and the dogs.

OLD AND NEW

Ranchlands' work methods are a combination of the old and the new. In the same ways that our predecessors used horses, we still do as well: for branding, moving pastures, sorting cattle, checking calving heifers, and doctoring sick cattle. As much as possible, we ride horses out to the pastures instead of trailering them out in a pickup truck.

Due to the scale of our ranches, the helicopter brings an efficiency component, saving a huge amount of time and the costs of vehicle wear and tear. We use it primarily for gathering

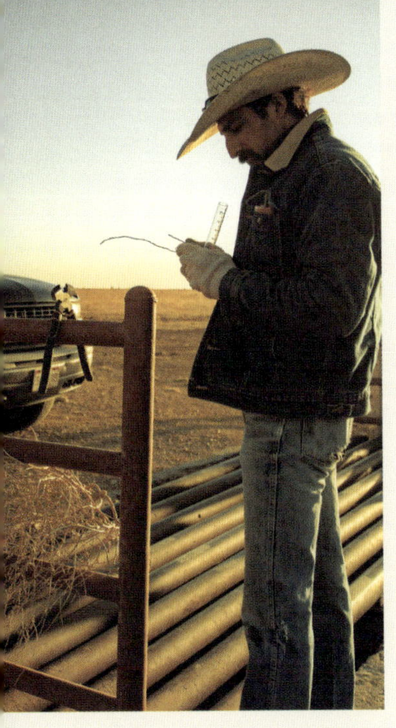

pastures and, at times, for checking livestock-watering points and rainfall on the back parts of the ranch when driving in the mud is not good for the roads.

DAY TO DAY

The seemingly endless number of moving parts and operations makes our business different from most. Principal activities are divided among livestock, land, and nonlivestock land-based work. Every day is different, and we spend time both inside and outside. A misconception about ranching is that we spend all our days on horseback working cattle, riding the range, or loafing on the side of a hill with our horses standing beside us. The truth is the majority of our time is spent on the management side of the business and on infrastructure. The most important element is the water system: pipelines, water troughs, wells, and pumps. On ranches without live water—such as springs, creeks, and rivers—the water has to be piped underground to livestock-watering points, switched from herd to herd when necessary, and repaired when broken.

Motorbikes, trucks, trailers, and riding gear must also be maintained and repaired when broken. Corrals, barns, and houses are on the list to be cared for. Administrative work is an important activity that brings us inside to work with maps, computers, tally sheets, grazing and monitoring data, and phone calls. We put a high priority on our weekly meetings so that everyone knows the plan for the week ahead. Also, scheduling yearly cattle works and marketing and selling cattle are important events in running an efficient business.

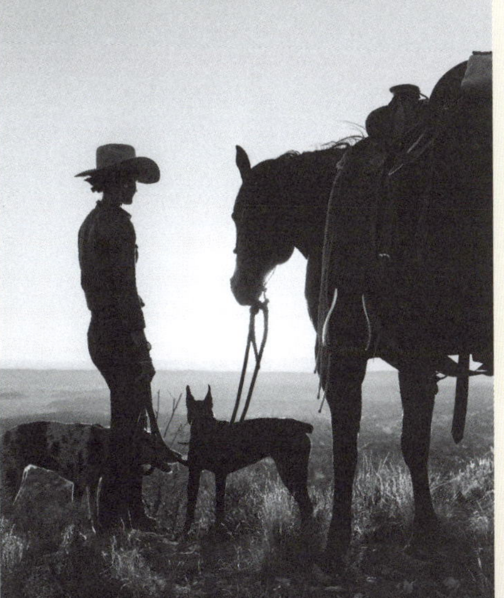

In addition to the ranch work, we have a large nonagricultural component to our business—hospitality—which is run by an entirely different crew. Guests from all over the world visit us and want to join our team's daily work, hundreds of school kids come to the ranch in the spring and fall to learn about the prairie, and visitors drop by through our open-gate policy. On this side of our work, there is a great deal of effort expended in marketing and sales, which continues to develop the Ranchlands brand across social media platforms and in the media. We also create leather goods that are sold in the Ranchlands Mercantile, and we host journalists, travel professionals, and friends.

INTERNS AND APPRENTICES

The Ranchlands Management Guild is a leadership training program dating back to 2000 that develops young people who want to learn how to manage ranches into the future. This apprenticeship program is based on learning by doing, by taking on a large amount of responsibility, and by being allowed to make mistakes—what we instead call "learning experiences." We believe that the best way to become good at managing a ranch is from the bottom up because there are so many little and big things to know and understand. Our apprentices begin as interns, and if they're successful, they can become apprentices. If they're successful from there, the apprentices can become ranch managers. Today, all of our ranches are managed by young people who were once in the apprenticeship program. They learn not only about handling livestock and horses, treating disease and sickness in cattle, reading the land, and managing the critical ecological resources but also about planning and organization, communication, working efficiently, and managing people.

SEASONALITY

The ranch work is heavily marked by the four seasons of the year. Baby calves

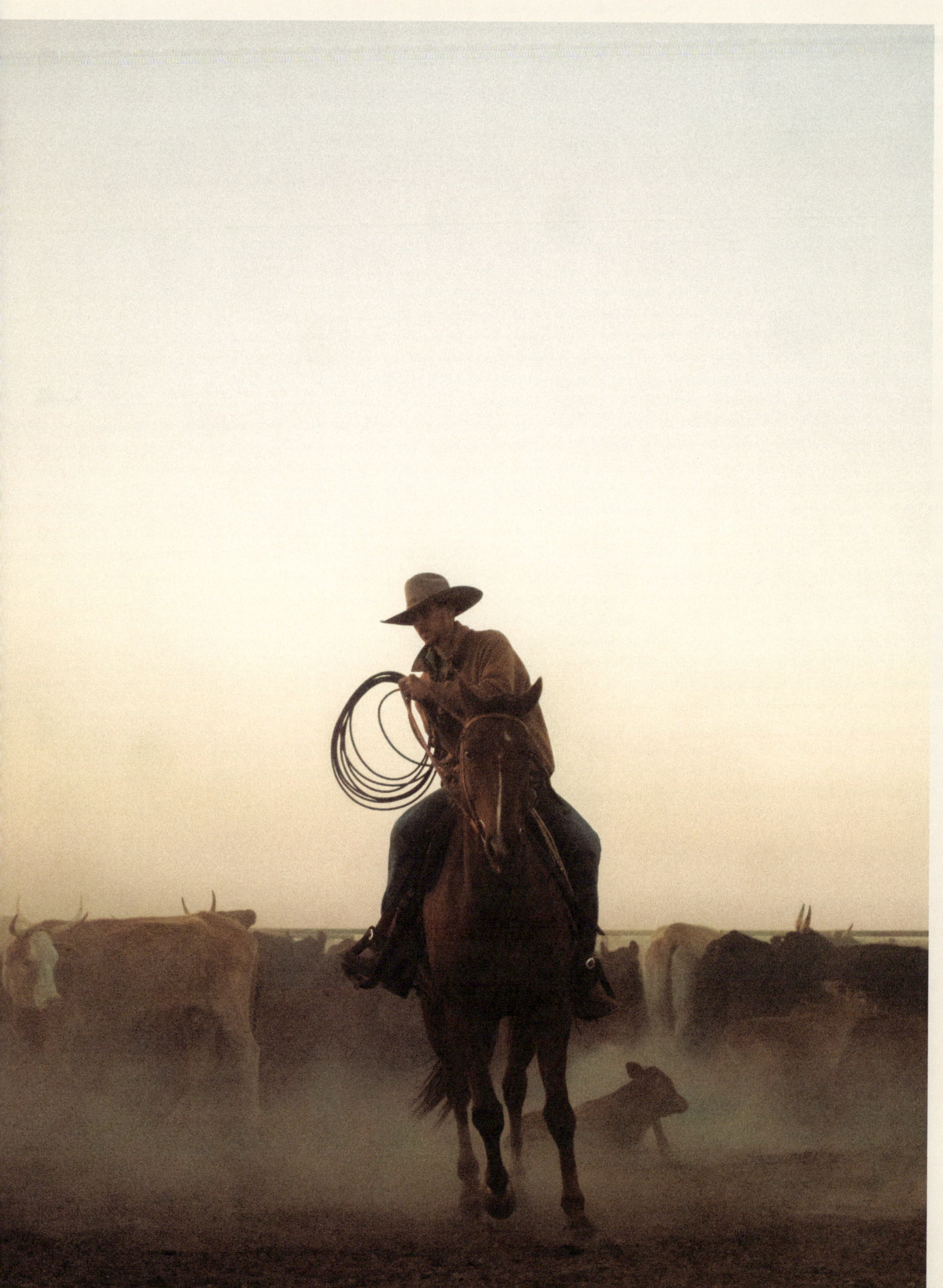

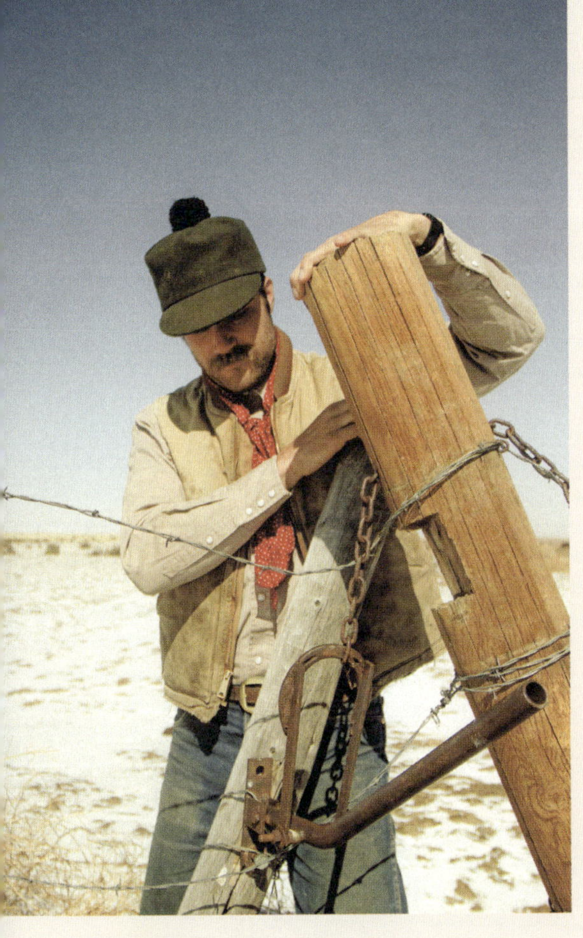

are born in the spring, and great care has to be taken to ensure the mother cows maintain their body condition in order to conceive during the breeding season. Fertility is mostly nutrition, which means we combine careful grazing campaigns with feed supplemented as needed. Branding occurs in late spring and early summer. Summertime's green grass provides the opportunity to raise the condition of the cows for the breeding season and maintain milk production to grow big calves. We wean and test for pregnancy in the fall, sending cull cows to market, and checking calves daily for sickness during and after the stress of weaning. In the winter, we plan, work on projects in the shop, break ice in the water troughs, and manage the cattle during the extreme cold that the season can bring.

BRANDING

Our favorite time of the year is branding season. We are coming out of winter, and if we are lucky, we've had rain and the pastures are greening up. The air is light and fresh; the full weight of summer heat is not yet upon us. Baby calves are everywhere, and they have to be branded before they grow too big. We trade work with the neighbors, helping one another brand. On branding day, which usually happens once or twice a week for about six weeks, everyone from Ranchlands travels to the ranch where we'll be branding. The hosting crew has to coordinate all the meals, have the cattle staged, and sometimes plan a fun day or an afternoon spent hanging out together after the branding.

In the dark of early morning, we trot out to gather the cattle into the corrals or around a watering point. We use long ropes from horseback to catch the calves in the herd and pull them to the branding fire. Their mothers follow and wait next to them as they are being branded. In teams of two, the crews wrestle the calves to the ground to be branded, ear-tagged, vaccinated, and castrated. The ropers are skilled at both moving through the herd with minimal stress to the cattle and catching calves by the neck or the back two legs.

The beginning is always chaotic, with calves, horses, and people going in all directions at one time. There are between two and six teams working simultaneously, wrestling calves to the ground and holding them. Everyone has a different job that requires a lot of movement—running back to refill the vaccine gun, ducking under the rope to the correct side of the calf for flanking, calling over the vaccinator, making sure the correct ear is notched and birthdate brand is applied depending on gender. If one aspect slows down, the whole crew slows down, with the calves waiting. There is no predictable path for the ropers bringing in the calves; they bring them to the crews however they can, with a lot of variability that depends on where in the herd and where on its body the calf was roped. But within the first 20 minutes, things start to flow smoothly. Everyone develops a sixth sense for what's needed where, the teams gel, and what appears to be bedlam is actually beautifully synchronized, controlled chaos.

LUNCH TIME

After a morning of branding, new energy arrives, as the traditional bean burrito lunch is started. By this time, everyone is covered with dirt, smiling, talking, laughing—and also feeling very hungry. The beans are smashed in a bowl that sits in the branding fire. People grab spoonfuls to lay into the tortillas that have been heated on the hot coals from the fire. Diced tomatoes, onions, and hot sauce go on top of the beans. Before you know it, legs can be seen sticking out from under wagons and trucks, where people have crawled to take naps in the shade. No one is talking, except for one or two people who are not sleeping. The cattle drift off slowly, and gangs of calves run around. The horses stand sleeping, resting along the fence. After a while, people begin waking, putting the kitchen and branding equipment away, readying themselves to go back to headquarters. If there is anything that comes close to a more beloved time than branding, it is the trip back to headquarters. We move in groups, strung out across the pasture, everyone talking and the horses moving easily at a slow gait. No matter how sore or tired we are, how many welts we have from kicks and hitting the ground, we ride together, happy to be exactly where we are at that moment. Although the work is long and tiring, work is not really work when you are branding.

The work we do requires so much more than just specific skills or being handy. It teaches all different forms of leadership: communicating effectively; making sure everyone knows the priorities; working in a tight-knit team to solve situations that are often intense, exhausting, and dangerous; and learning how to talk about difficult things. Practicing these skills will not only make us better ranch managers but also better human beings, better community members, and better citizens of this planet.

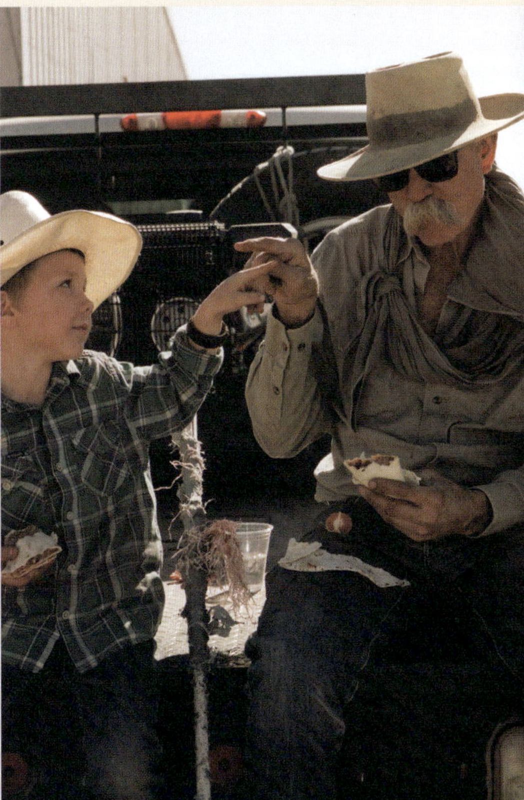

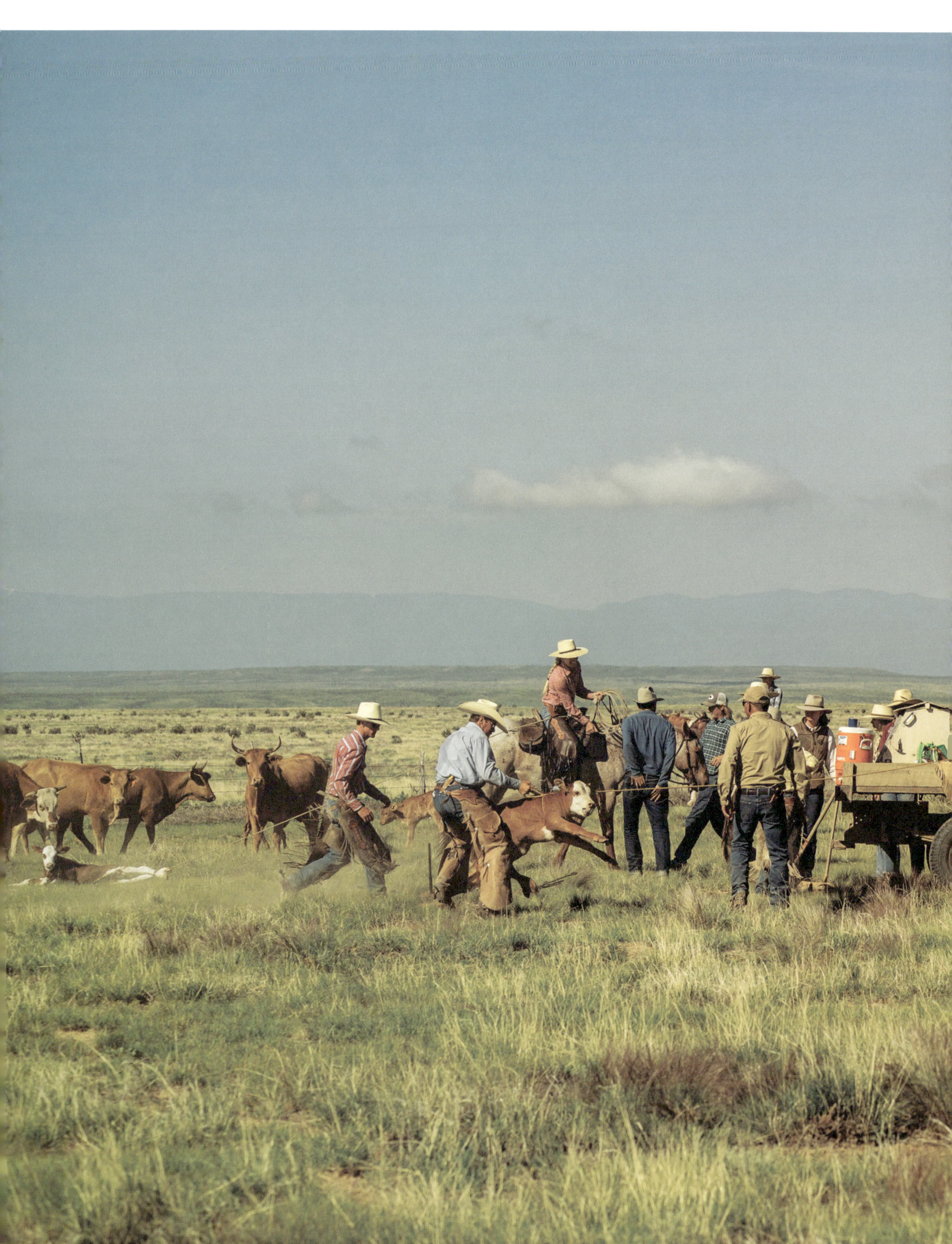

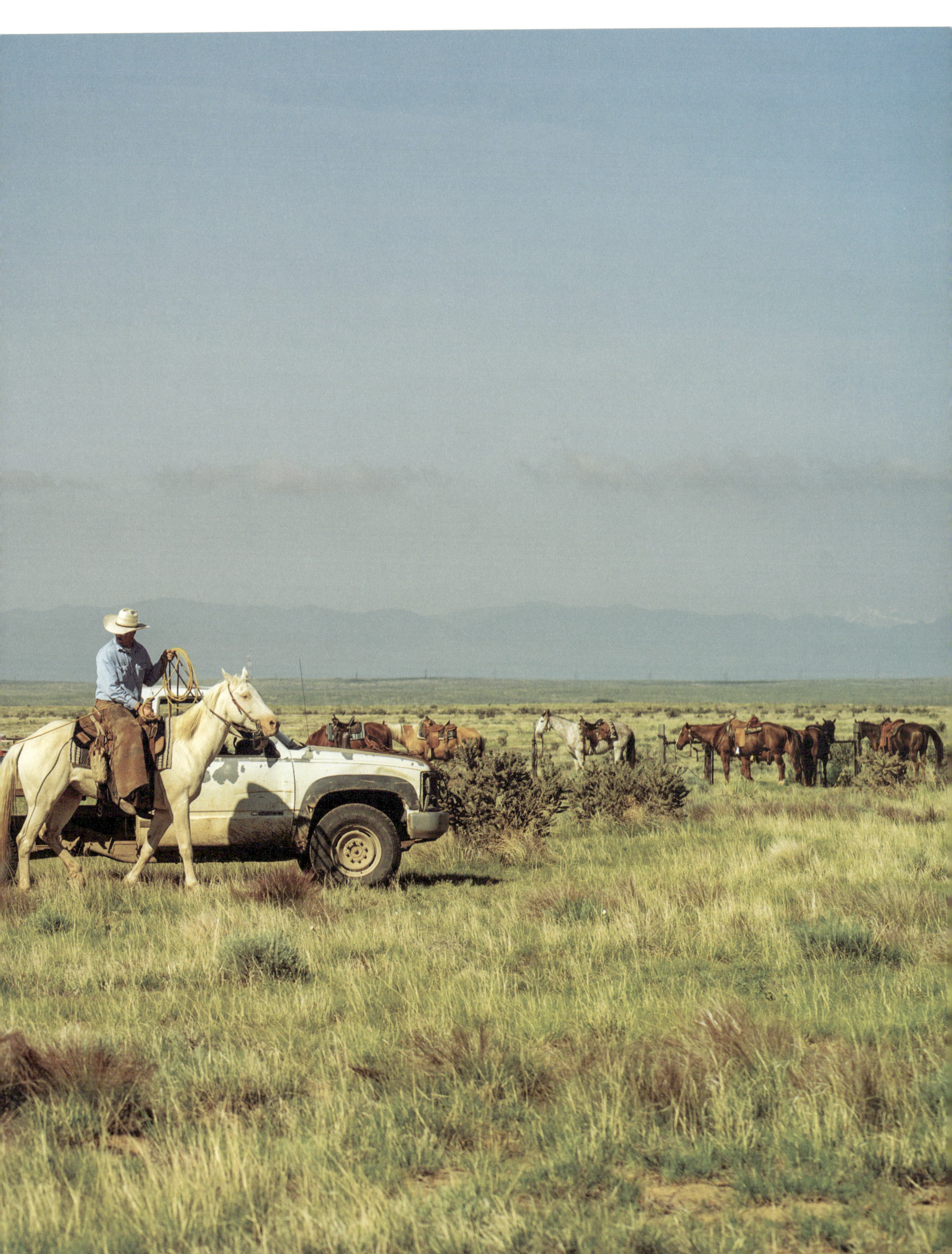

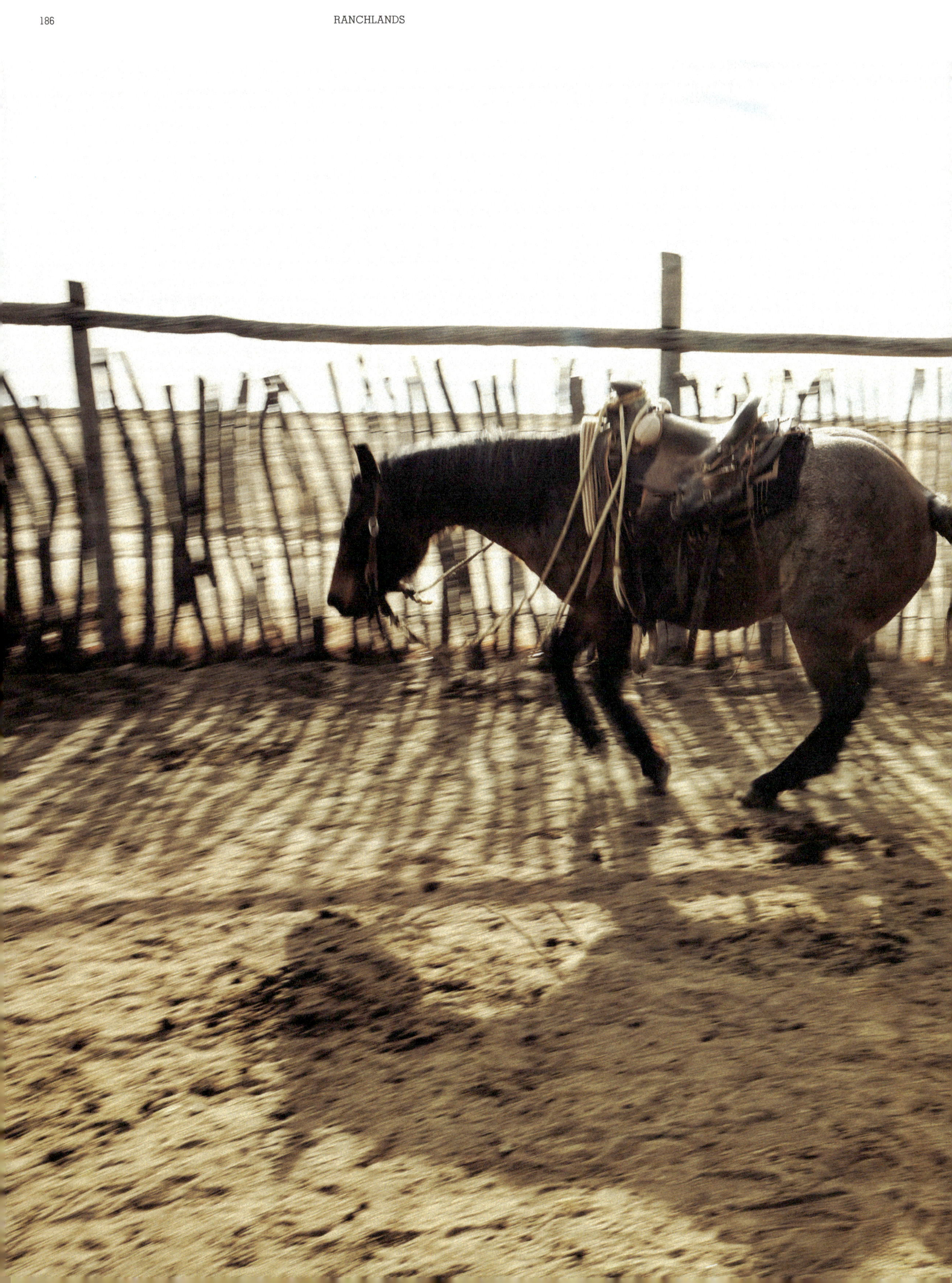

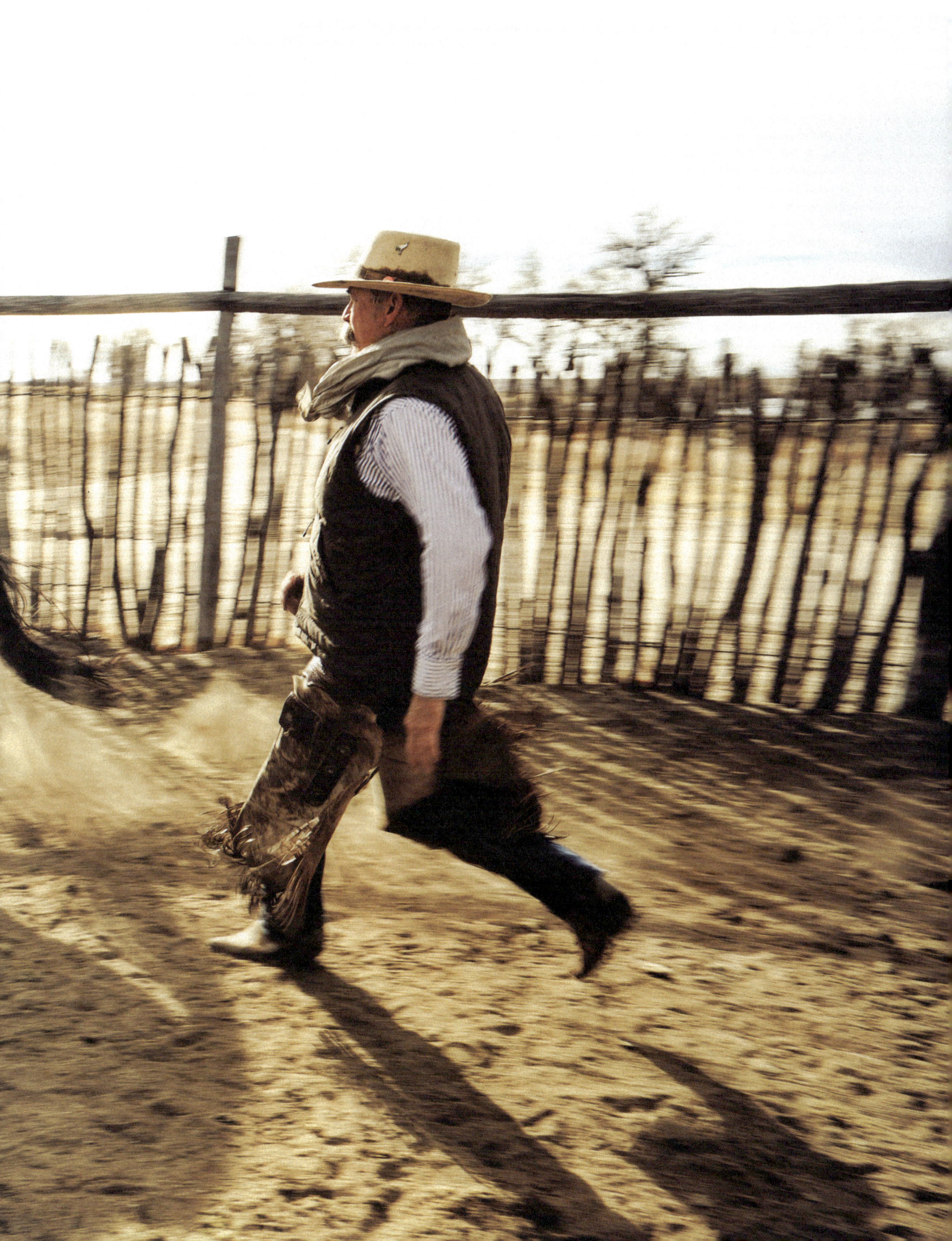

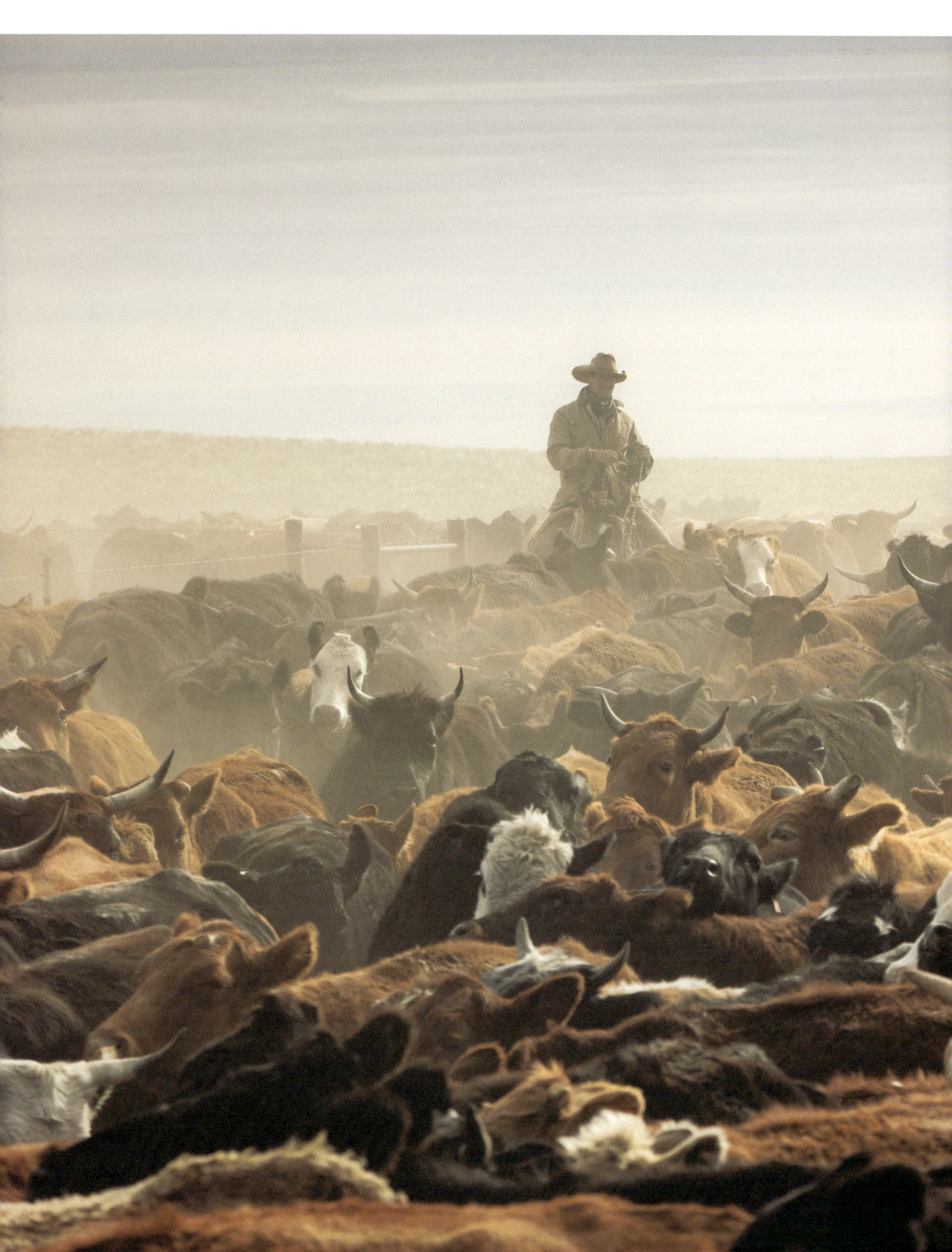

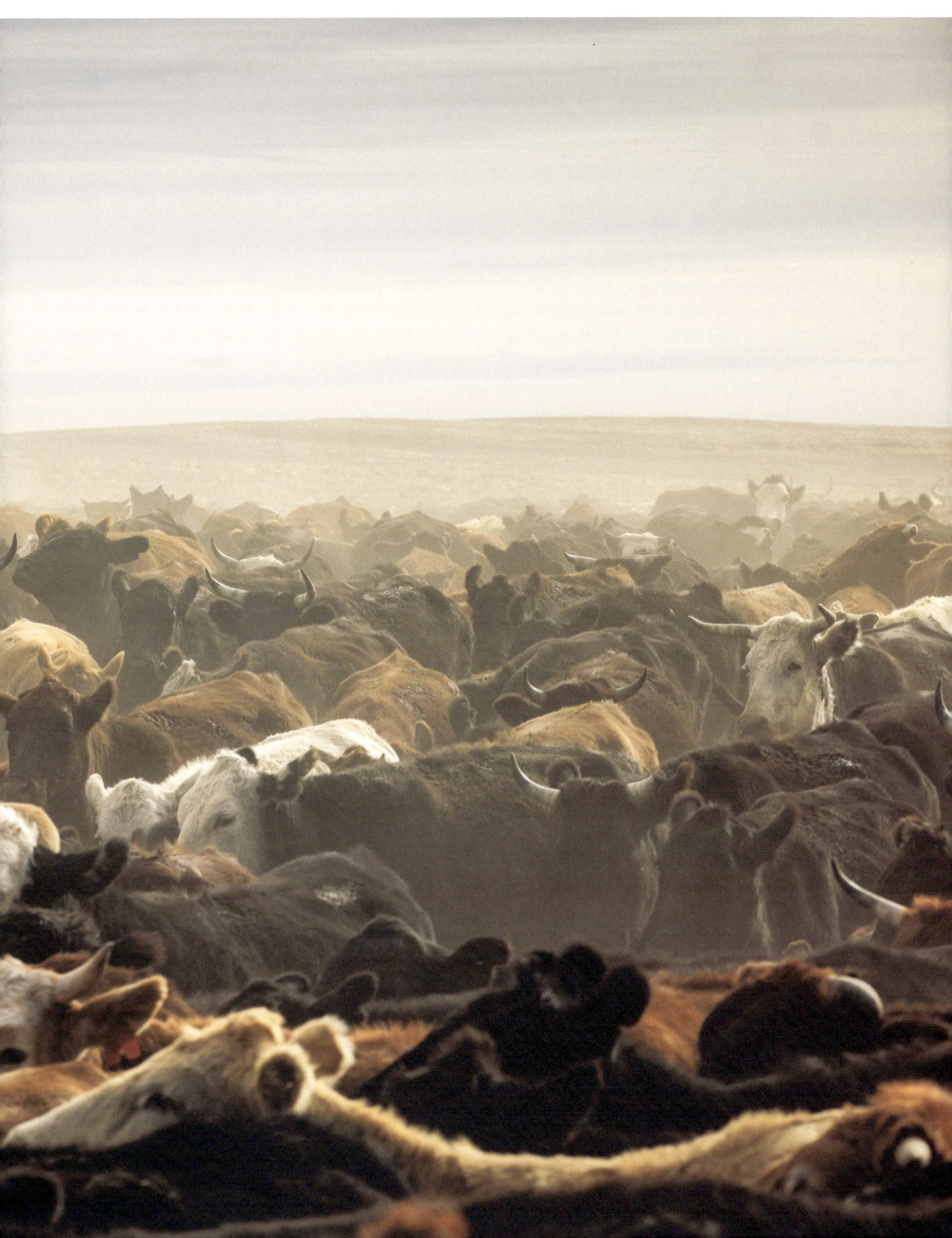

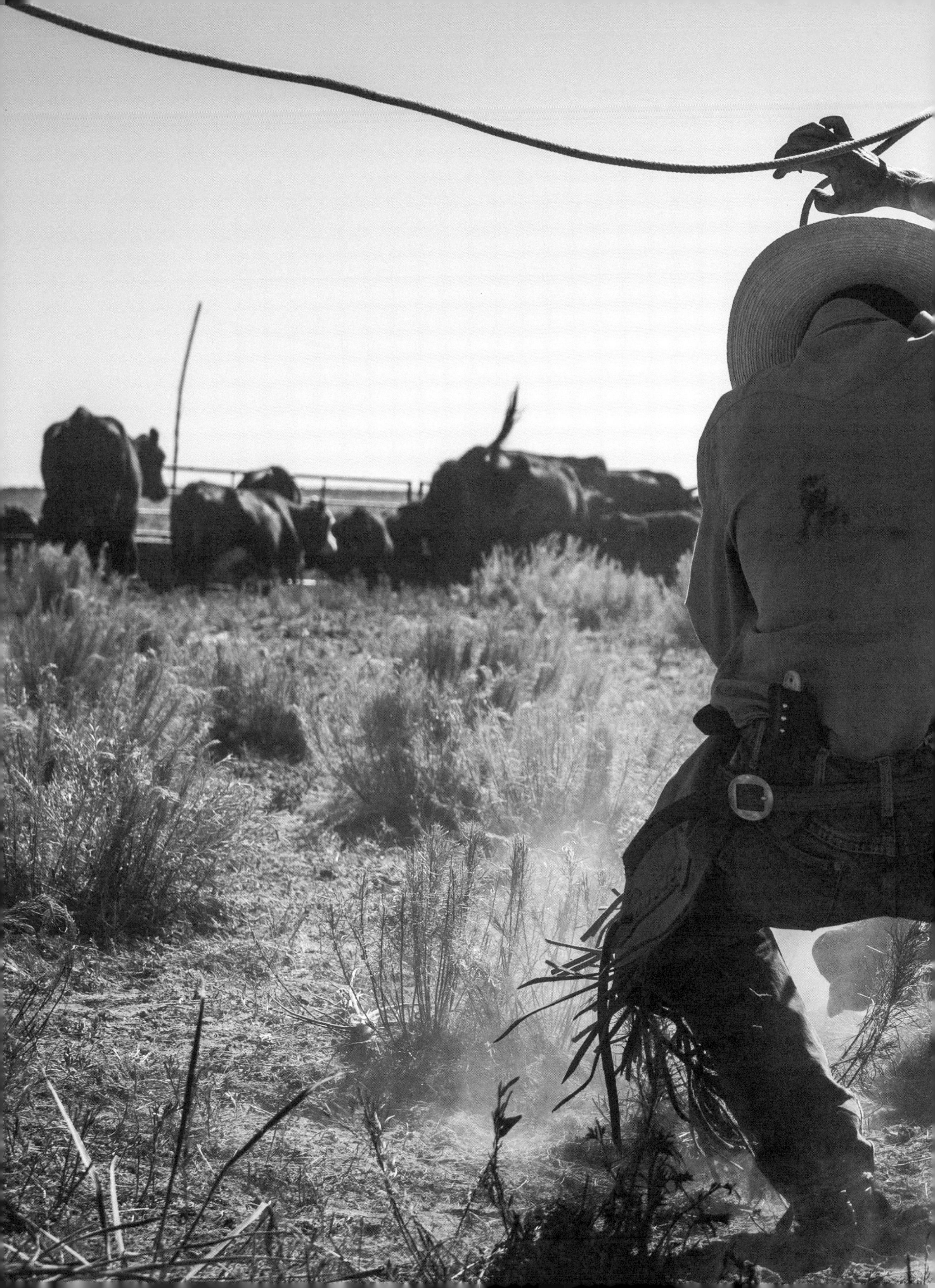

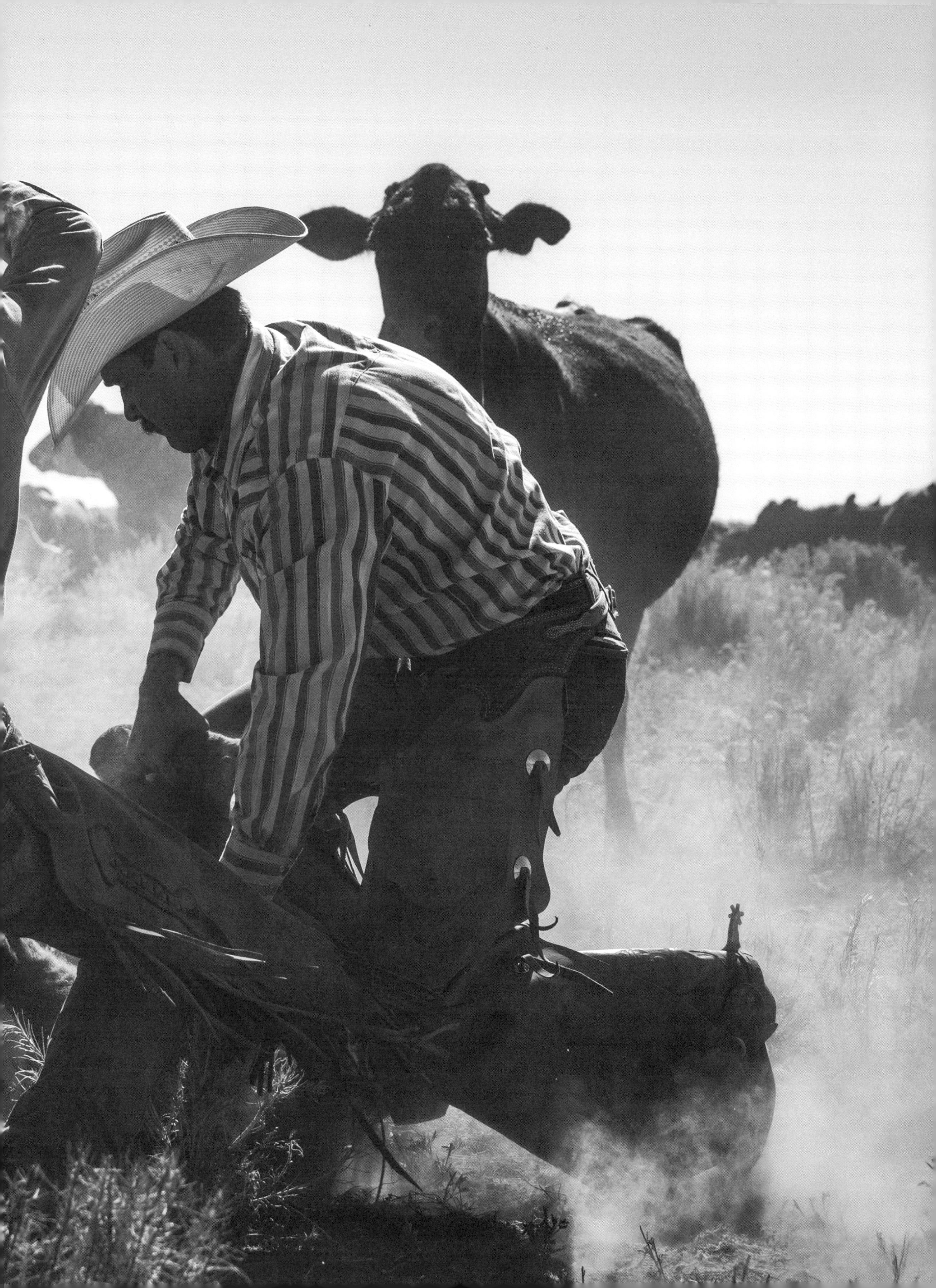

CONNECTION

Much of our work revolves around developing, maintaining, and enhancing connections. We work in the currency of continuity. When a fence breaks, we mend it. When a water pipeline leaks, we patch it. When a cow strays from a trailing herd, we ride in and push her back to her place. It's our job to restore connectivity to these lines of wire and PVC and rushing water and cattle. The ranch is blanketed by a scaffold of such fences, pipelines, and cattle trails, layered over one another and imprinted upon the land, above and below ground. These systems allow us to manipulate how our cattle behave, where they graze and when, like force fields pushing and pulling them across the land.

What binds us all to the land and cattle, the wildlife, the plants and rocks, and one another is that we are members of a shared community. The constant work of maintaining these systems connects us to the cattle and the land, and it sanctifies our mutual dependence. All of us at Ranchlands are connected to one another by our commitment to learning this work. Wherever we are, however different the climate might be, we all have cattle, land, and one another to take care of, and we share the same daily rituals and responsibility to maintain our webs of continuities. It is made known to us that decisions we make on a daily basis, both large and small, may have a direct impact on the nonhuman beings around us and the landscape we share. Perhaps we will observe the effects of our decisions tomorrow, or next week, or next spring. Or we might never get the chance. These stakes are high, but they are more energizing than paralyzing. It only deepens the mandate of stewardship to be humbly aware that our connection to the land spans large temporal distances as well as physical.

Imagine nourishing drops of monsoonal rain that fall on a neighbor's pasture. You can follow that water as gravity pulls it from higher to lower ground, following continuous flow lines that converge into the Chico Creek. It meanders downstream, cutting across silt bars, plunge pools, and cattail stands. It gains momentum, carrying sediment and debris further down while rich soils and roots—made strong and vigorous by a cow—filter and try to slow the water as it passes, before it inevitably finds the Arkansas River. These seasonal floods carry the effects of our soil and management to the world beyond, connecting our creek basin to a larger river basin and our small crew to a larger community of humans.

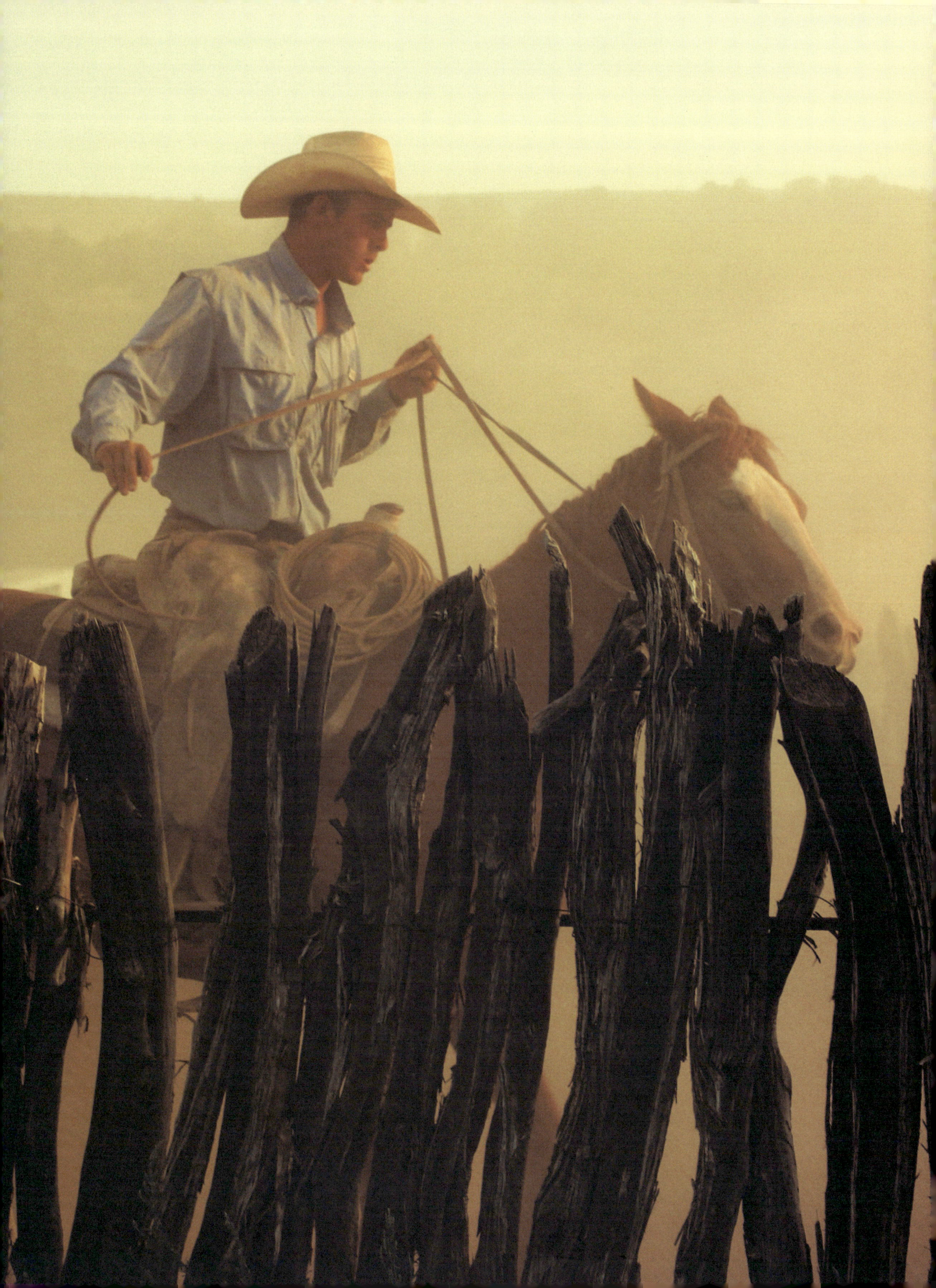

TOOLS OF THE TRADE

The equipment used by Duke III has become a part of his everyday life on the ranch. From the spurs handcrafted in remote Mexico to the belt his father passed down to him and then to Duke IV, all of these items hold significance and purpose.

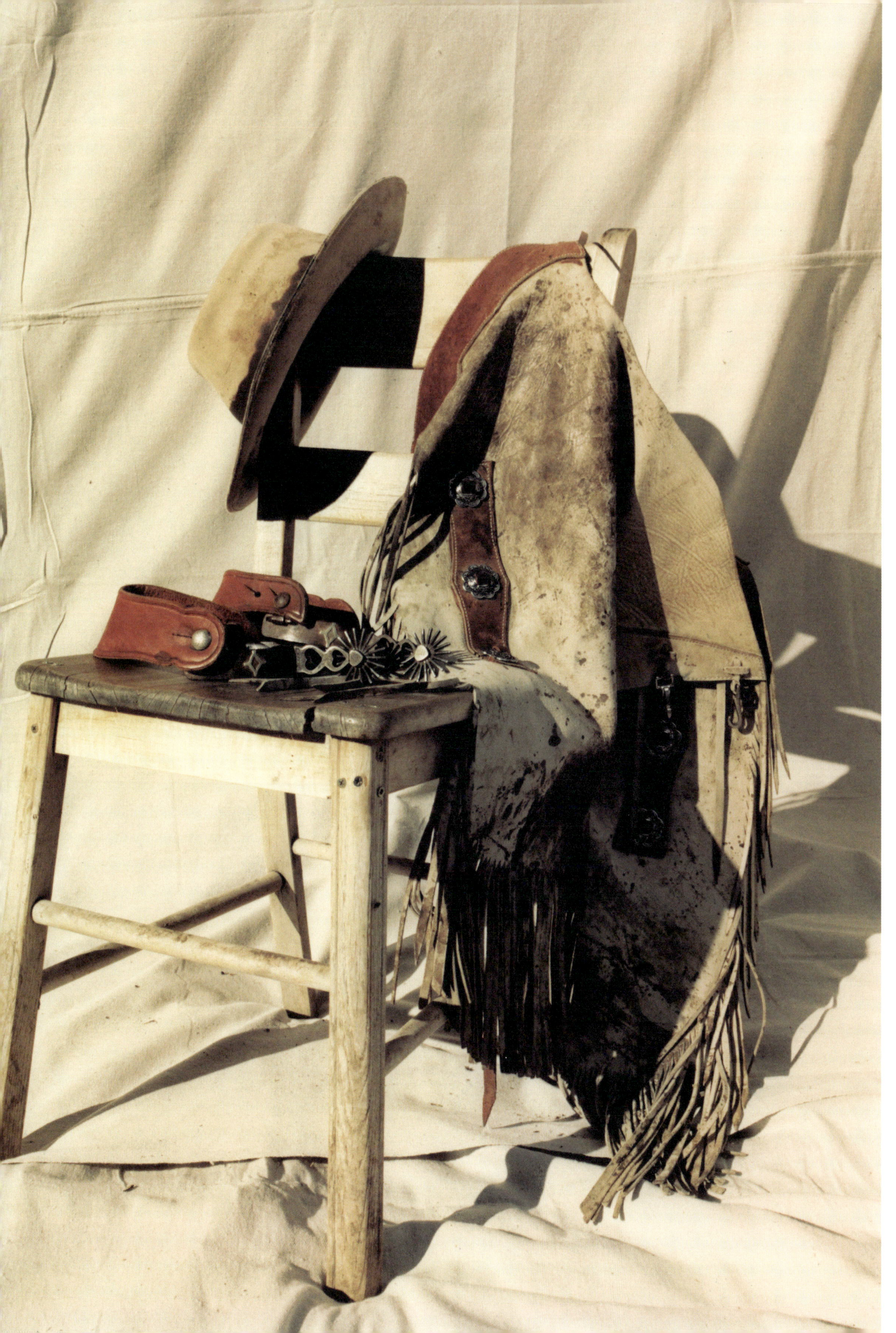

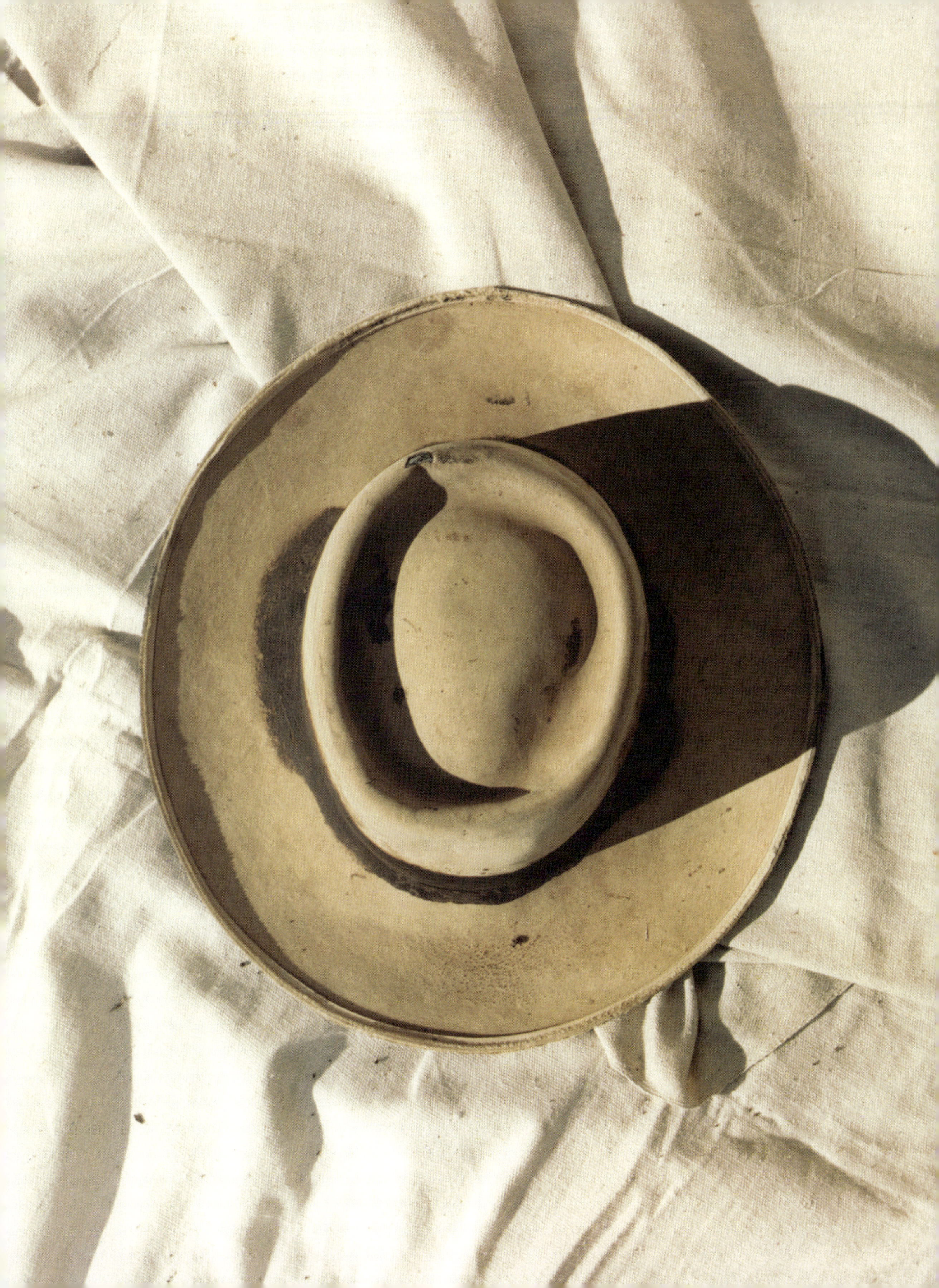

HAT

This is Duke III's handmade beaver hat that he has worn since the beginning of the Chico Basin Ranch lease almost 25 years ago. Throughout all the seasons, this hat accompanies him in his daily tasks. The brim is flat to provide maximum shade coverage from the sun and protection when running through thick brush. When he's not wearing it, the hat is used for many other things, such as a pillow or a bowl to provide water for the horses to drink.

BOOTS

These boots belong to Duke IV and have been worn by him nearly every day since they were made.

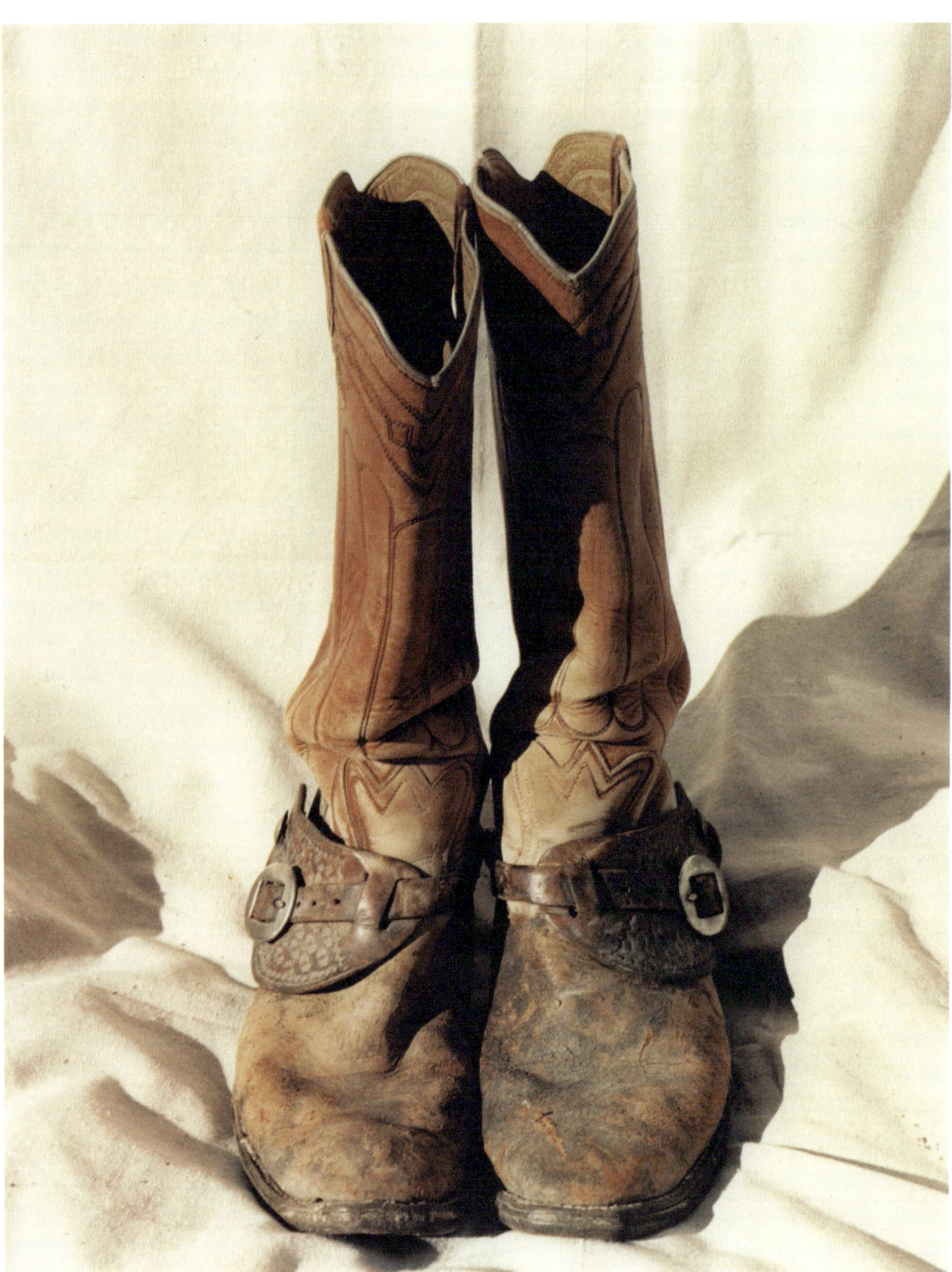

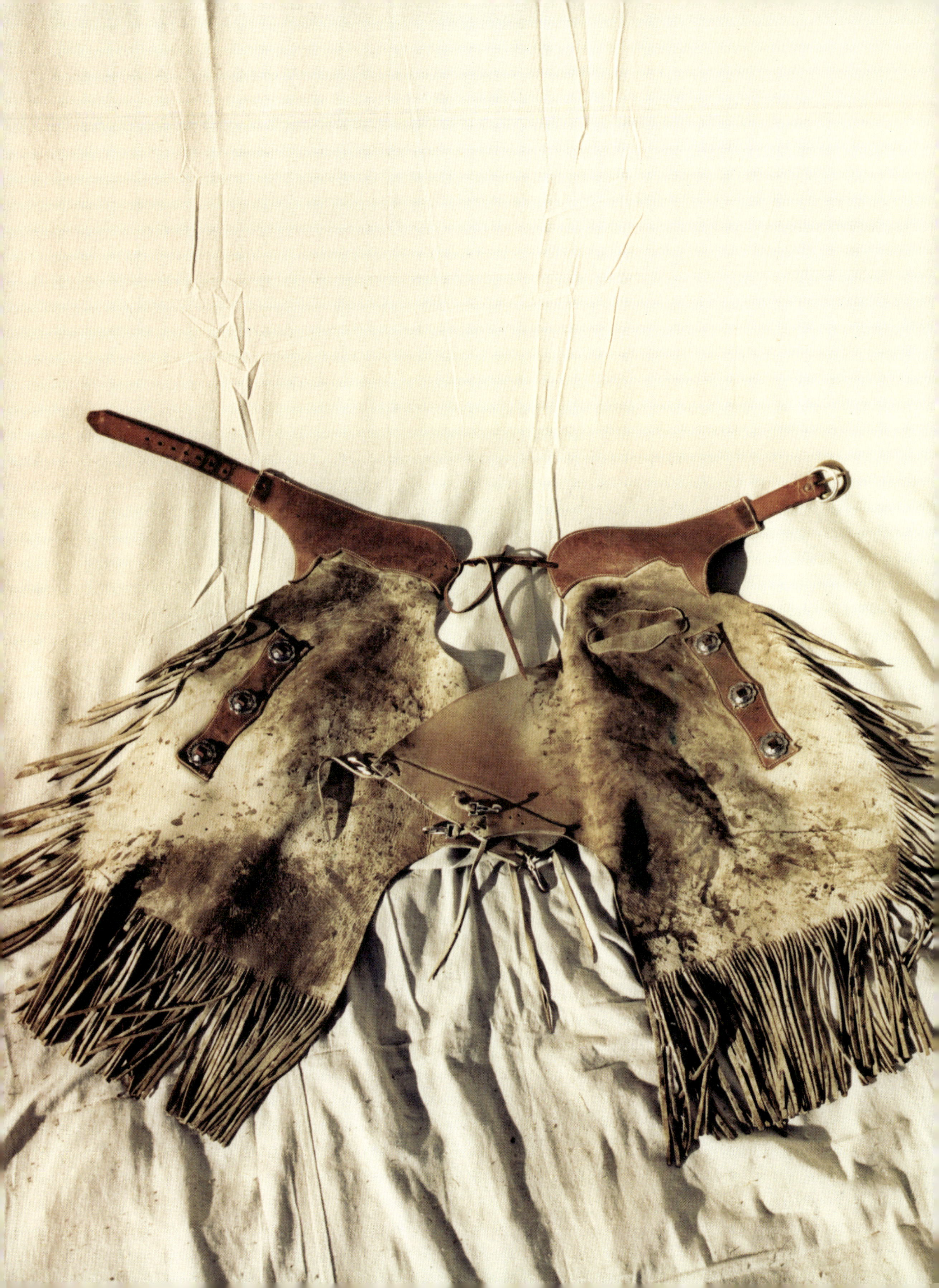

CHAPS

These chaps were handmade in Ranchlands' leather shop in 1999, and Duke III wears them almost every time he rides. The large batwing design keeps brush off of the legs and provides a sleeve against the saddle when riding. At the hip, they protect the sides of the body from rope when pulling something with weight from the saddle horn. Duke has also been known to use them as ground cover for napping or protection when caught out in a storm.

BELT

The DP initial belt clip is a family heirloom and tradition that has been passed down for generations. This one first belonged to Duke II and was later given to Duke IV. The leather portion was crafted by Duke III in the Ranchlands Mercantile. Duke IV wears this belt every day.

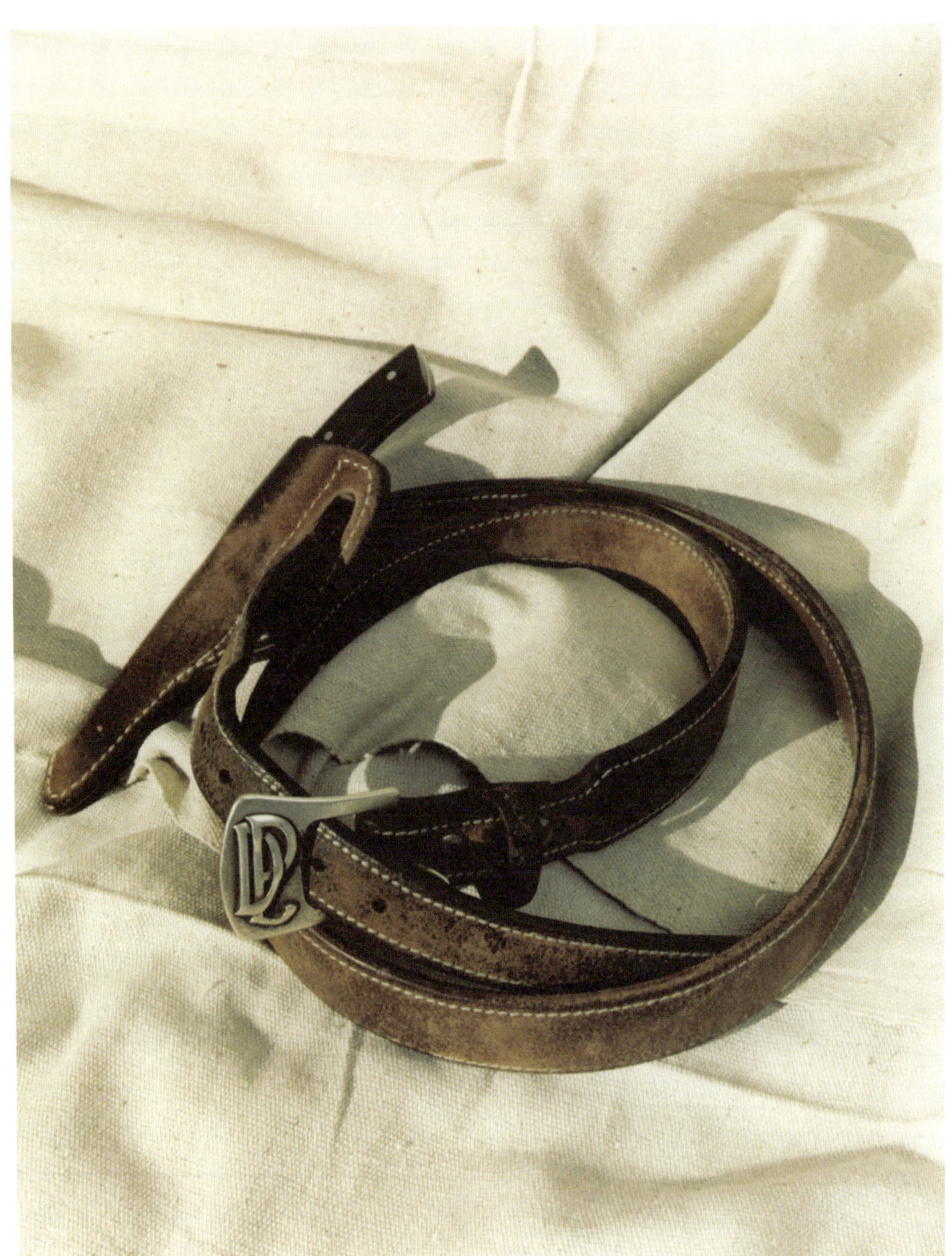

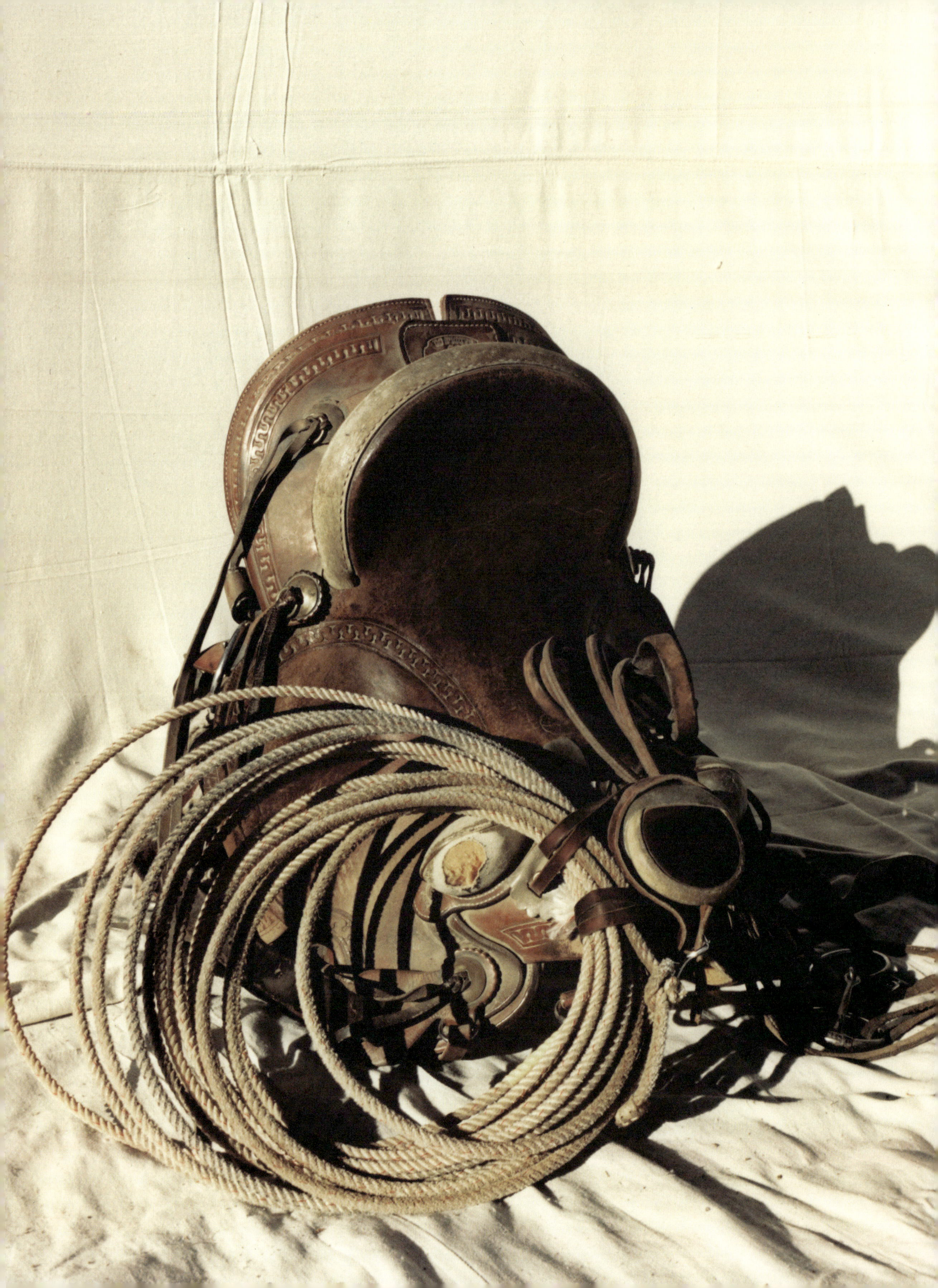

SADDLE

Duke IV has used this Wade saddle for nearly 20 years. It has taken him thousands of miles on horseback.

SPURS

These spurs are a one-of-a-kind pair that Duke III wears whenever he rides. The design was handcrafted in the backyard of a Mexican blacksmith, who used to make them after melting down old pesos. His shop was made of bare-dirt ground, situated under the shade of a huisache tree in his yard. He used car springs to shape the spurs on a small handmade forge.

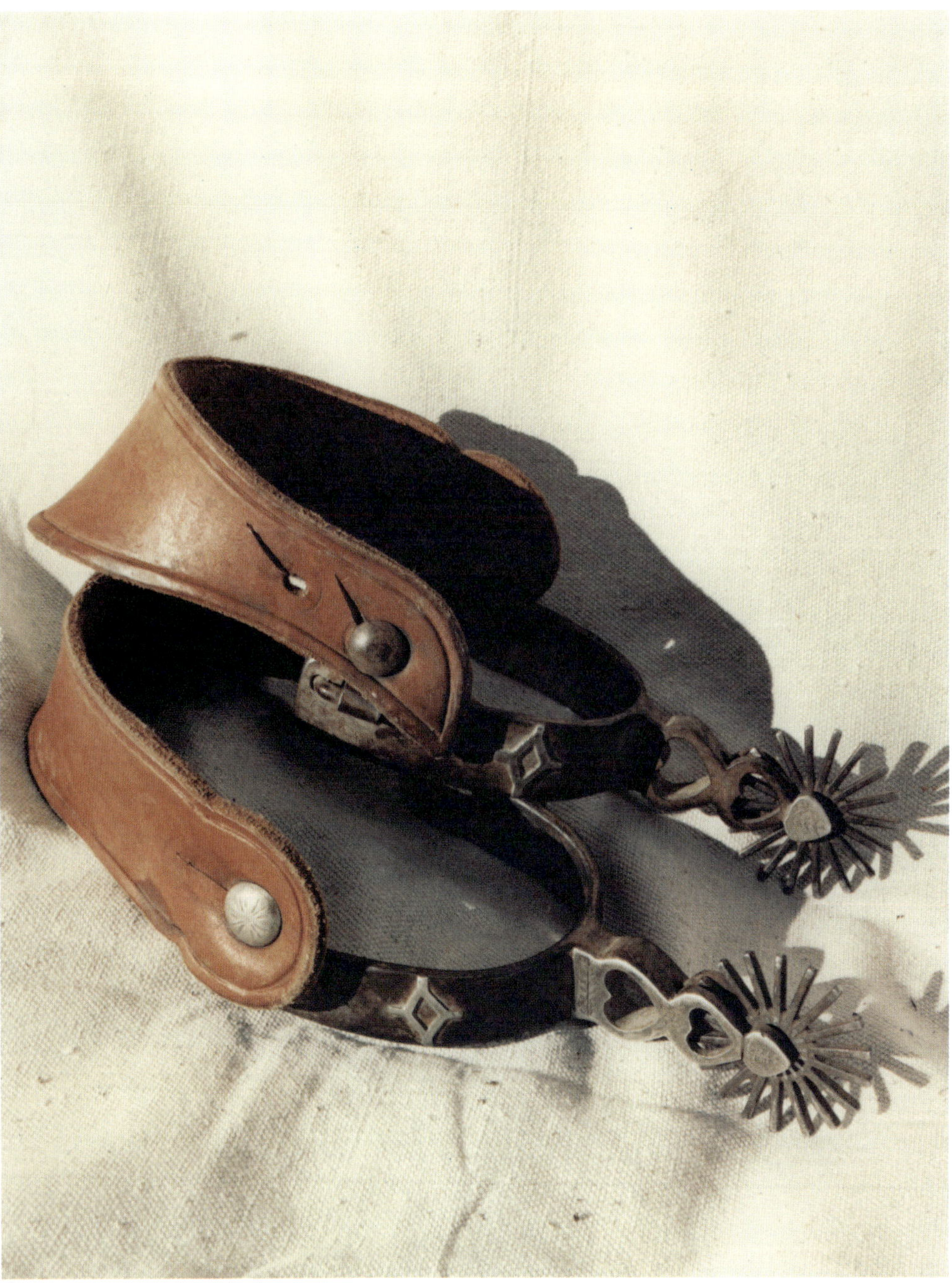

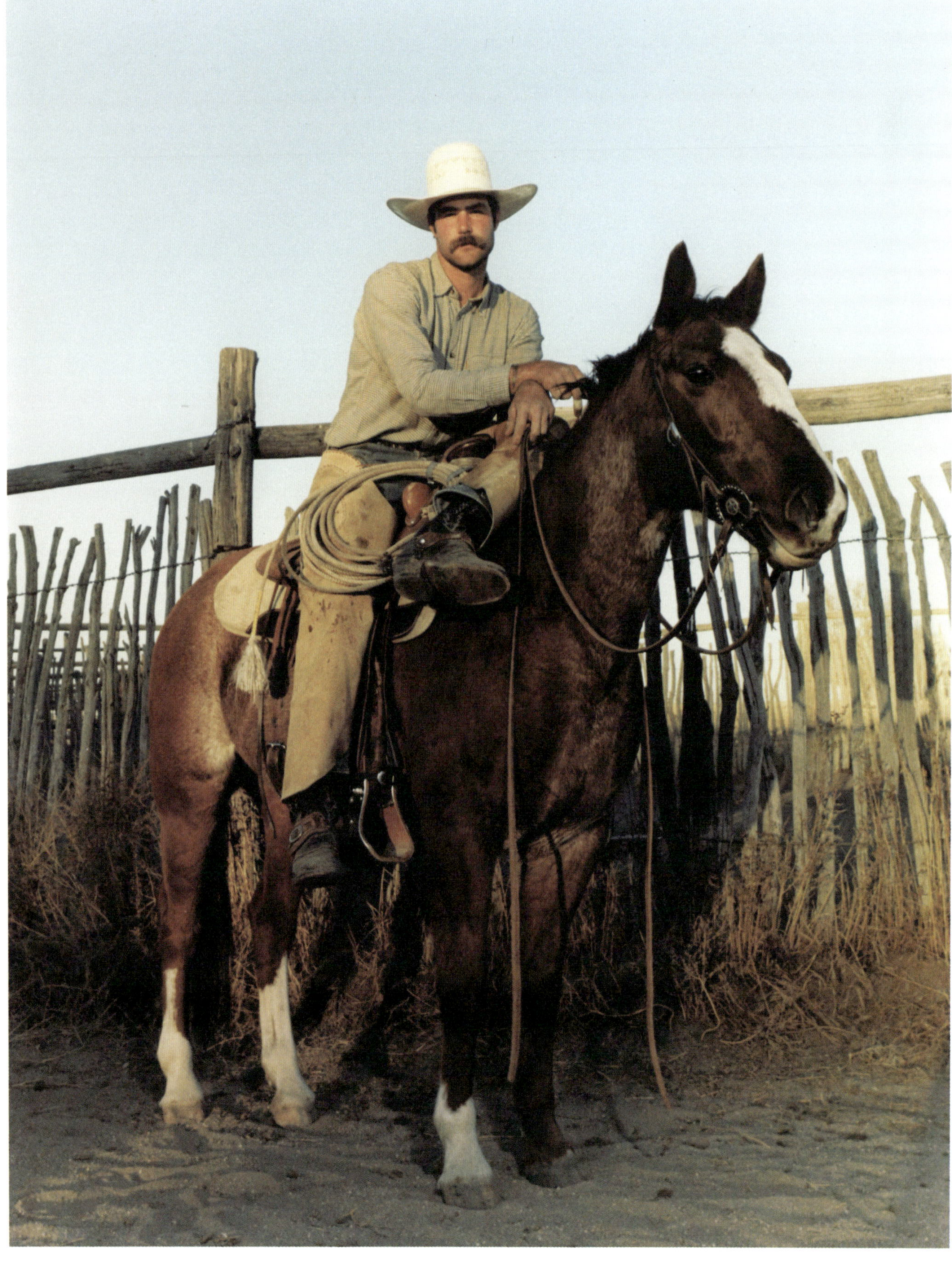

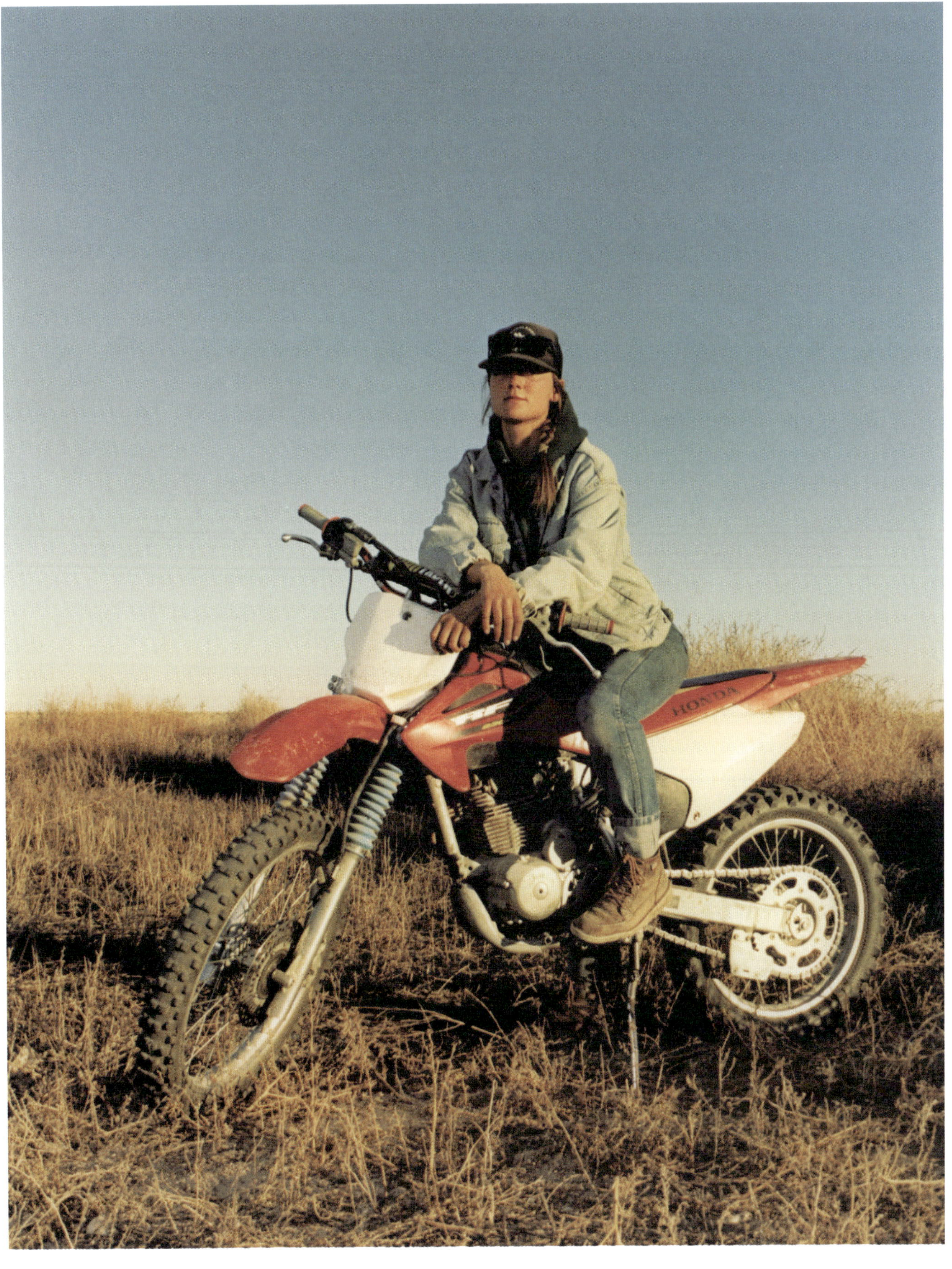
HONDA

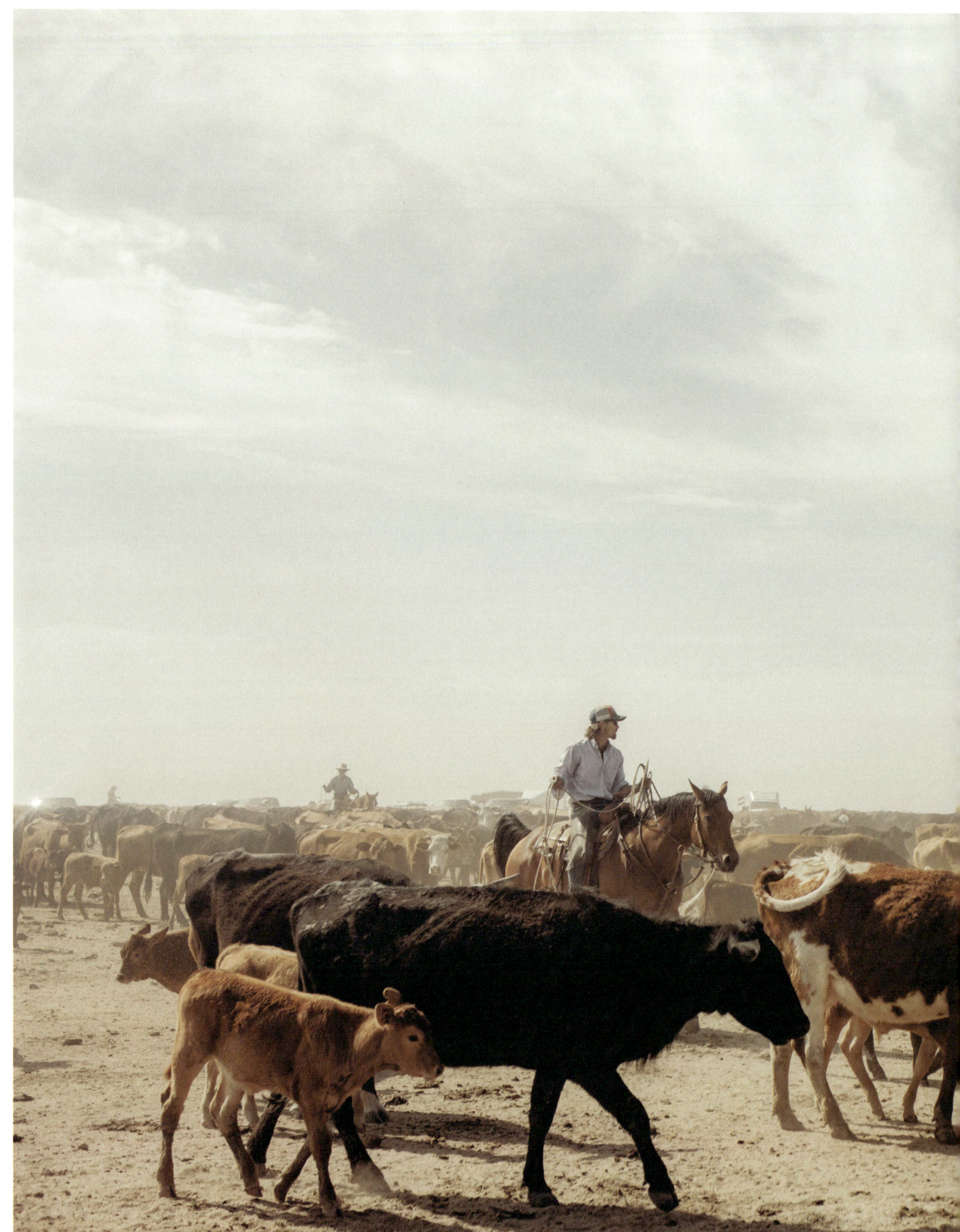

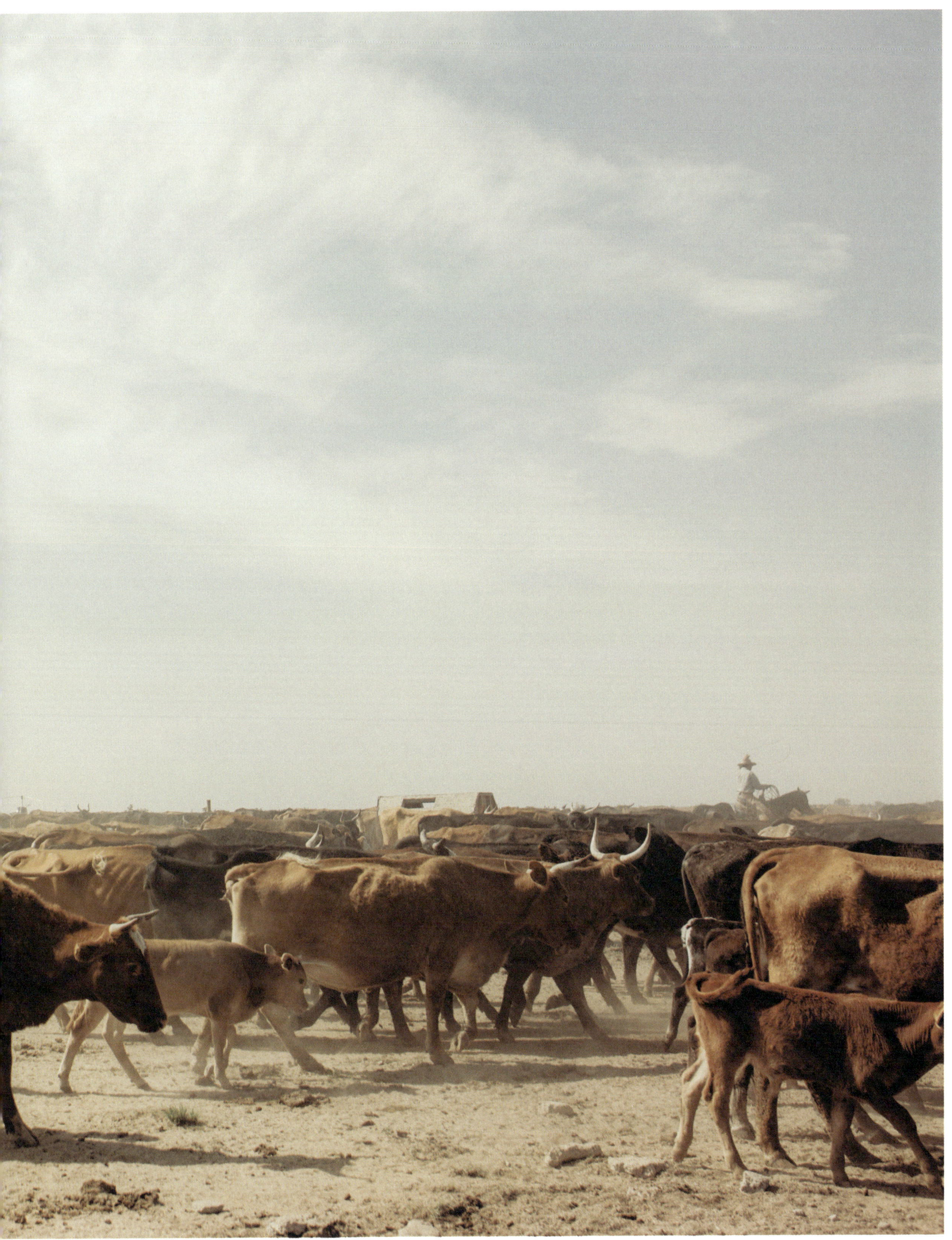

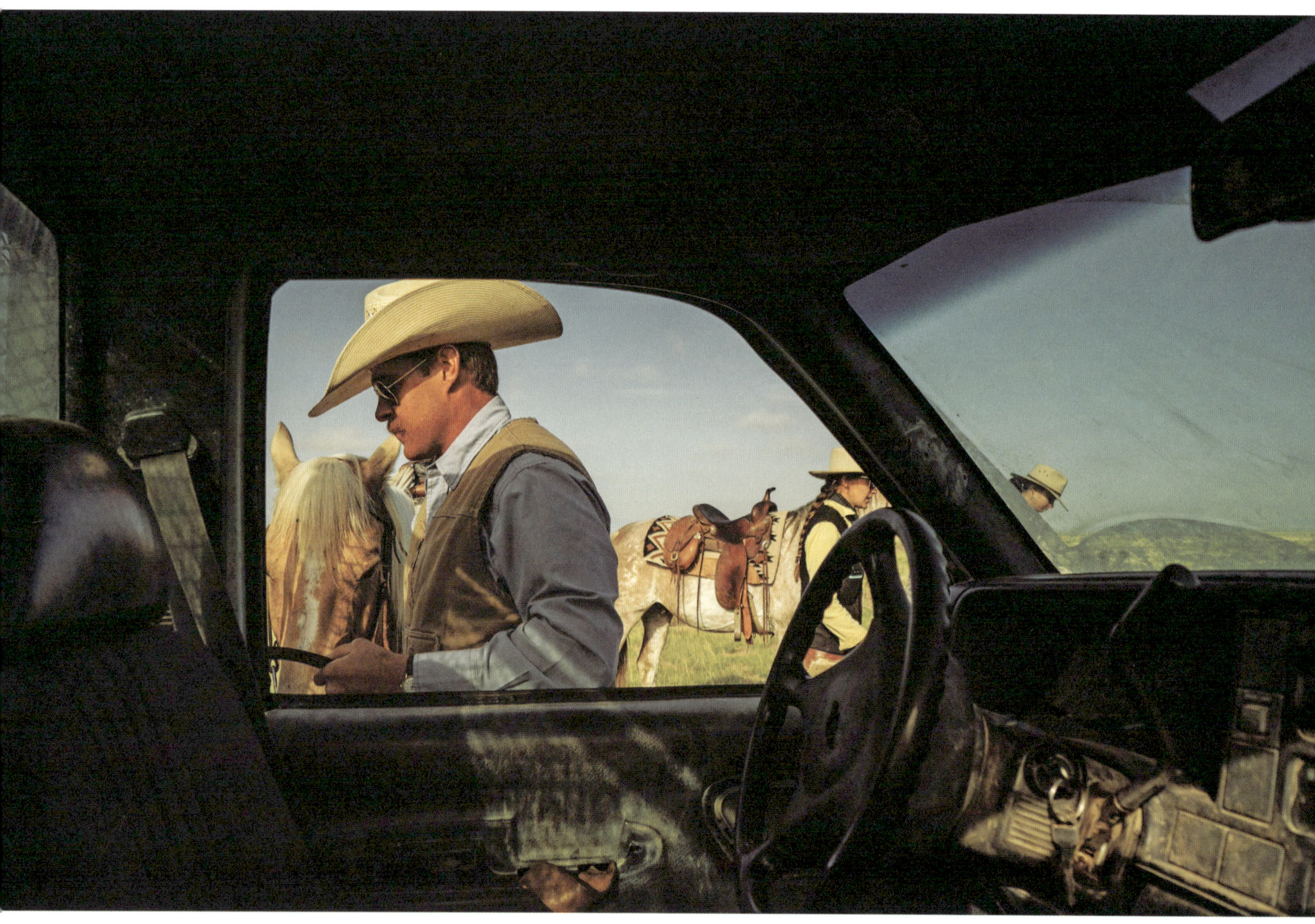

We use helicopters, dirt bikes, and horseback riders to gather herds of cattle and move them to new pastures according to our grazing plans.

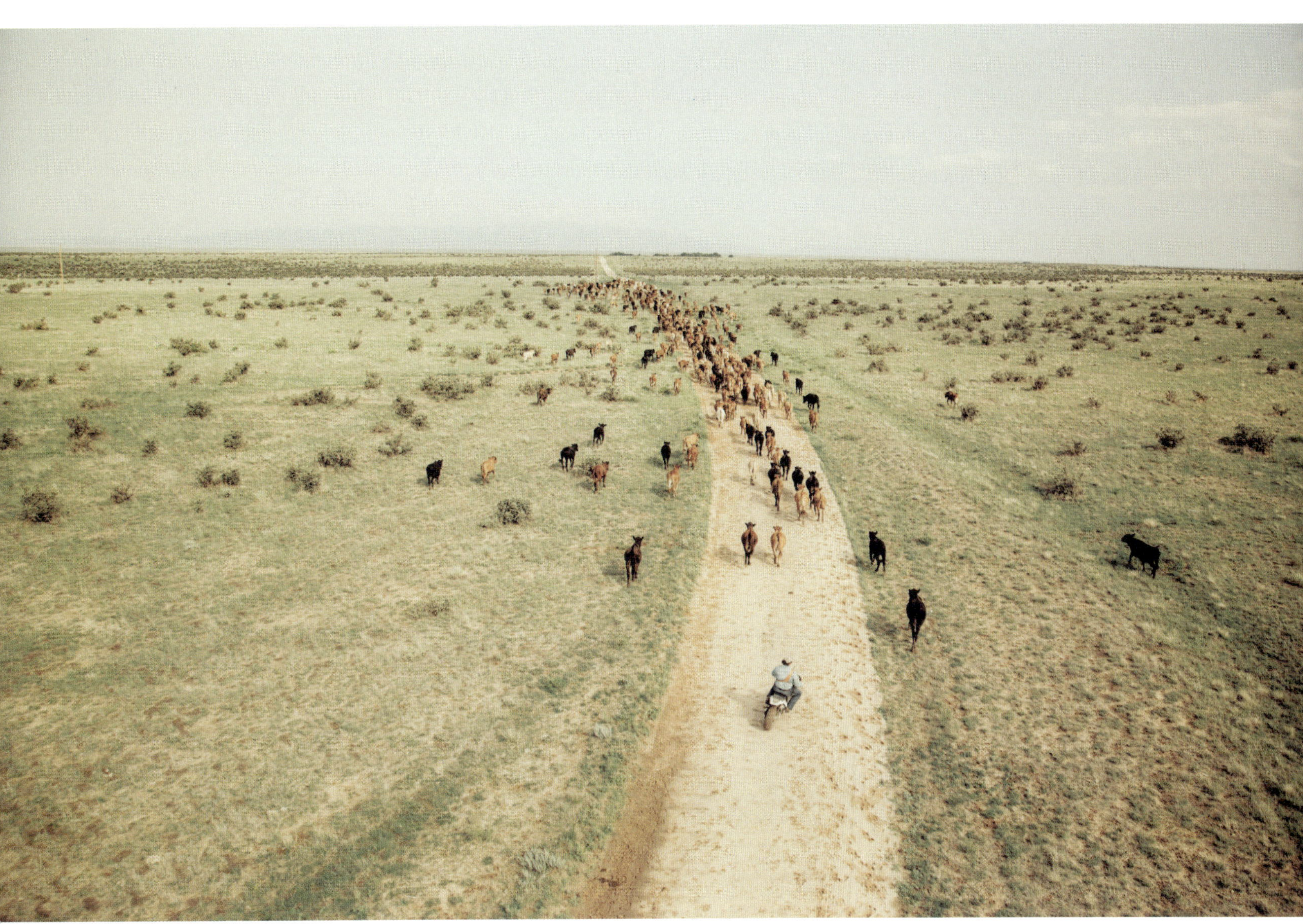

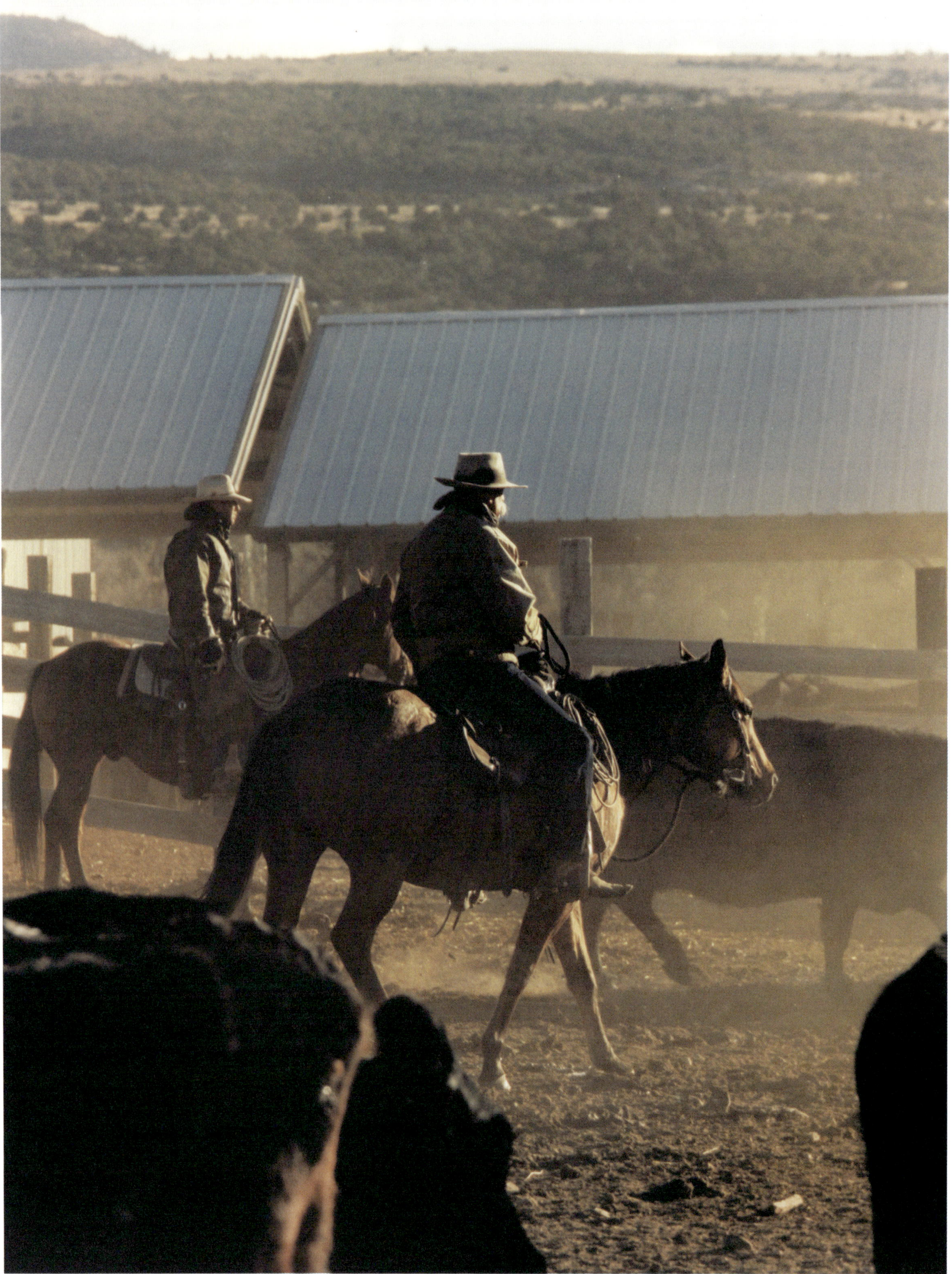

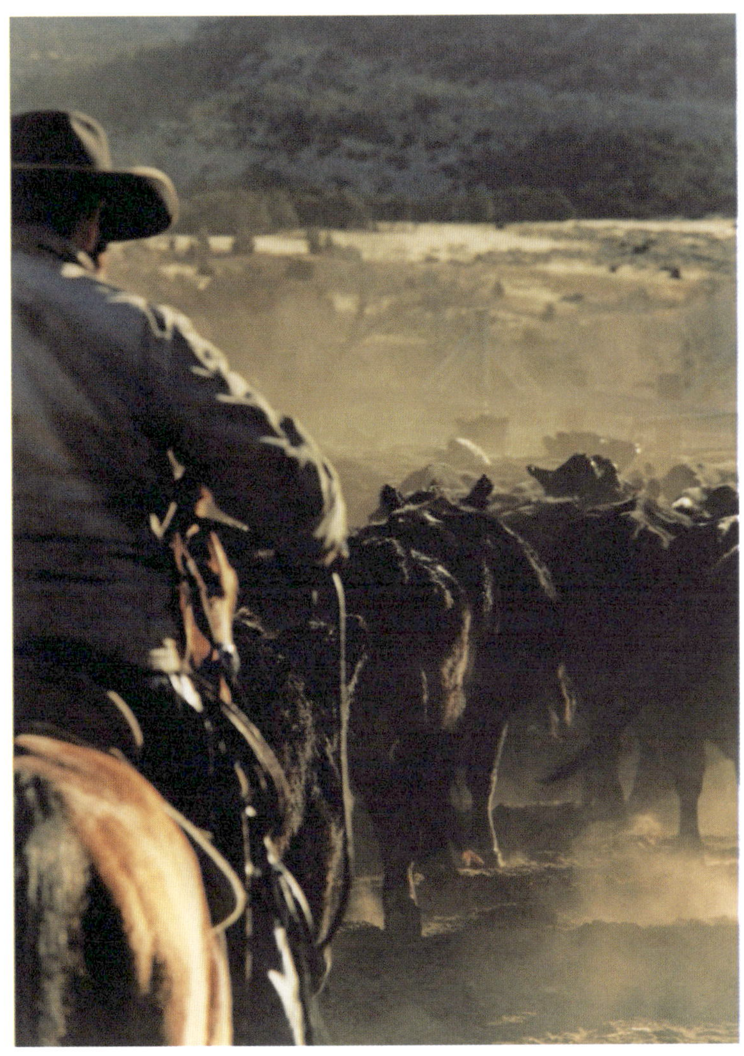

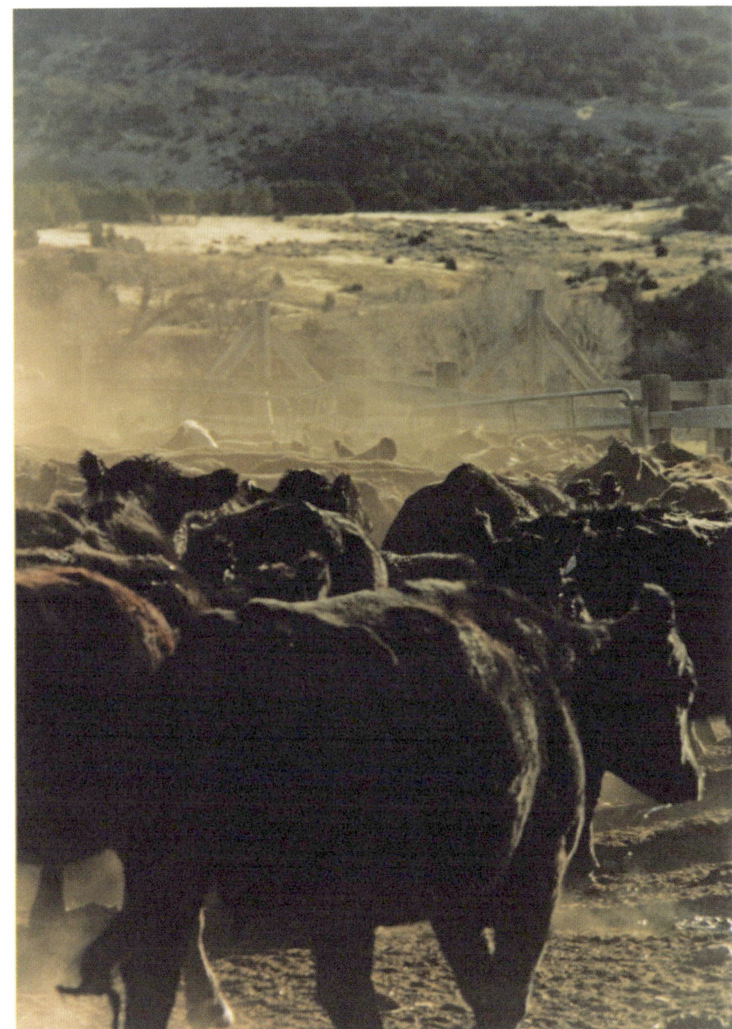

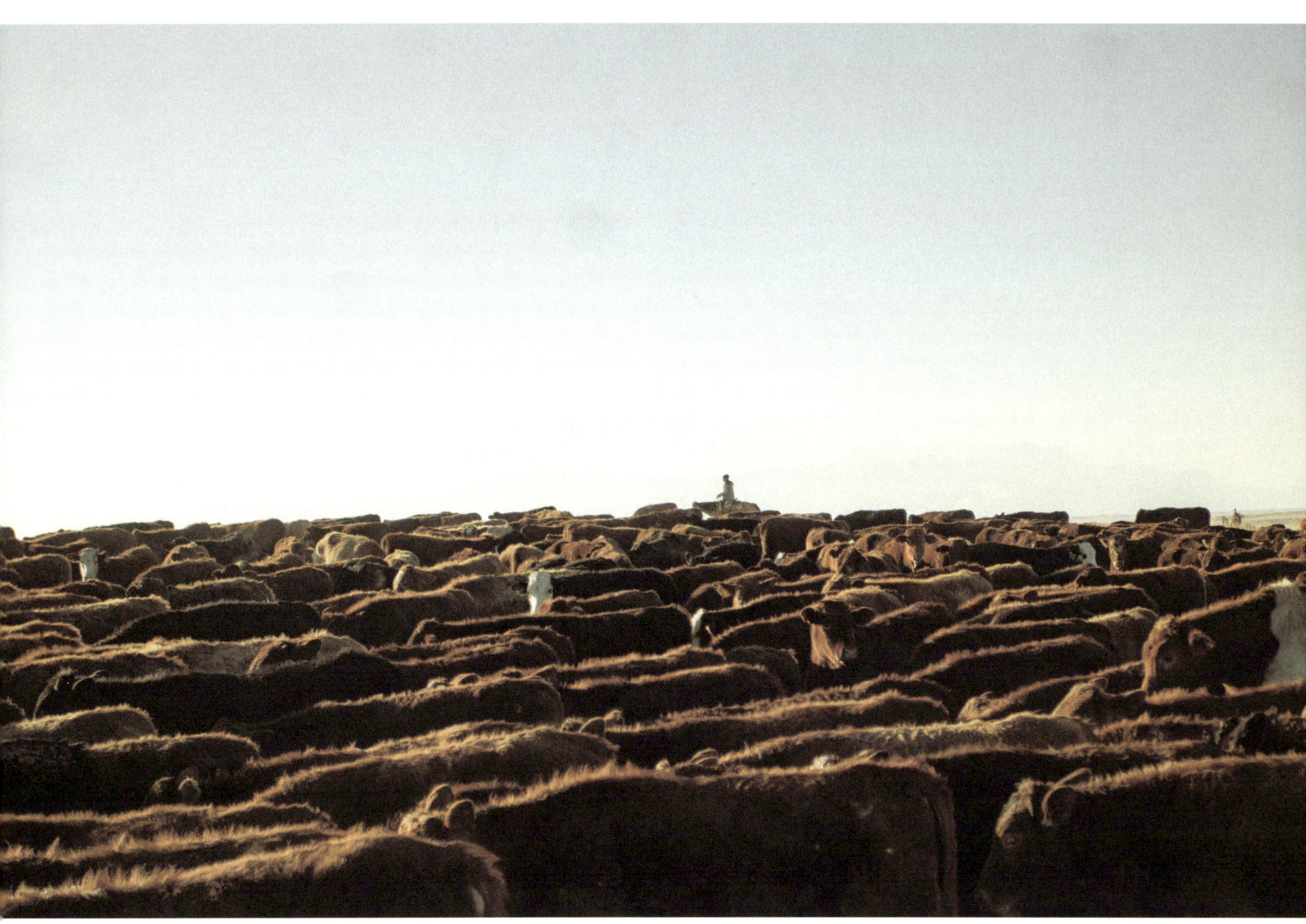

MIKE GIORDANO

"One time, I was gathering on the MP Ranch with Duke IV. He was in the helicopter, and I was on a bike. The pasture was just under 8,000 acres of absolute wilderness—canyons, mesas, rocks, juniper, cholla—nasty, thick, beautiful New Mexico country. It was incredibly physical work maneuvering the motorcycle around those rocky mesas, and it was extremely hot. I also had no water, and this gather was taking up a good part of the day. I knew I was in a pretty dicey spot. That was probably the thirstiest I've ever been in my life. I was lightheaded by the time we got the cattle to the drinker. I stuck my head under the water and began pulling in gulp after gulp, fully aware of the amount of algae at the bottom, the bugs swimming around, and all the times I had seen cattle pissing and shitting in the drinker. I had even pulled out a dead oryx right before. That water is also heavy with gypsum and sulfur, but none of that mattered—it was the best drink of my life. I was definitely sick later on, but it kept me going long enough to finish the gather. There's something rewarding about being pushed uncomfortably far and then having to improvise in order to keep going."

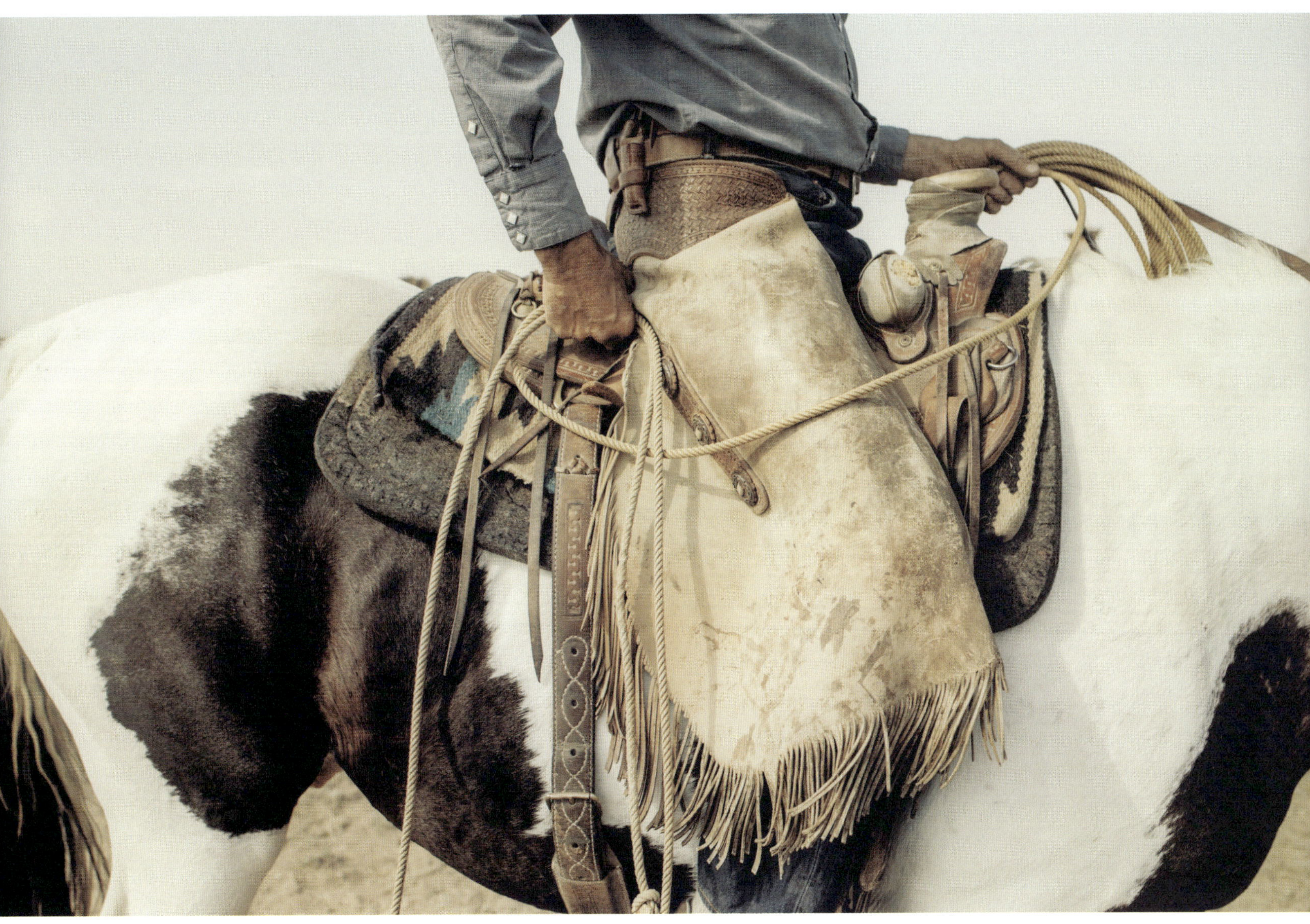

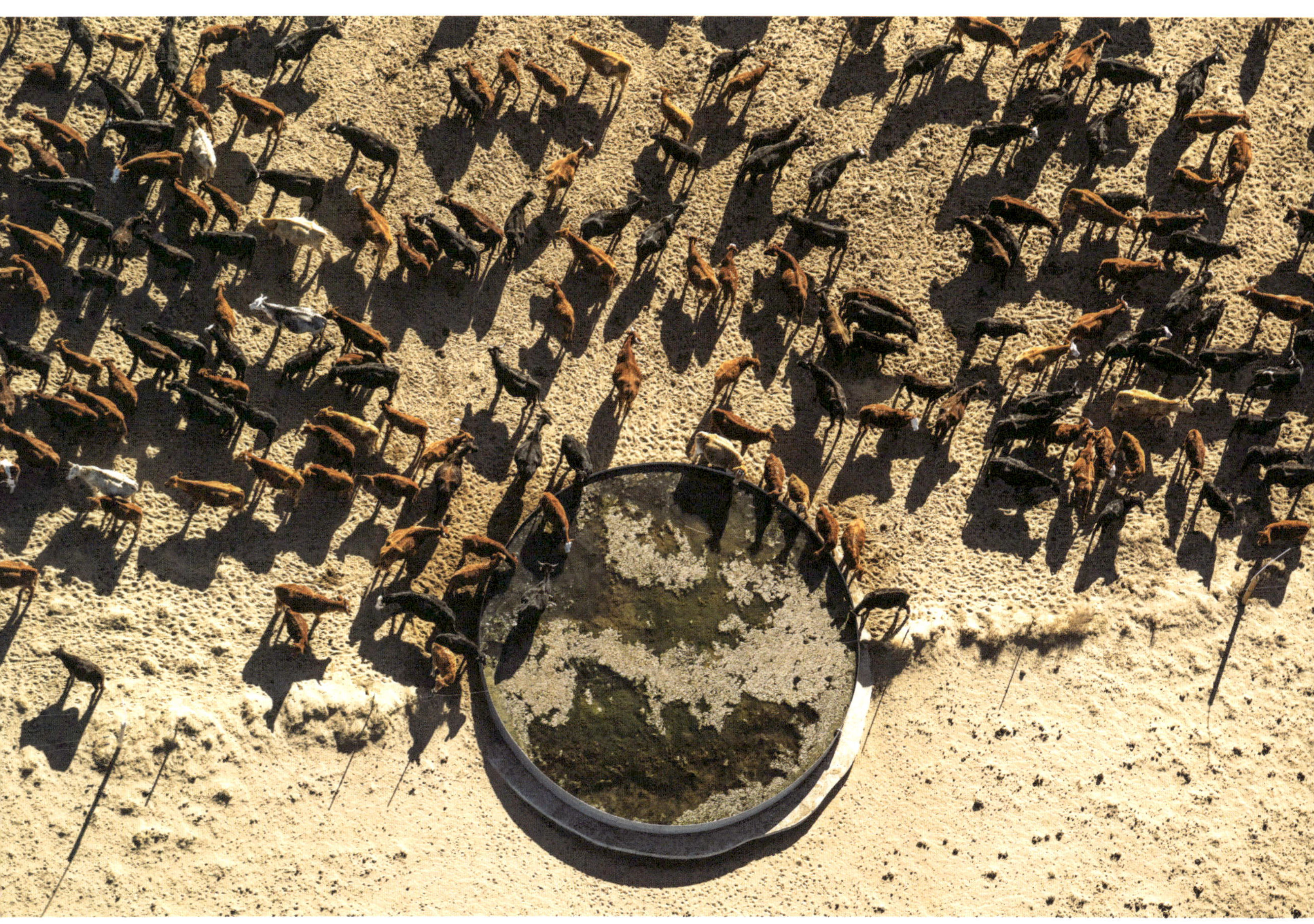

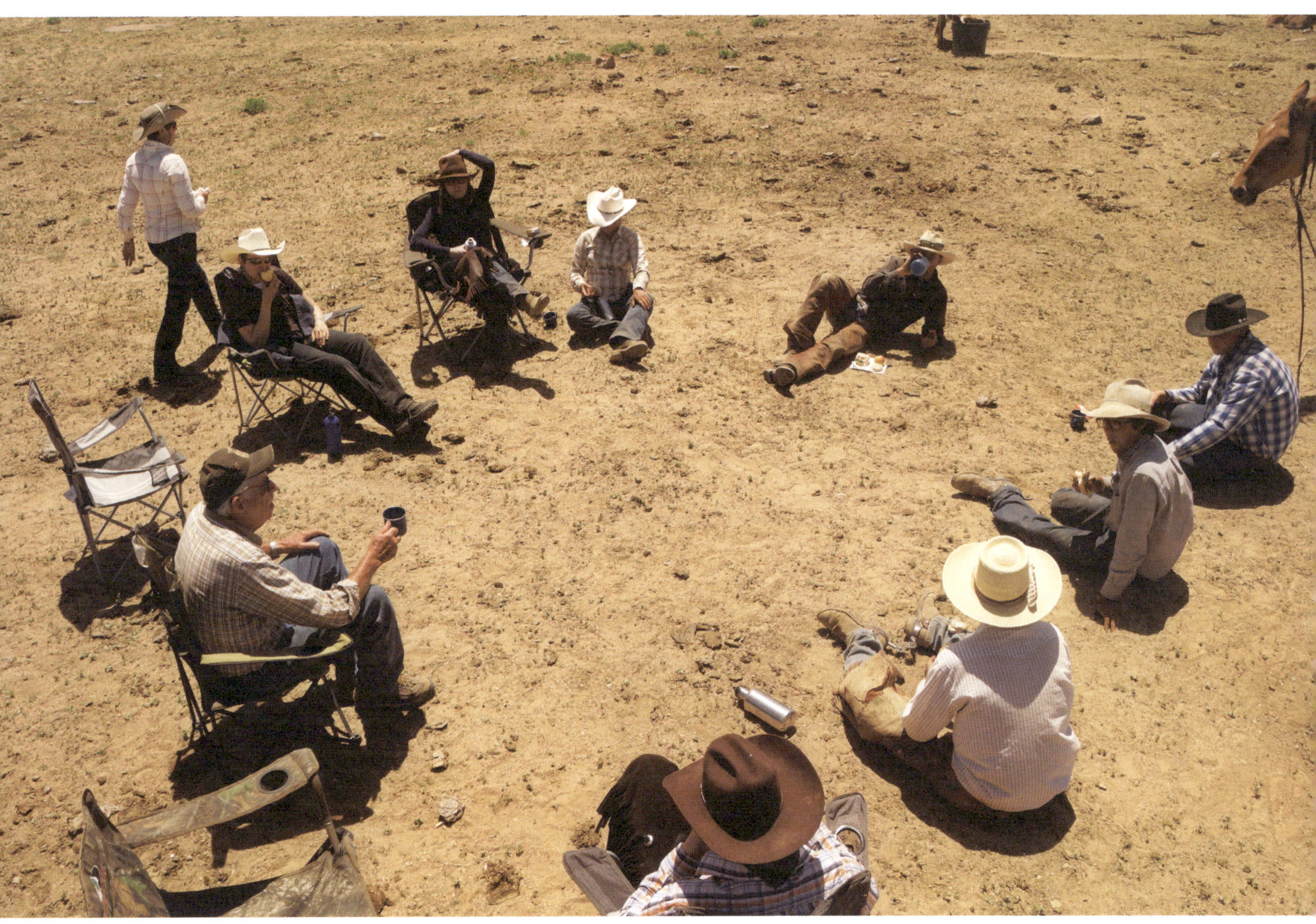

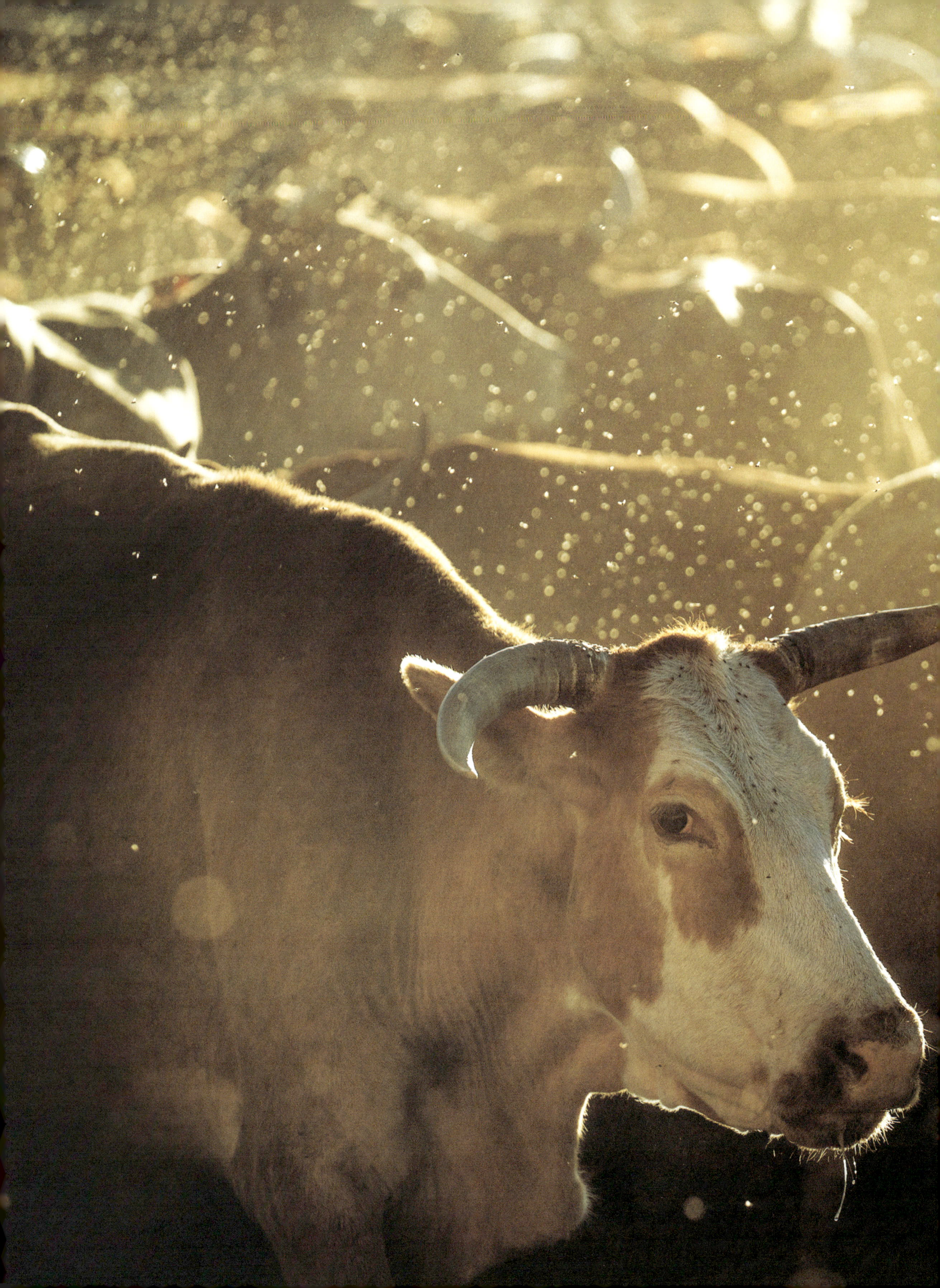

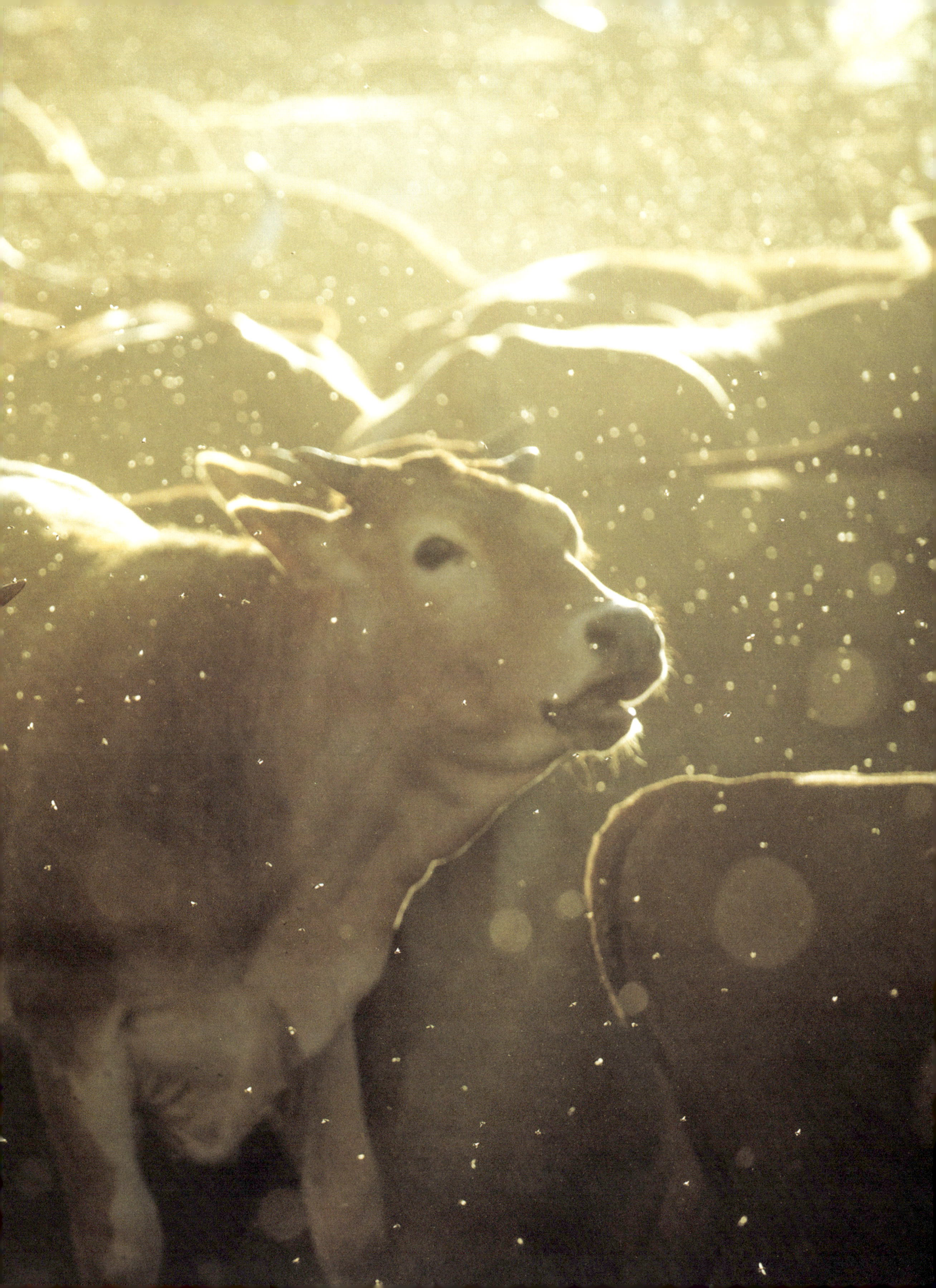

MEAT PROCESSING

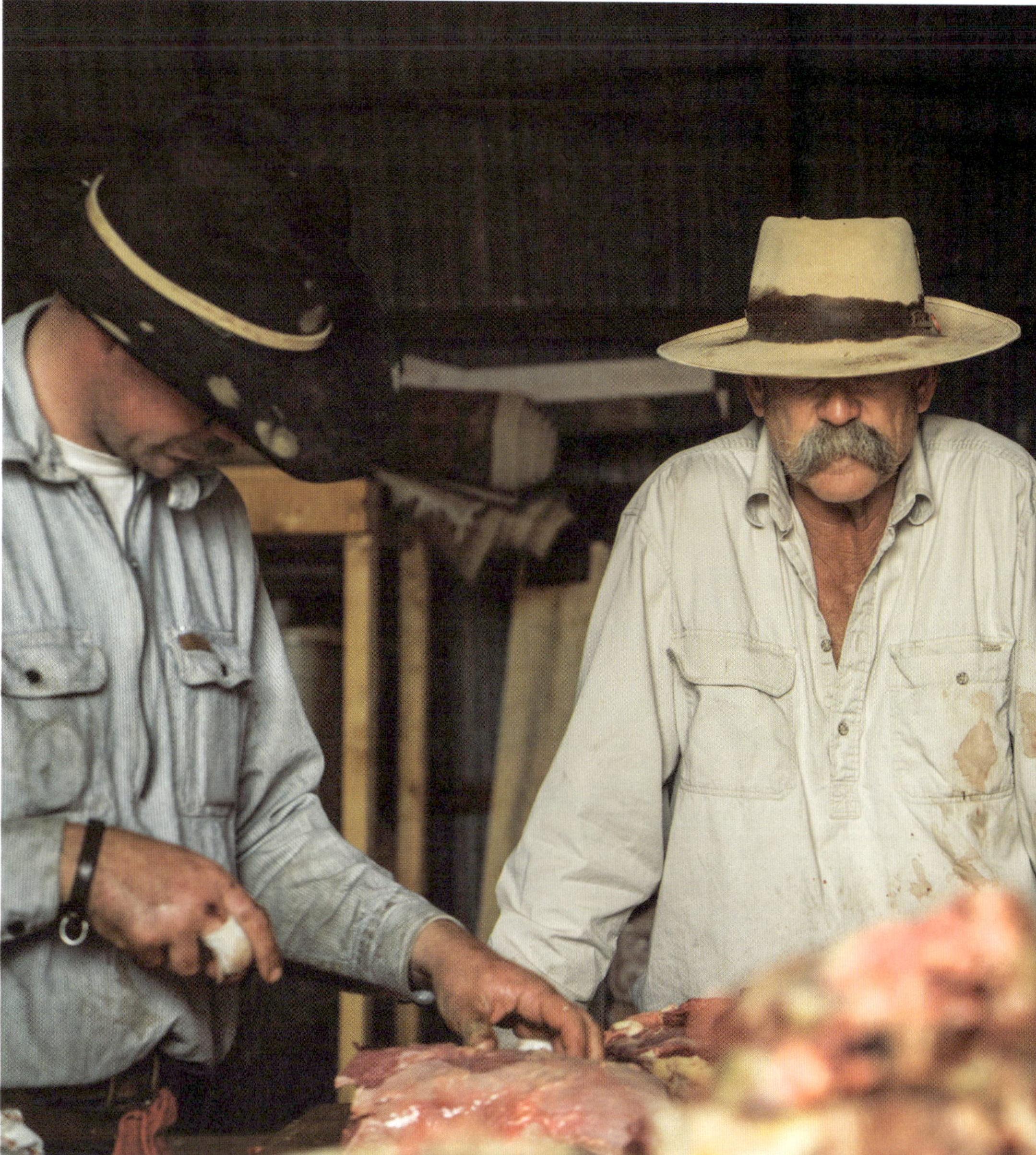

Most people are initially drawn to Ranchlands by the more visible, romanticized aspects of ranching: horseback riding, the dramatic openness of prairie landscapes, trotting out before the sun rises, and cattle roaming what remains of the great American grasslands. Once all the romantic dust settles, however, what's left is the core of our lifestyle: producing food. The job of any rancher is to convert grass into meat protein, all the while using cattle as a tool to manipulate the grass life cycle, manage whole watersheds, and build healthy soil. But making meat from the natural resources the land provides will always be at the heart of our work at Ranchlands and all of the properties we manage.

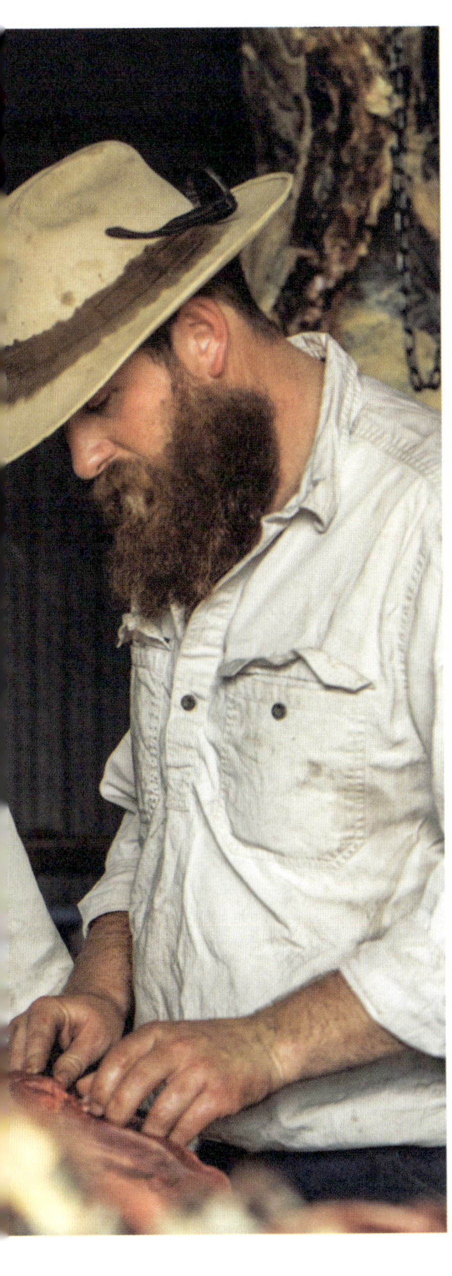

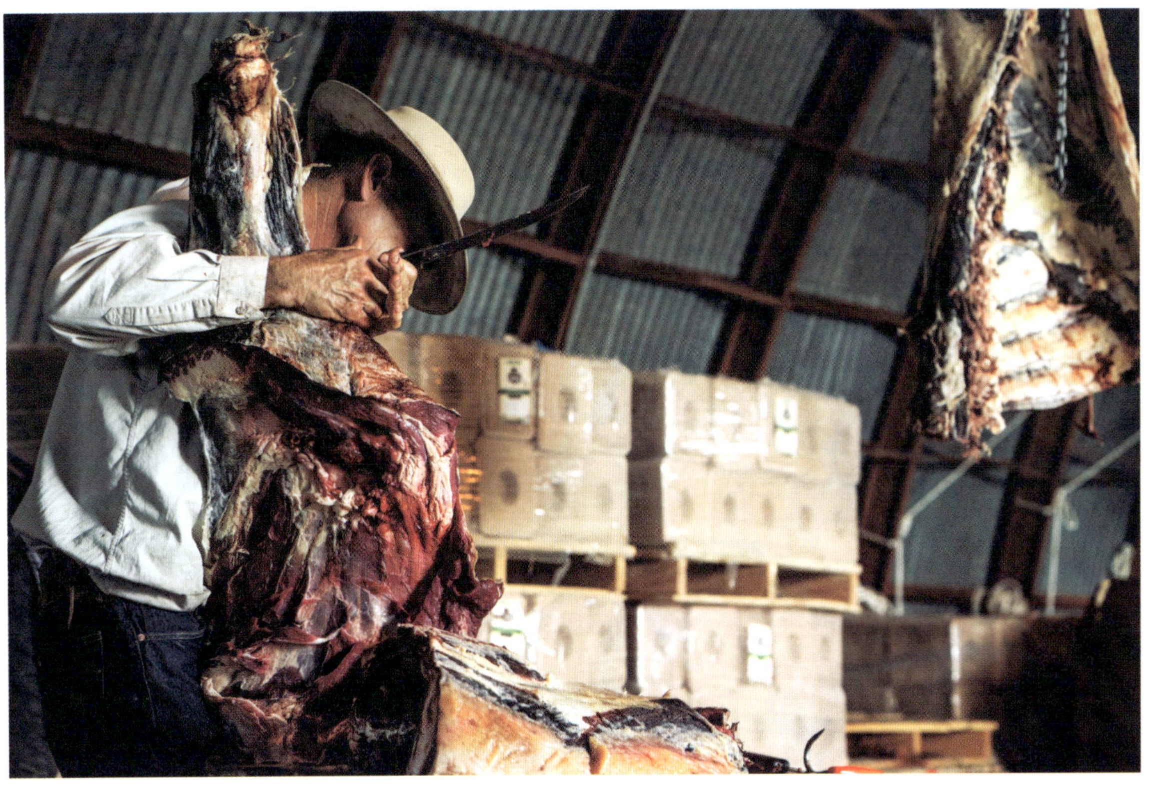

THE MERCANTILE

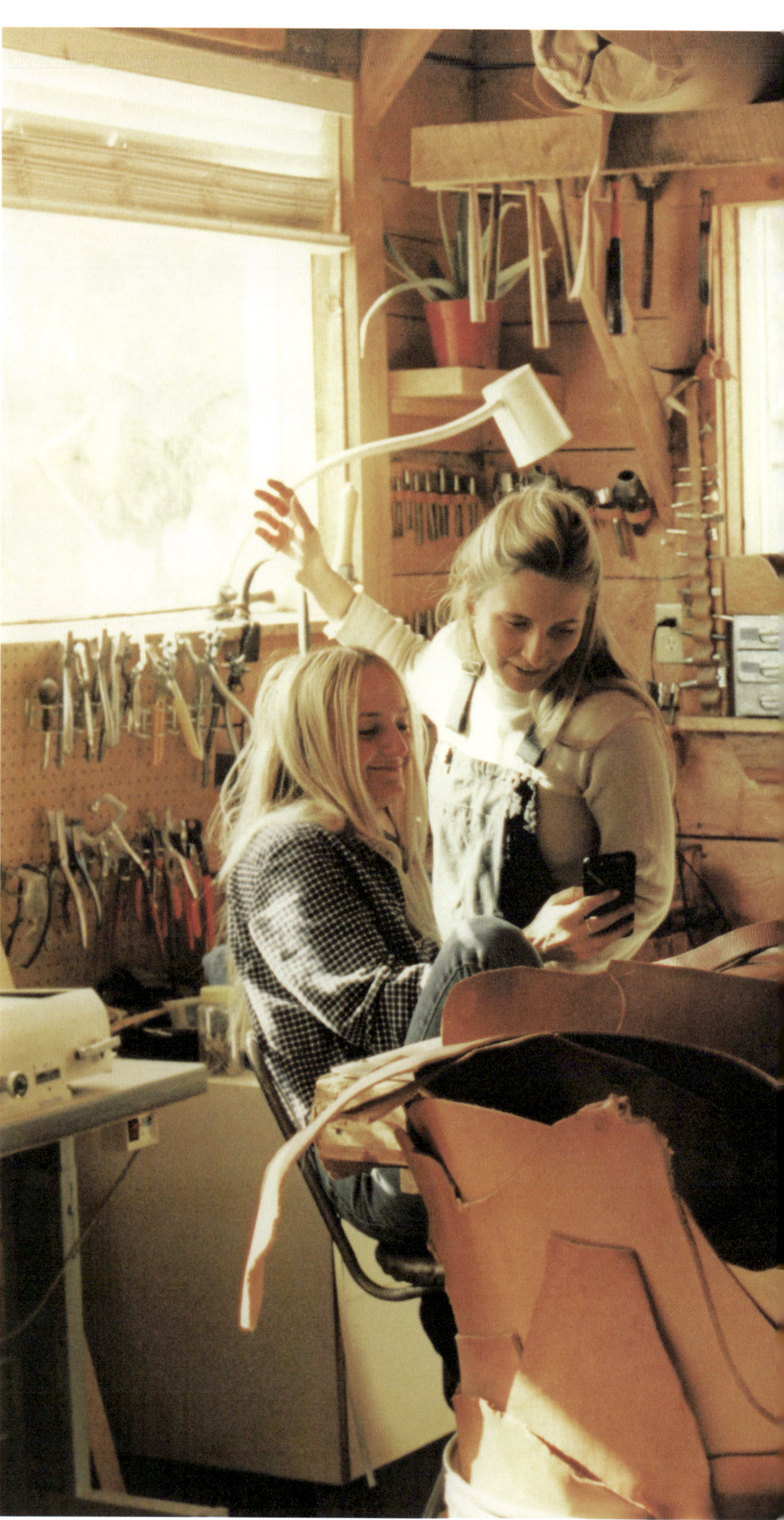

The Ranchlands Mercantile was first born in Mexico on an isolated ranch, where Duke III was raised. Each month, a saddlemaker would come to fix their tack and make any new bridles or reins they needed. Duke became interested in the trade and began a collection of his own tools that lived in a small leather bag on his saddle. Years later, as he began managing several ranches, he always created a space for leatherwork. When he moved to the Chico Basin Ranch in 1999, he selected the old woodshop for the leather room. He first started making leather bags for his three daughters, but soon more of their friends and neighbors started wanting them too. Thus, the mercantile was born.

Today, the leather shop is led by Madi Phillips and Tess Leach, Duke's daughter-in-law and daughter. The artisans who join the team often arrive without prior leatherworking experience, and training them is part of Ranchlands' mission to keep the traditional methods of leatherwork—an essential part of ranching heritage—alive.

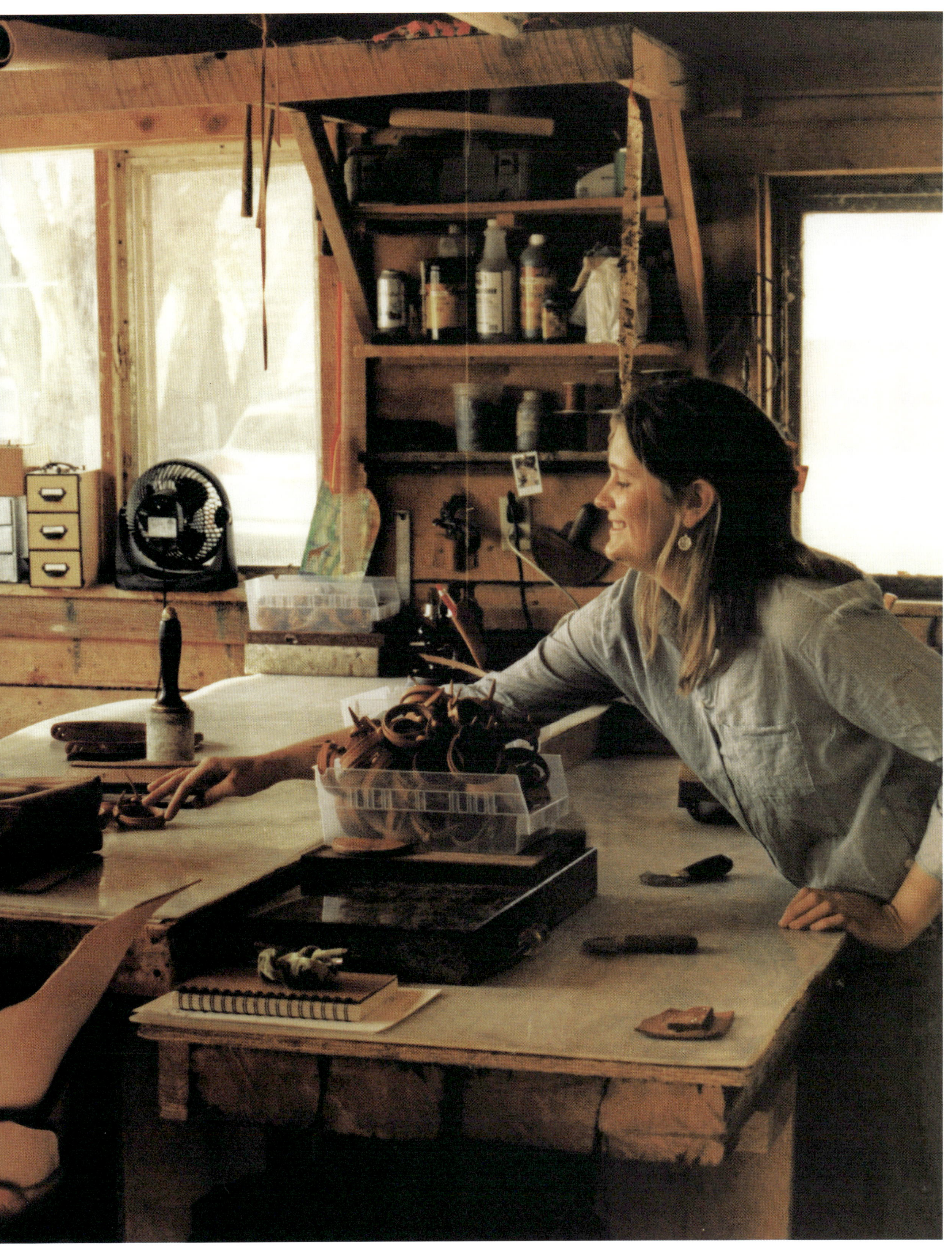

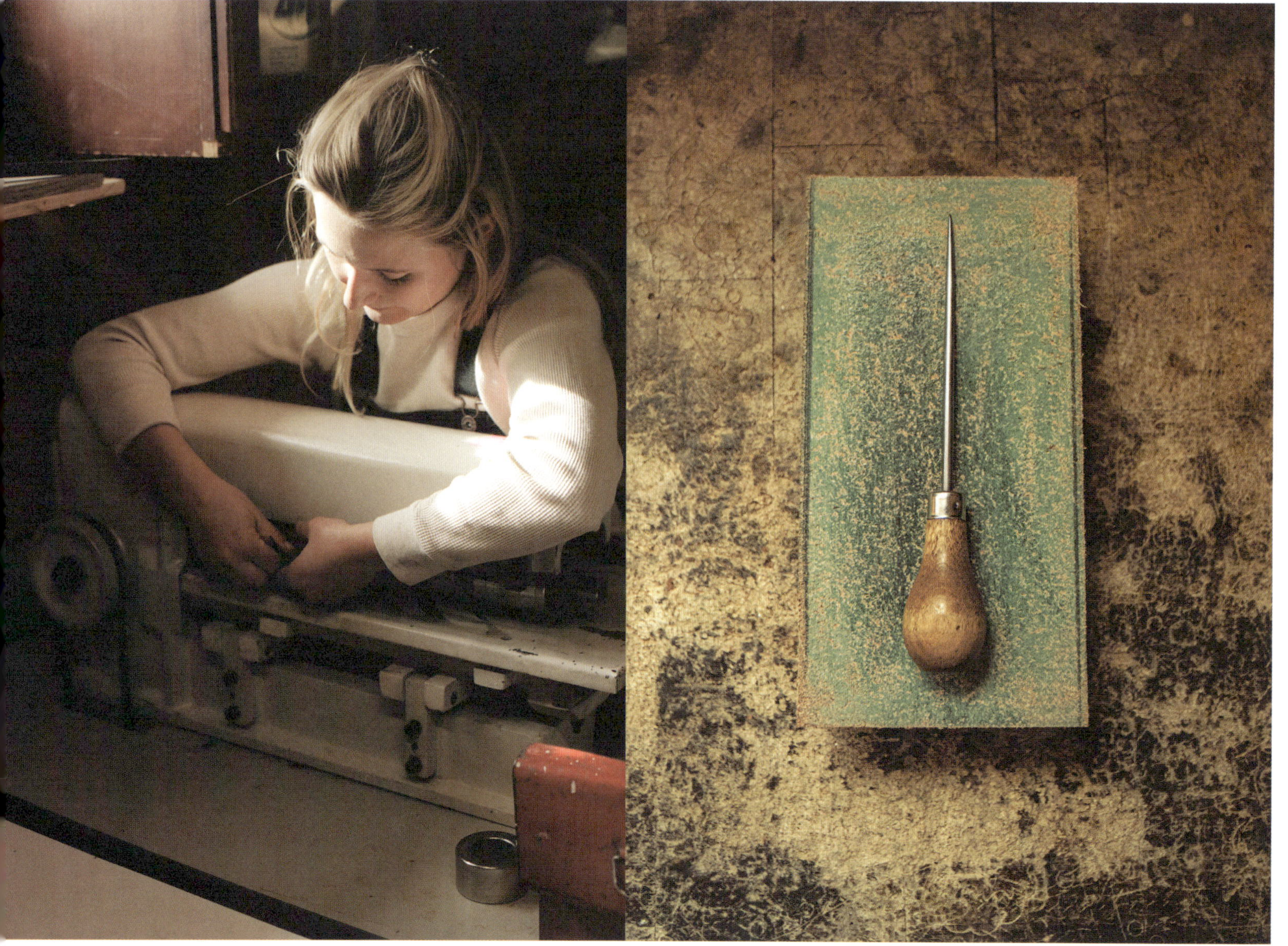

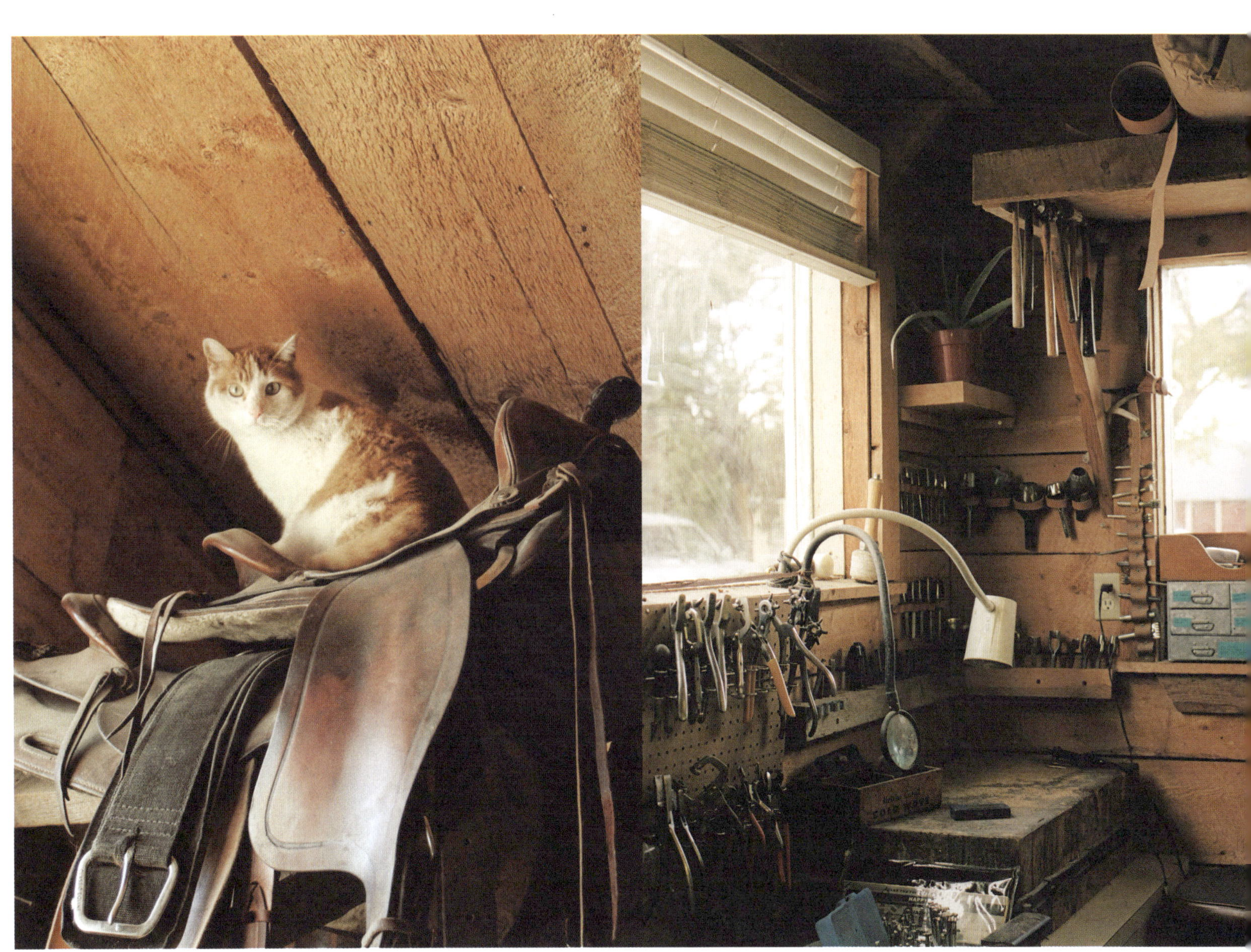

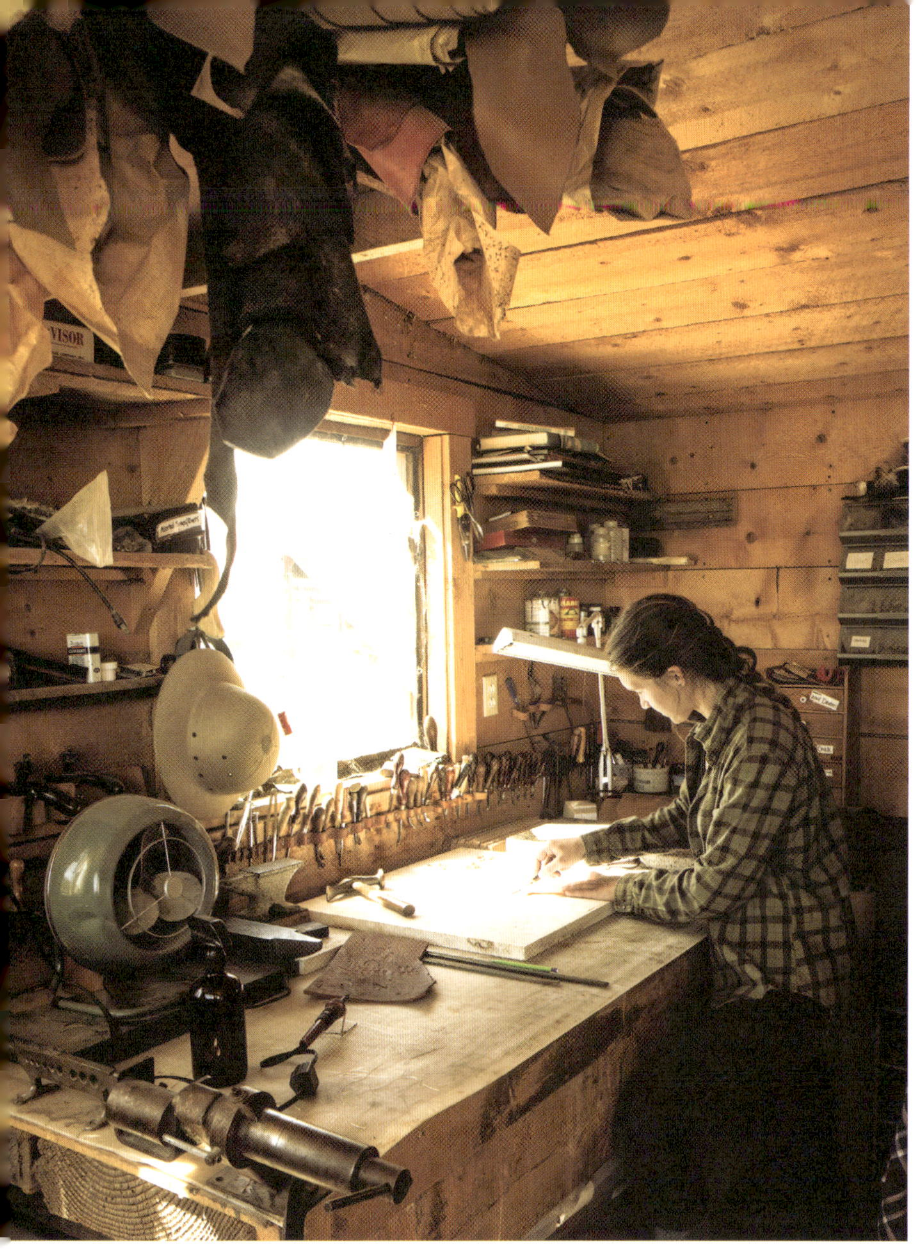

MADI PHILLIPS

"We want the shop to be an arm of the business that's diversifying income. But it has also become the heart of the ranch. One day, Duke III went into town and came back with a big sound system for the shop. Anytime there's a cattle move or a branding, the crew shows up here. It's the gathering spot because the leather shop is home to the saddle house, so everyone comes to do last-minute repairs, fix gear, and saddle up. We go out to brand or move the herd, and then trot back here to unload gear. And then we turn on the music, crack open some beers, and hang out in here together."

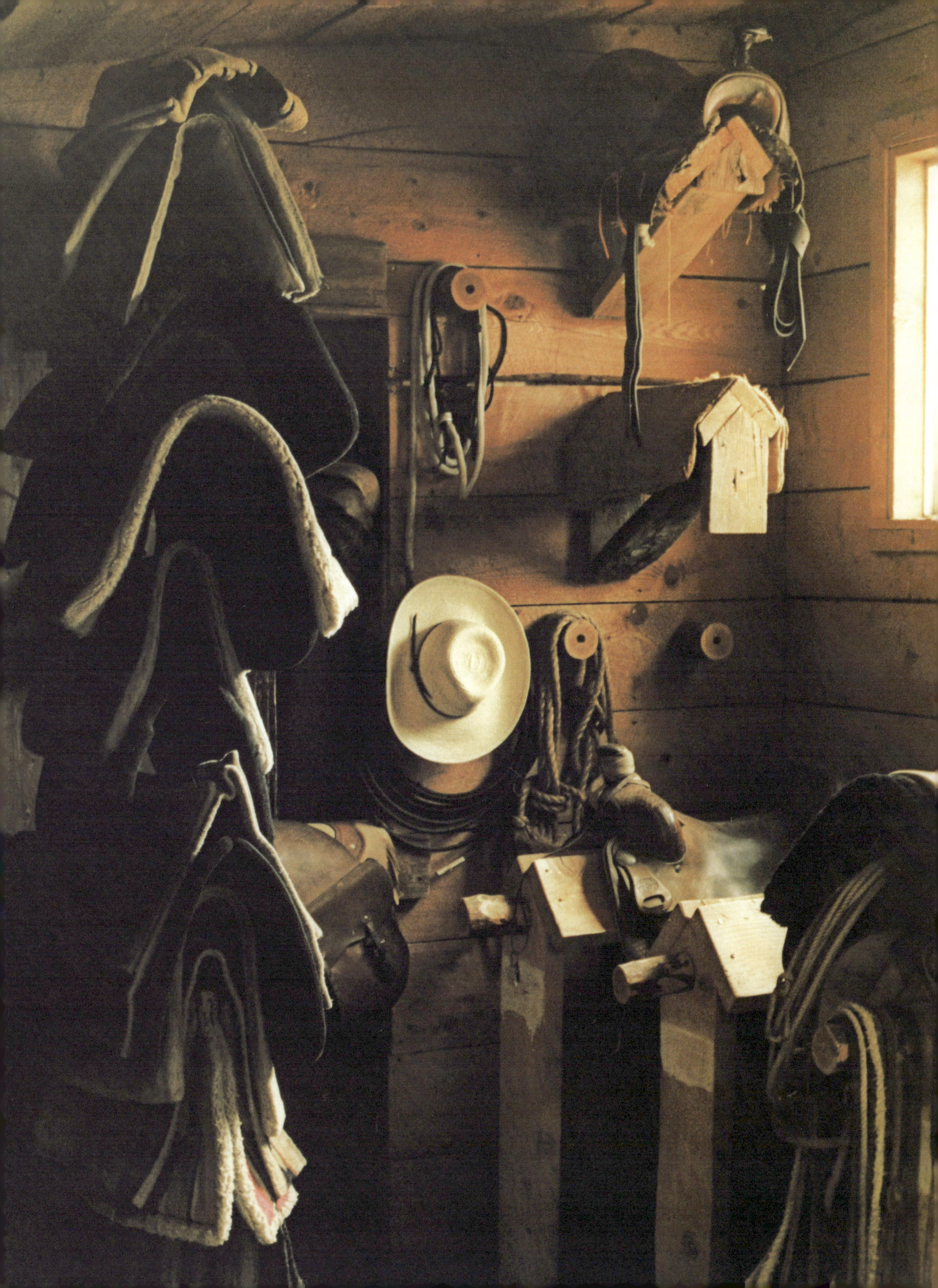

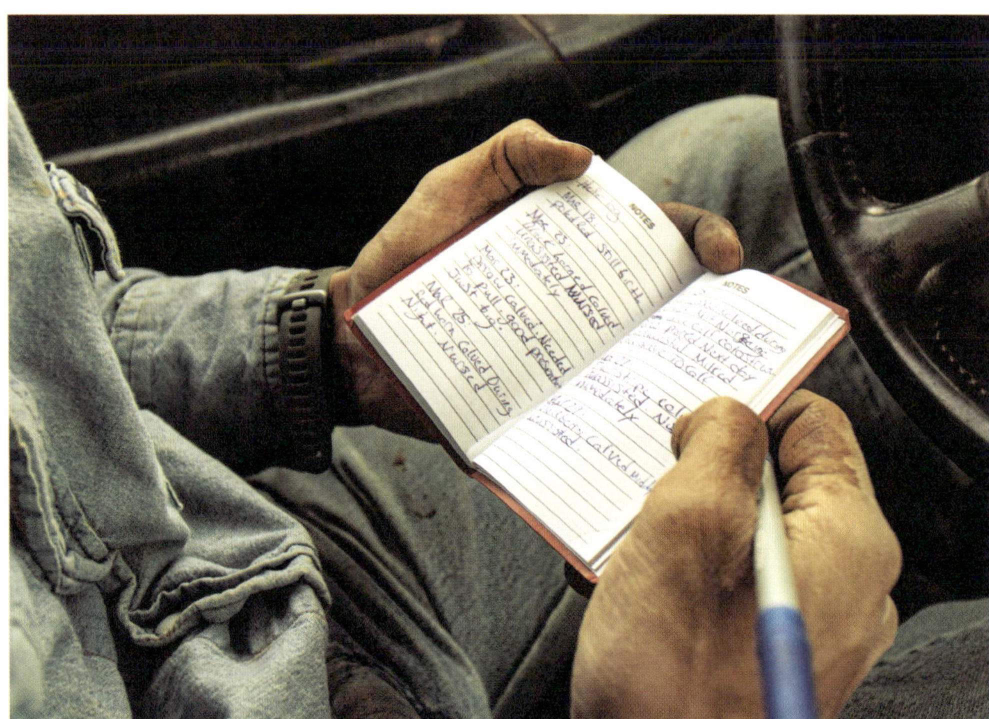

A NEW ENTRY

Calving heifers is the busiest time of the year on the ranch. We check our herd multiple times a day and throughout the night, looking for any signs that a heifer might need help delivering her calf. Statistically, one in five first-calf heifers will need some sort of assistance. Most of the time it goes well, but when something goes wrong, it's hard not to dwell on what we could have done differently. We keep logs of what has worked, what hasn't, and any insights that might help us improve. It's rewarding when you're able to assist a heifer that really needed your help. When you pull the calf and it goes smoothly, the mother turns around and begins licking her new calf, and then the calf gets up and starts sucking.

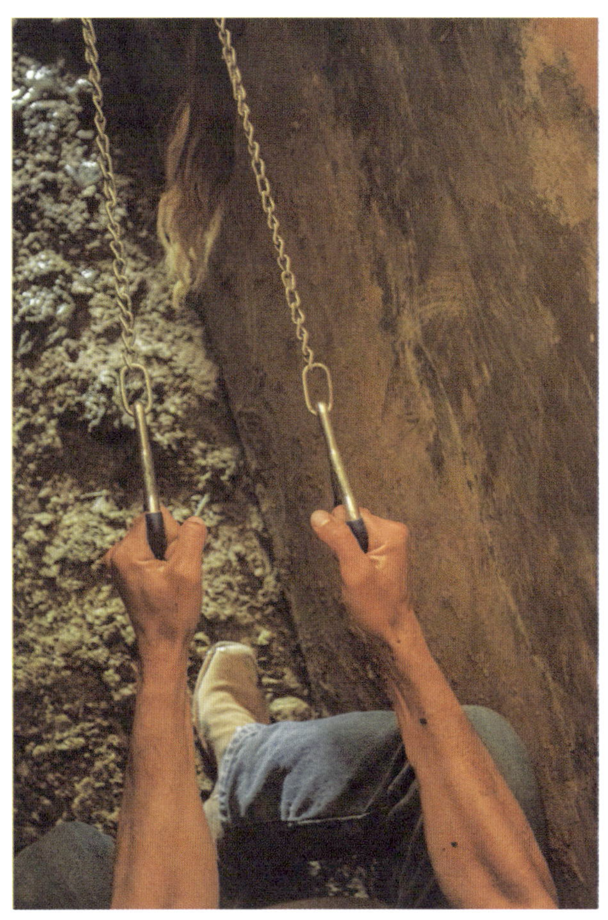

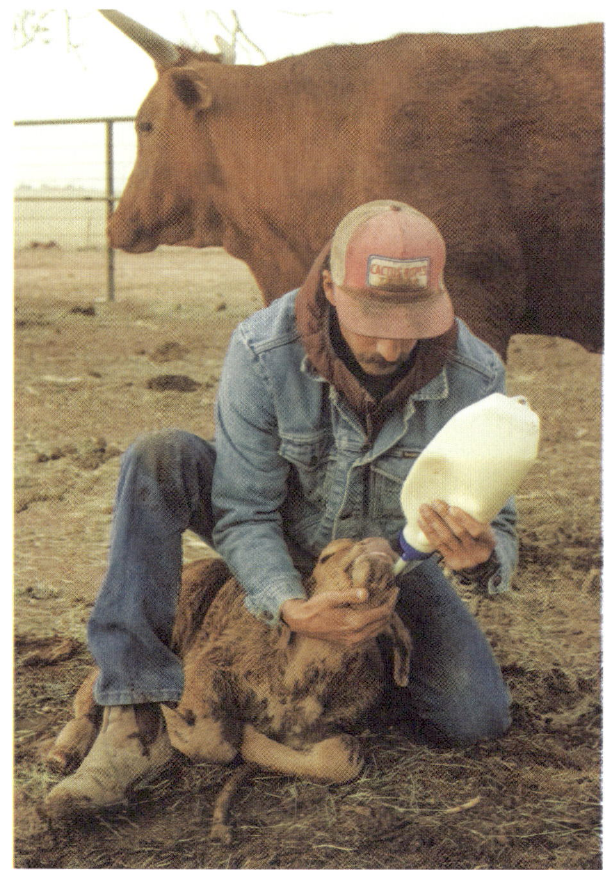

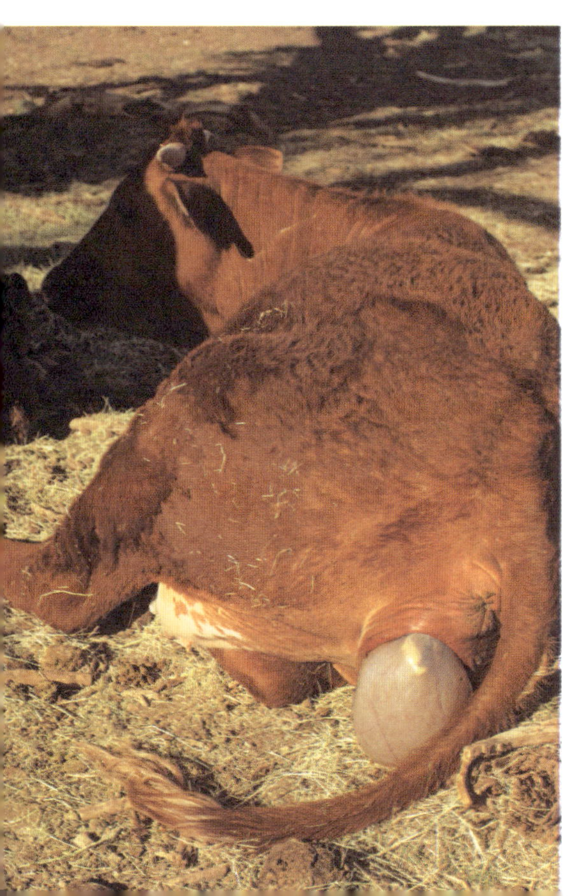

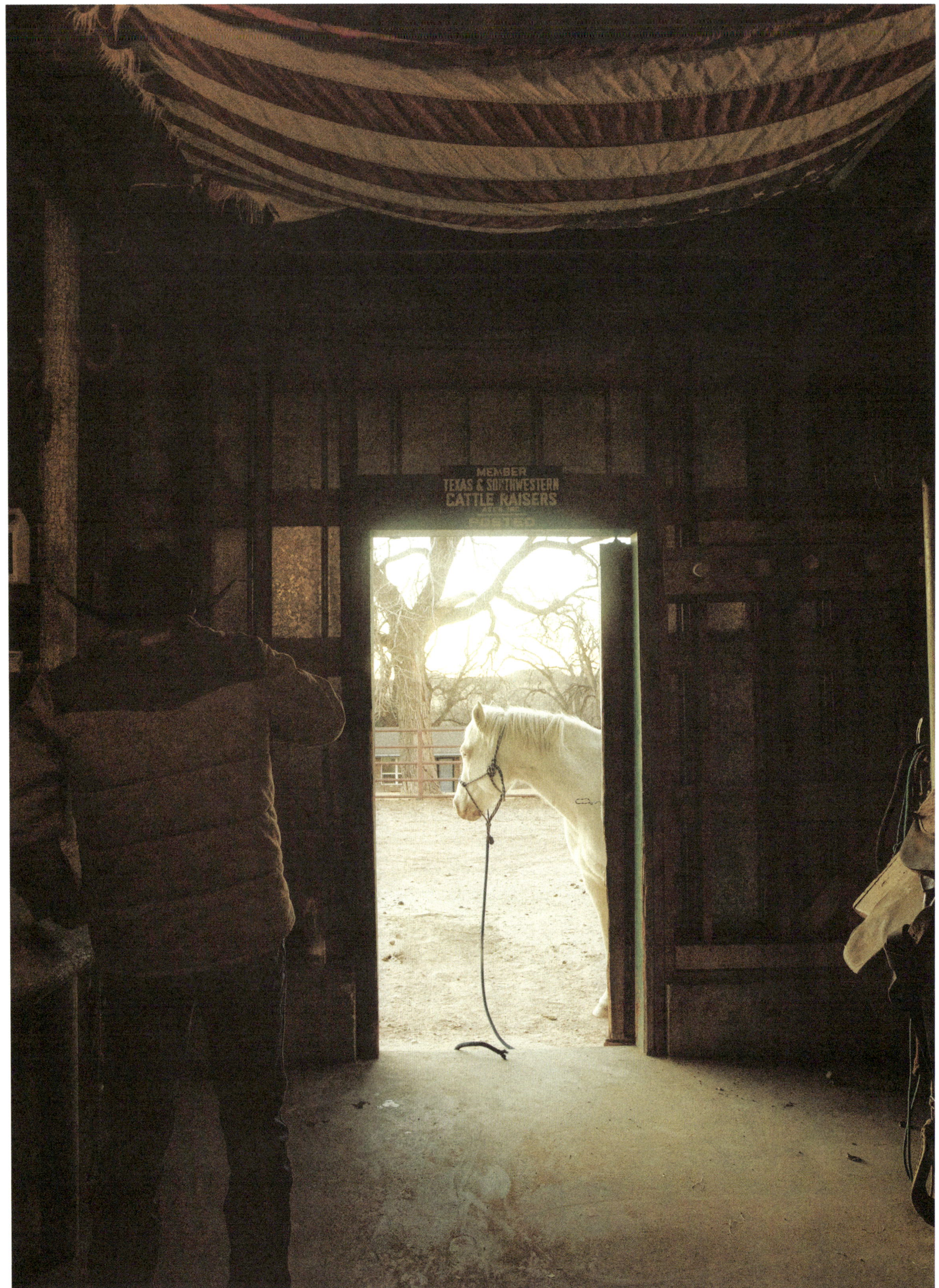
MEMBER
TEXAS & SOUTHWESTERN
CATTLE RAISERS

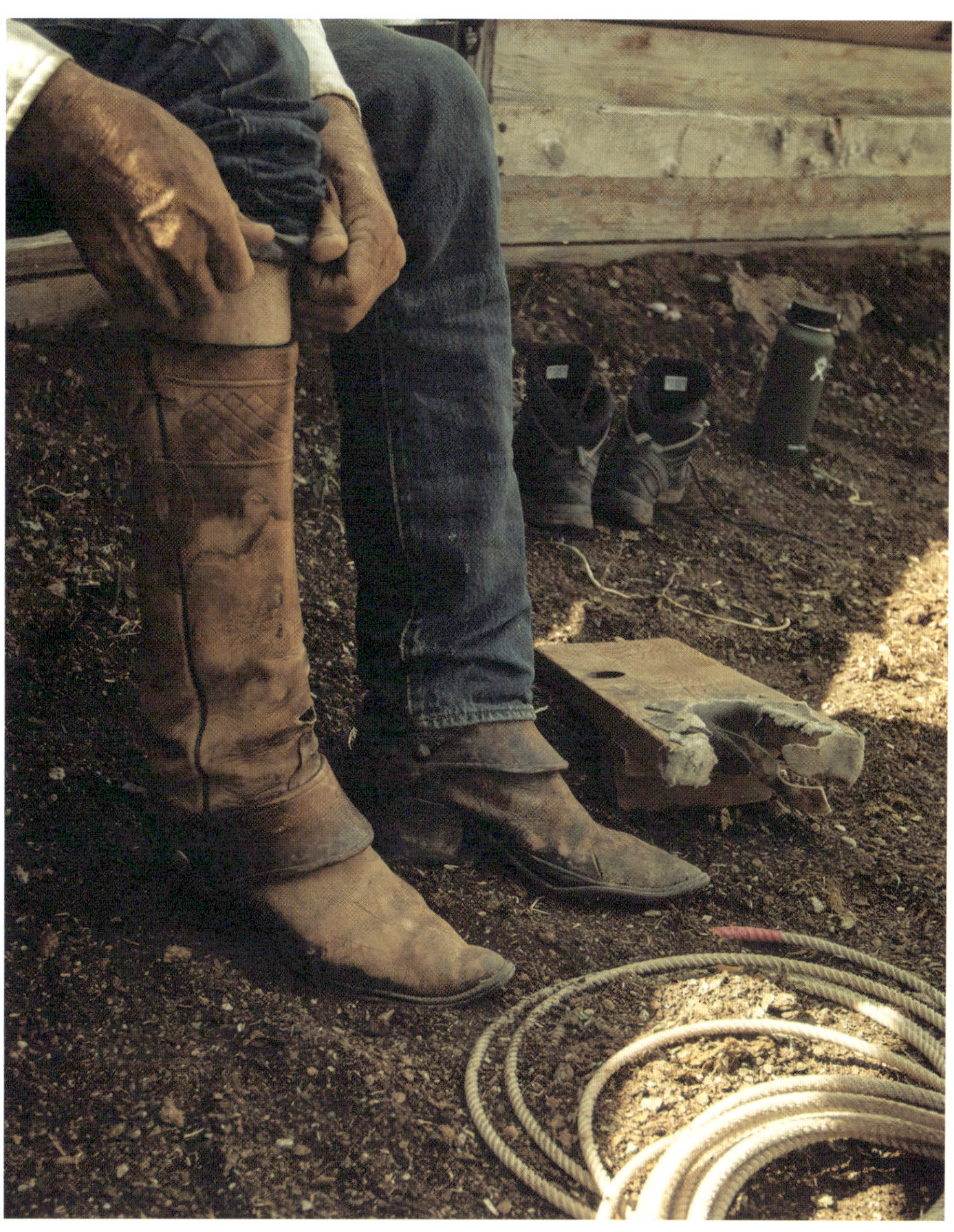

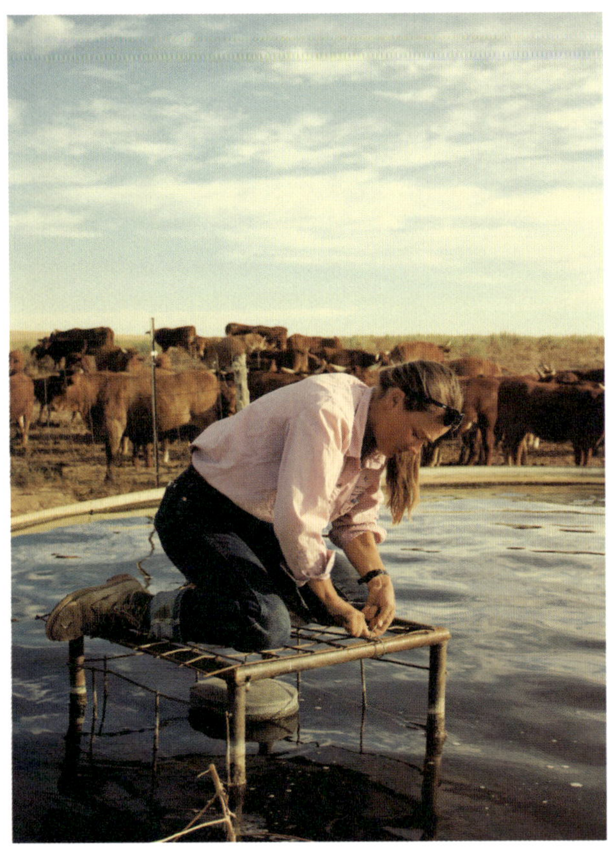

ANJA STOKES

"Different areas of the Chico Basin Ranch come with their own sets of challenges related to fencing and water infrastructure. I oversee the eastern part of the ranch, where the deep, sandy soil causes challenges over shorter periods. Buried or uprooted fence lines, altered soil depth, and erosion around stock tanks are the most common challenges. Most of the internal fence on this part of the ranch is electric, with a single strand of high-tensile wire subdividing a 22,000-acre pasture into nine smaller ones, which makes the frequent improvements more manageable. As apprentices, we're given freedom to be creative in our problem-solving when dealing with infrastructure challenges."

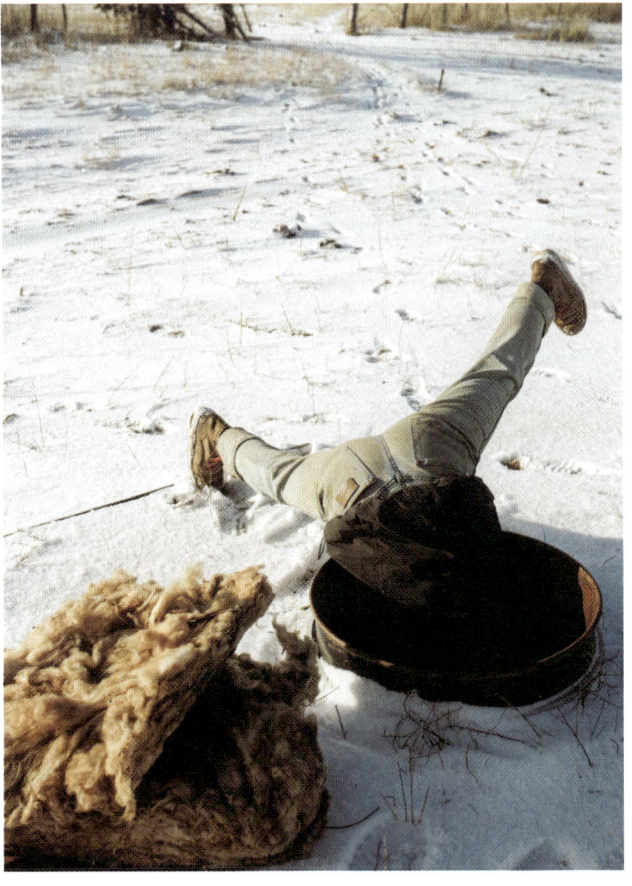

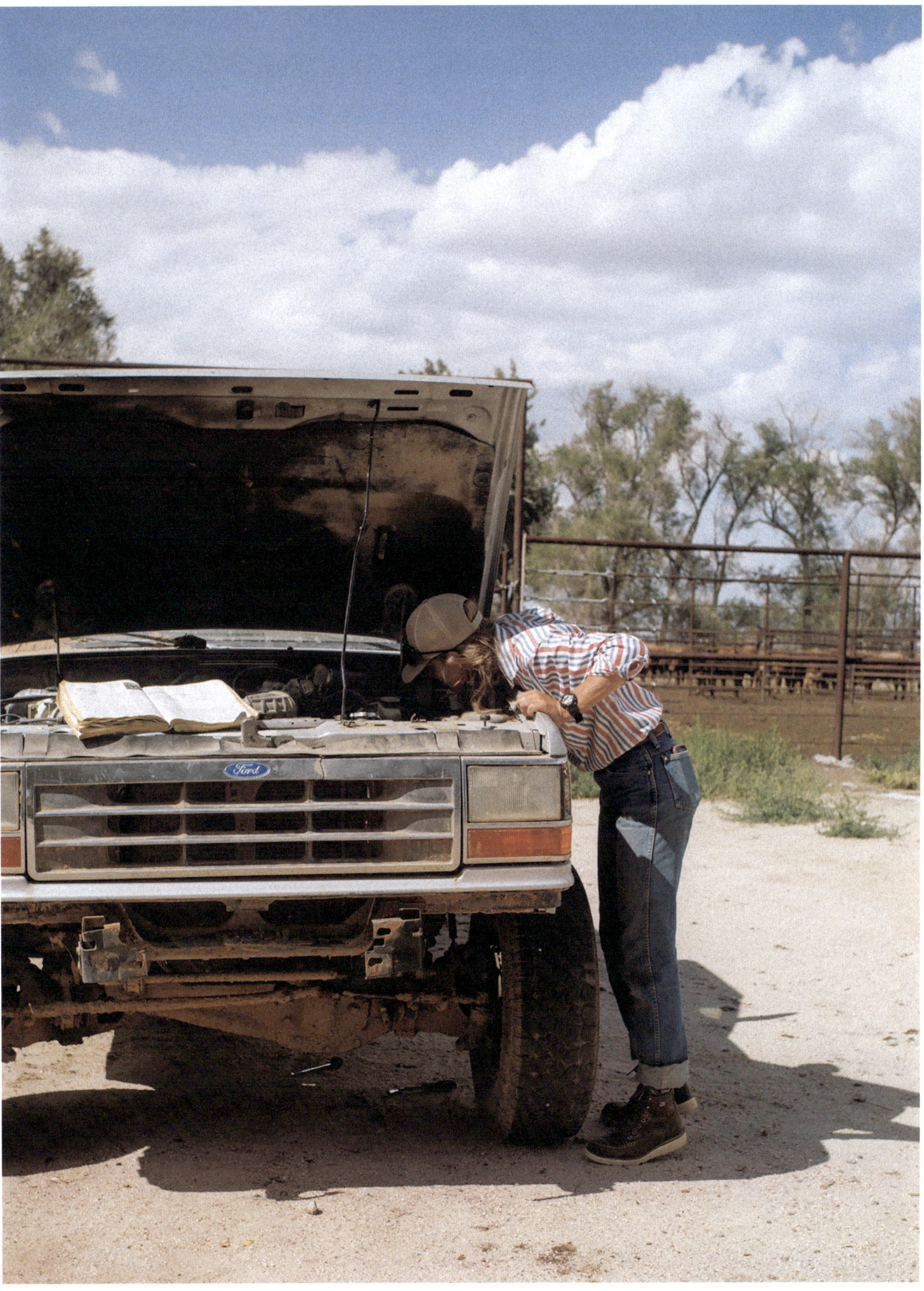
Ford

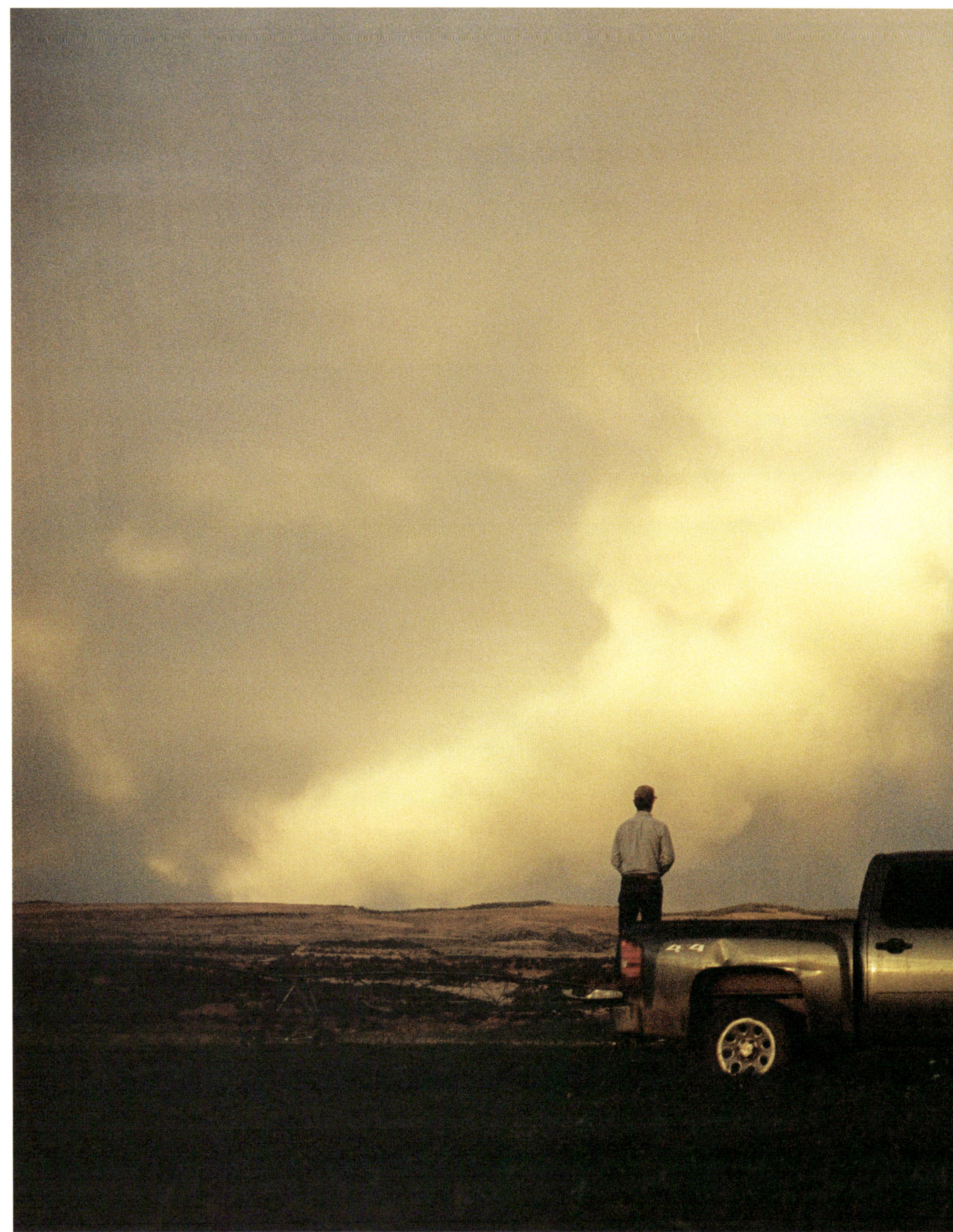

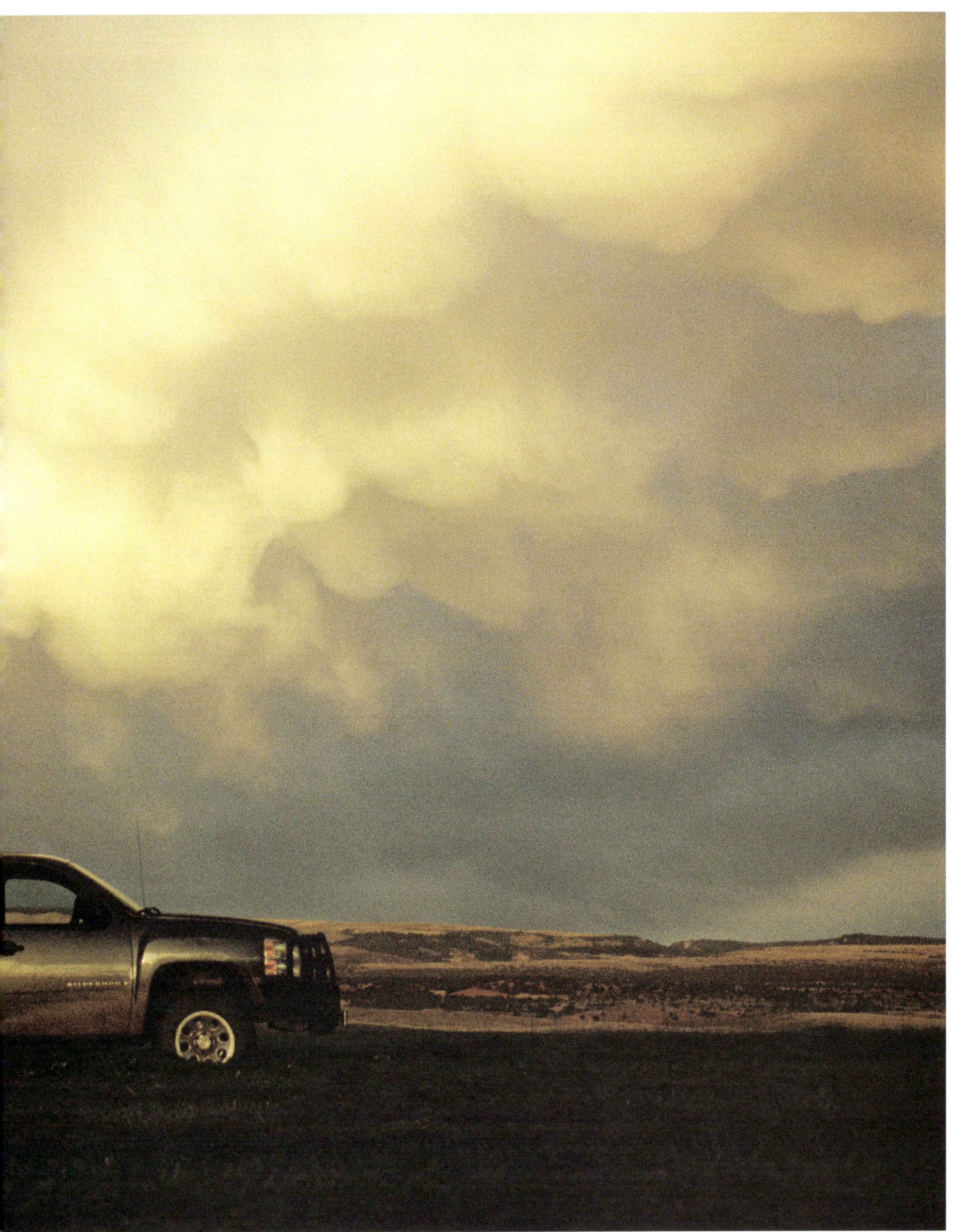

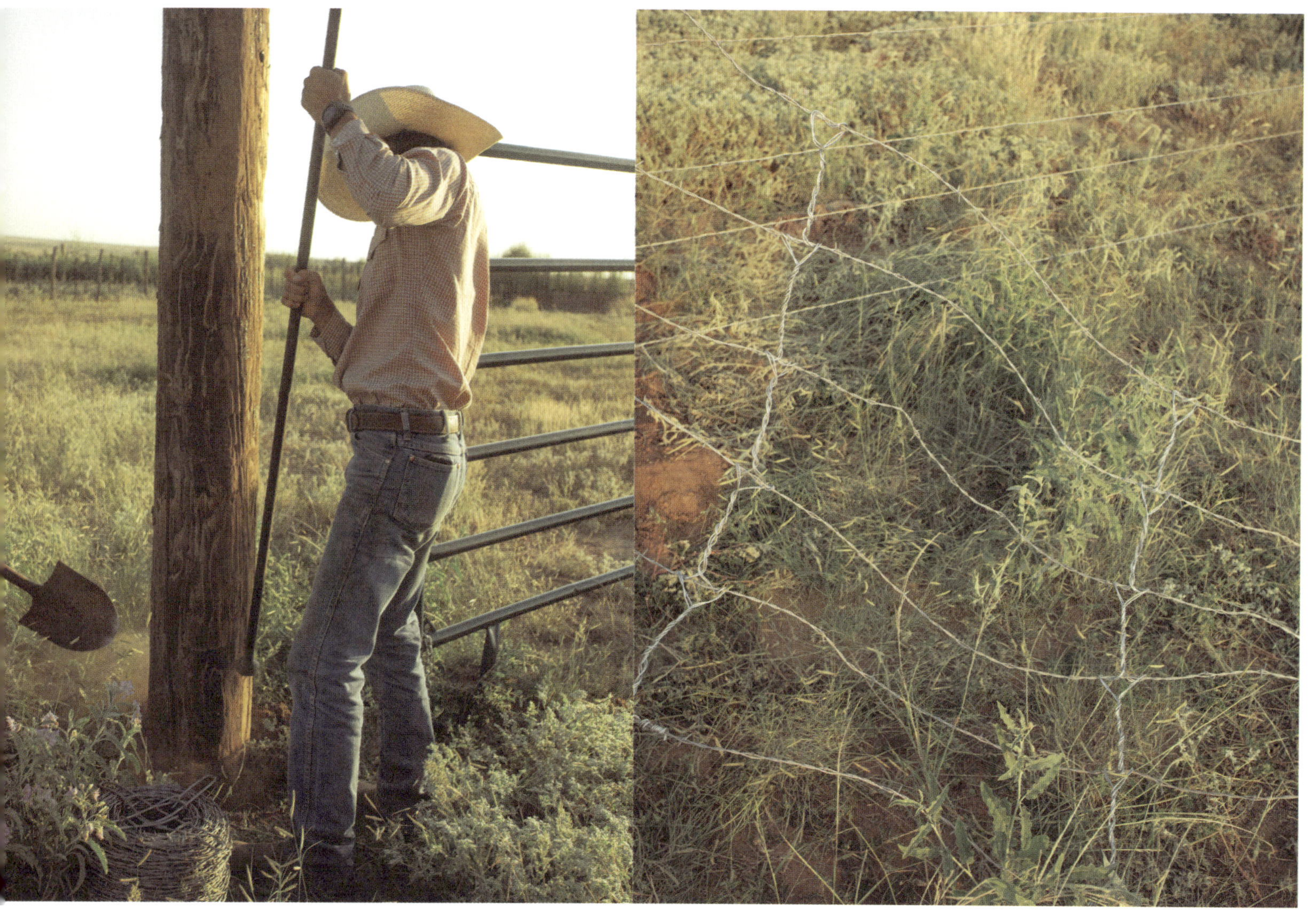

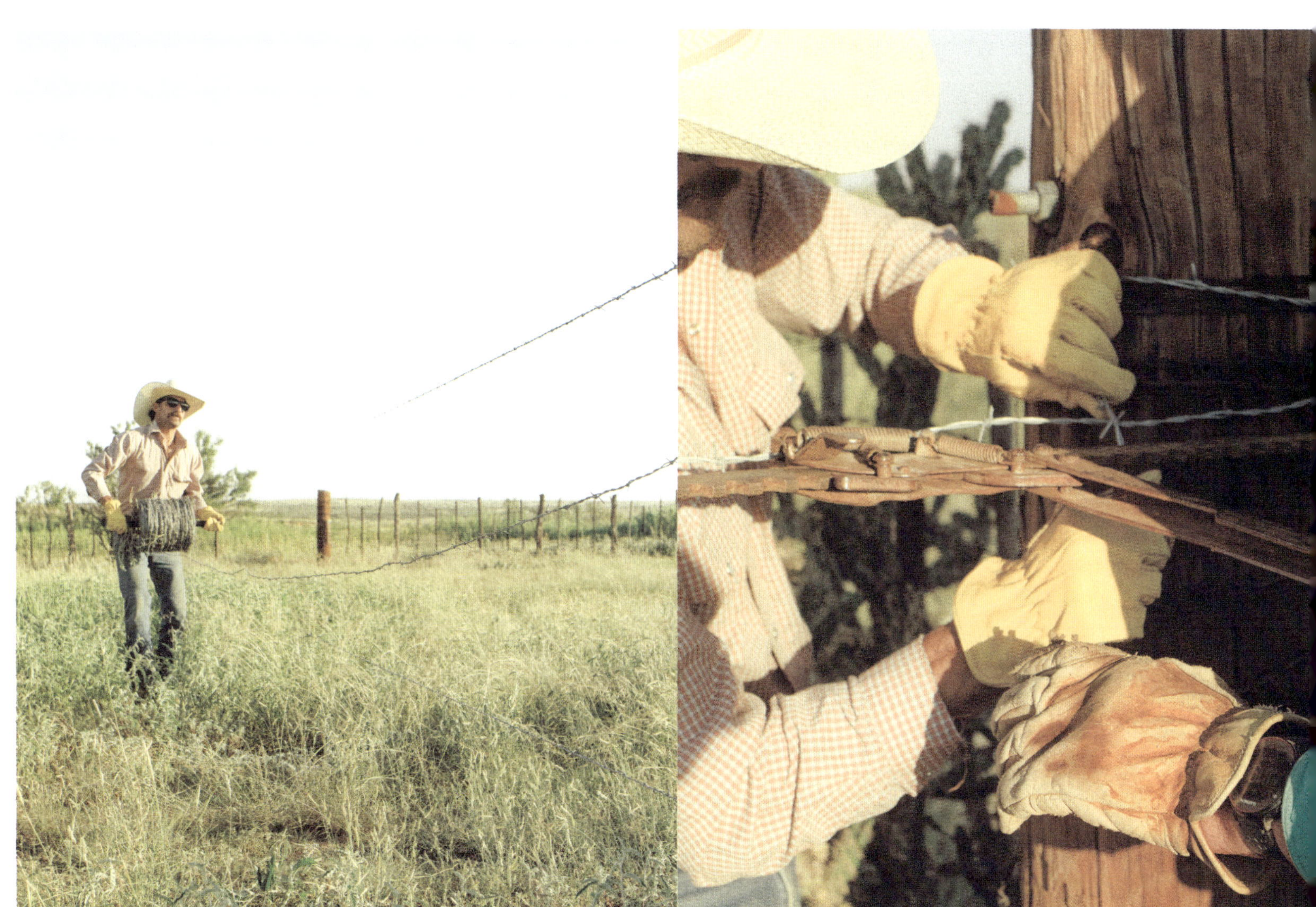

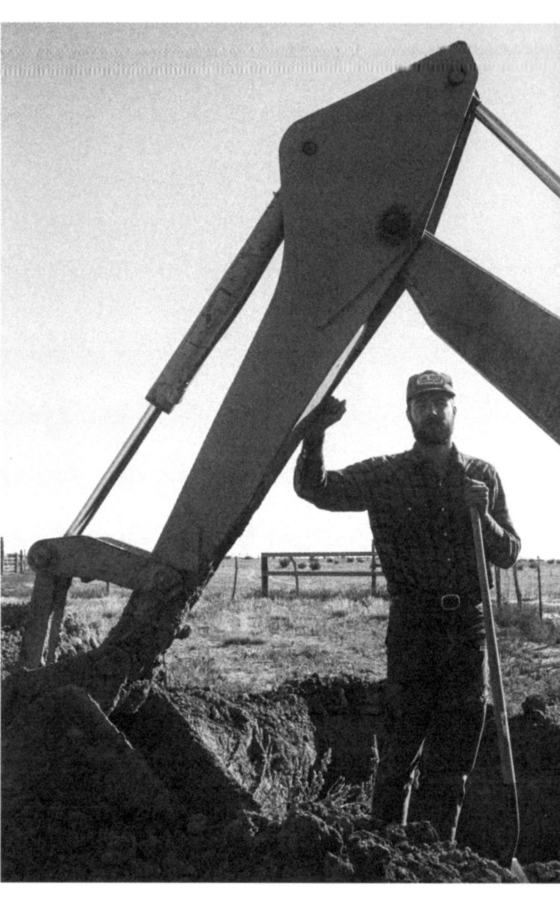

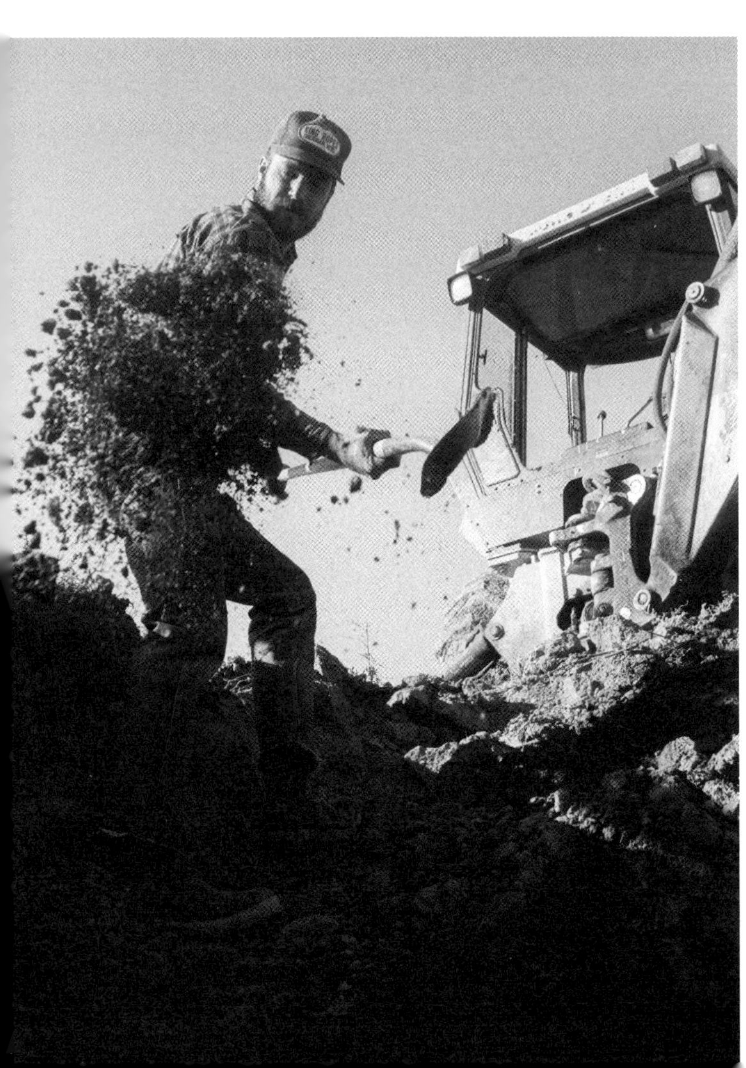

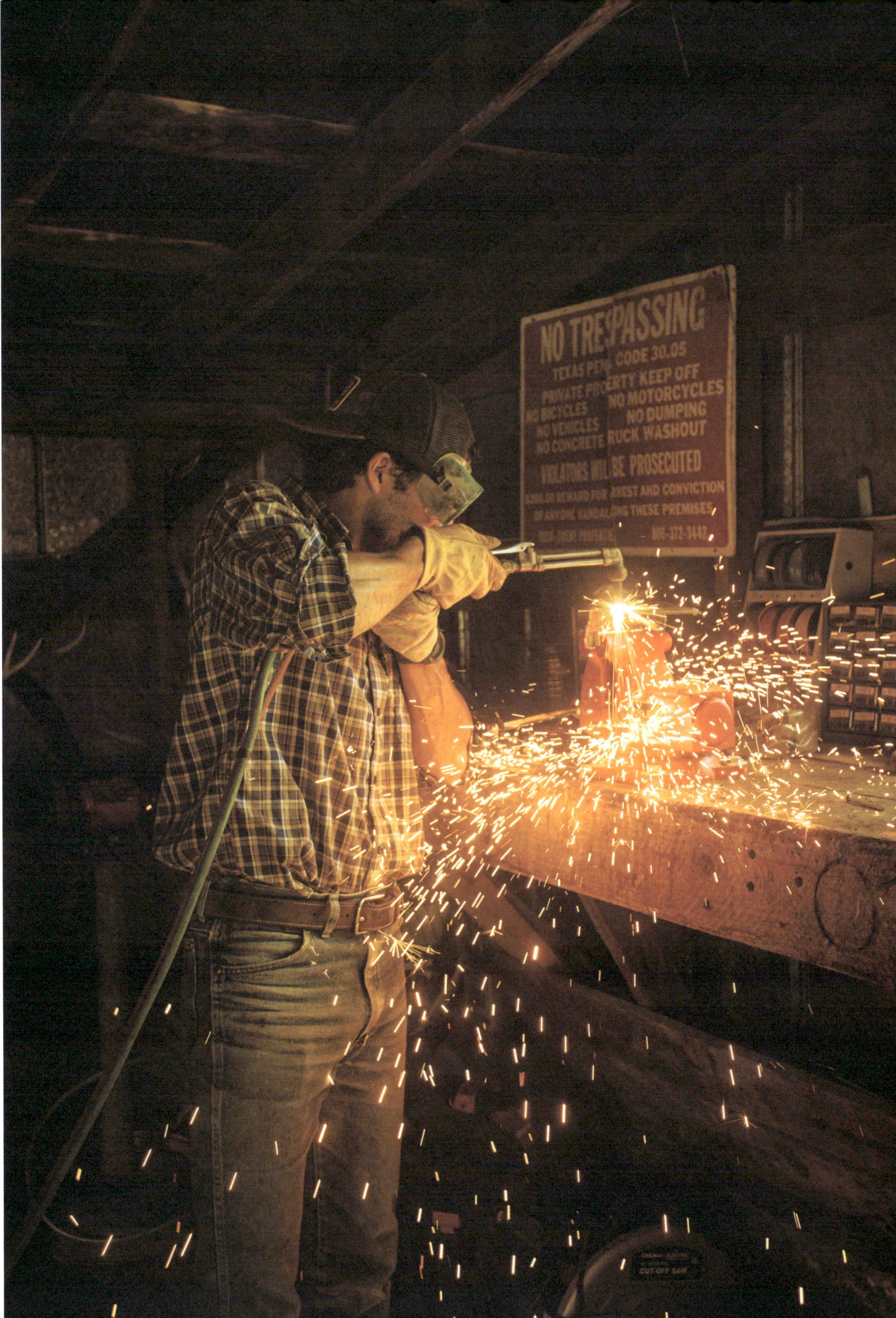
NO TRESPASSING
CODE 30.05
KEEP OFF
NO MOTORCYCLES
NO BICYCLES
NO DUMPING
NO VEHICLES
NO CONCRETE
RUCK WASHOUT
BE PROSECUTED
ARREST AND CONVICTION
THESE PREMISES
CUT-OFF SAW

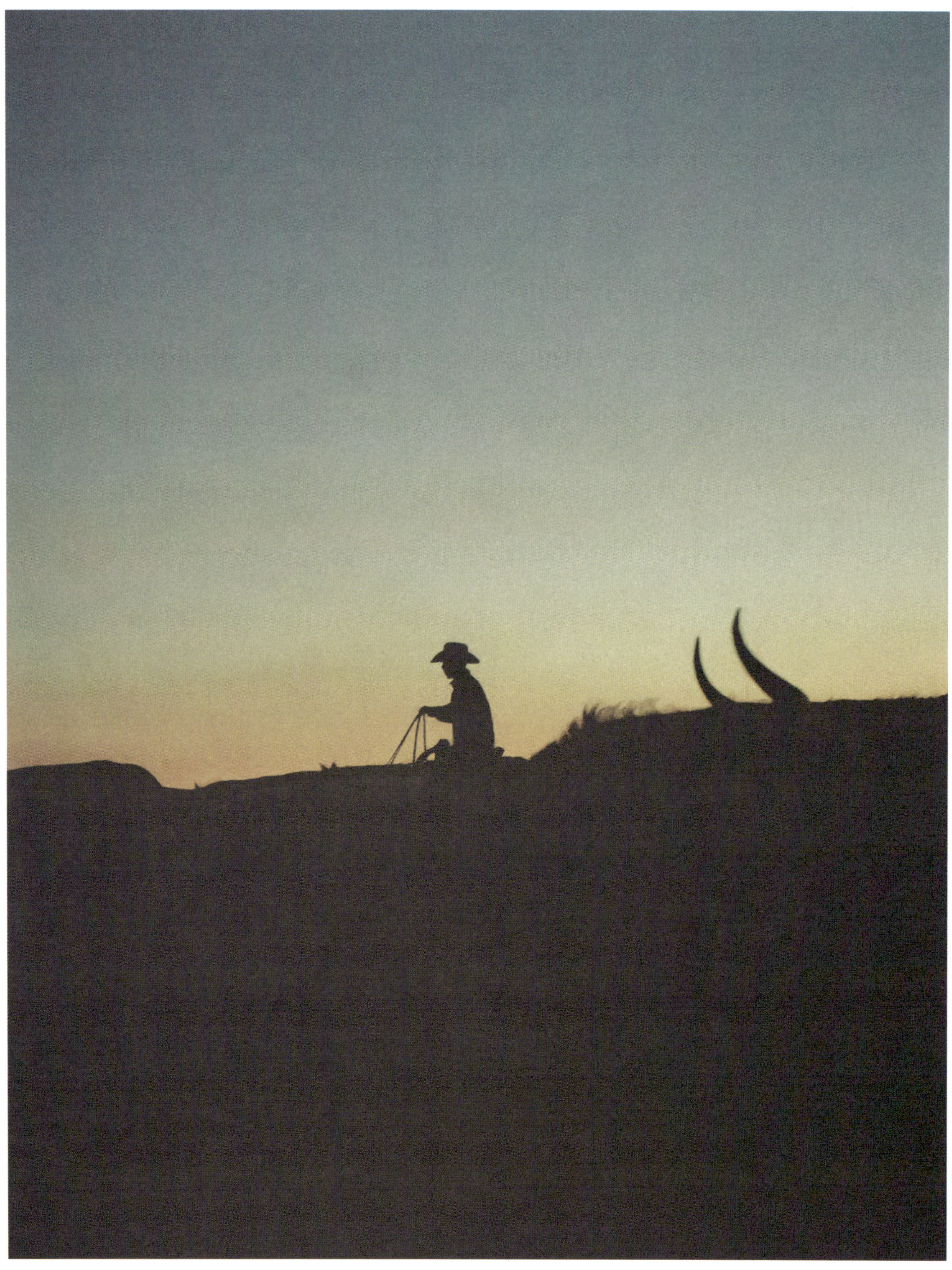

FROM DAWN

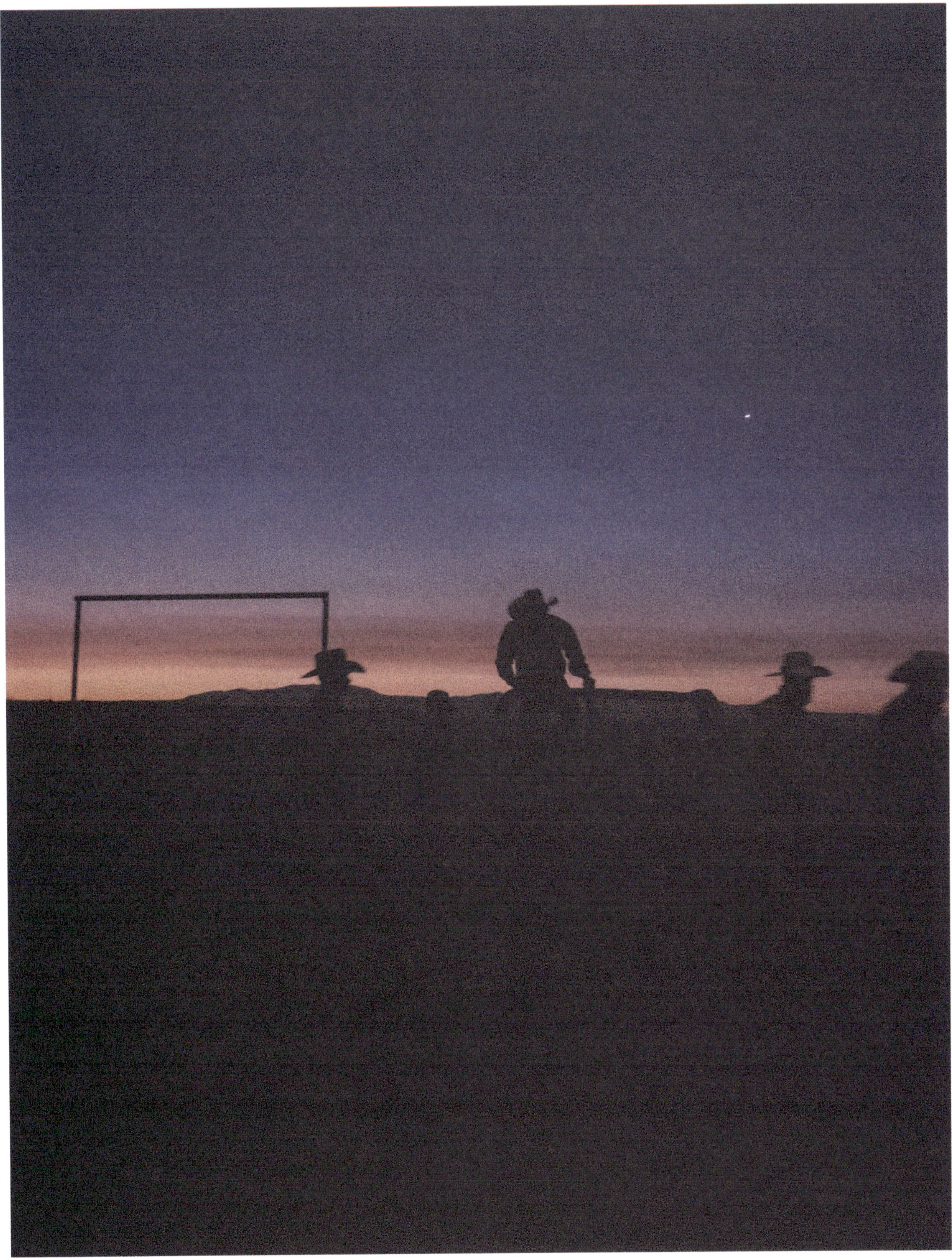

TO DUSK

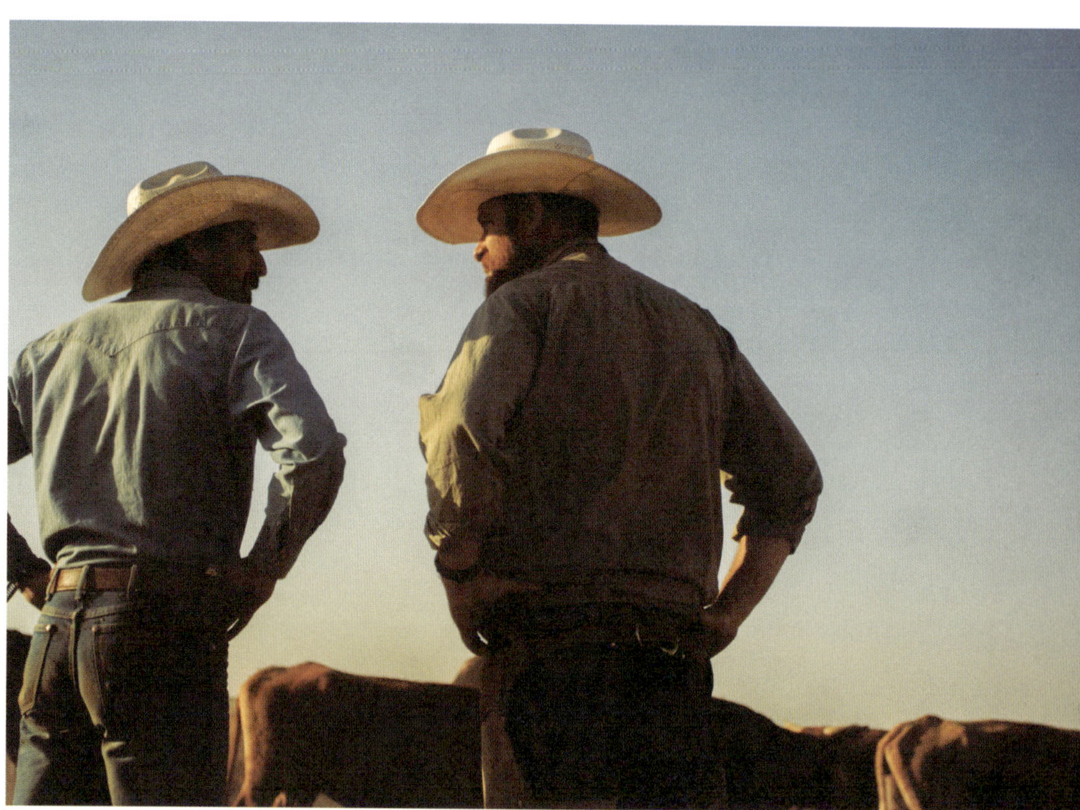

BRANDING

At Ranchlands, we practice a traditional style of branding. Branding season begins at the end of April and lasts until the middle of June, as calves are born and reach the age at which they can be safely branded. A branding day often starts around 4:00 a.m., with the cows penned by 6:00 a.m. Once the first calf is roped, everyone is ready to work. The calf is then flanked by the ground crew, branded, vaccinated, ear-tagged, and the males are castrated.

The process is done in less than two minutes per animal, and low-stress stockmanship methods are employed. The more relaxed and efficient the process is, the better. The calf-cow pairs are kept in the same pen to minimize their time apart, which in turn helps alleviate stress. After the branding process is complete, the calves are immediately reunited with their mothers. Branding allows calves to be vaccinated against diseases, provides data on herd health and numbers, and brings communities of ranchers together.

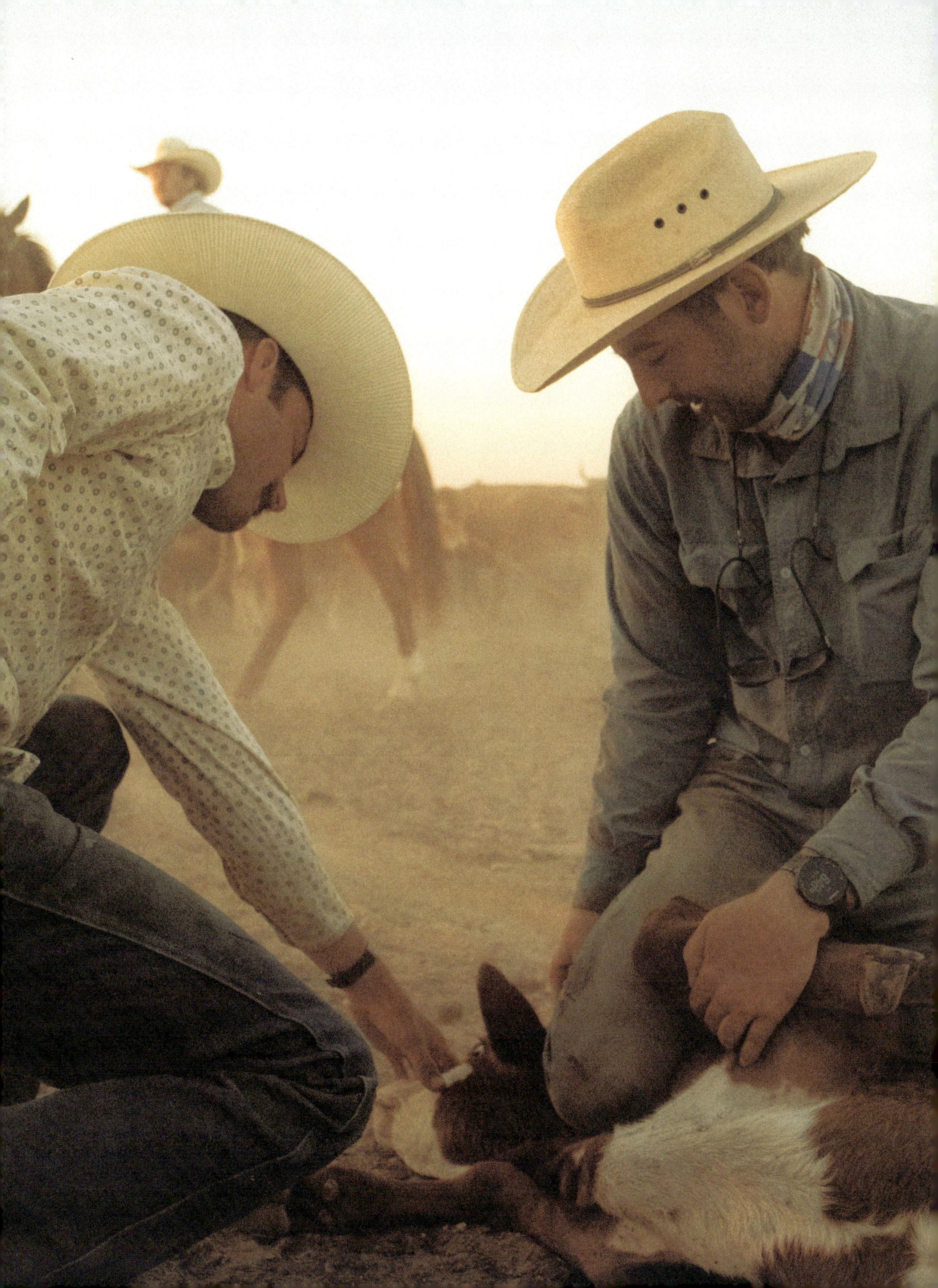

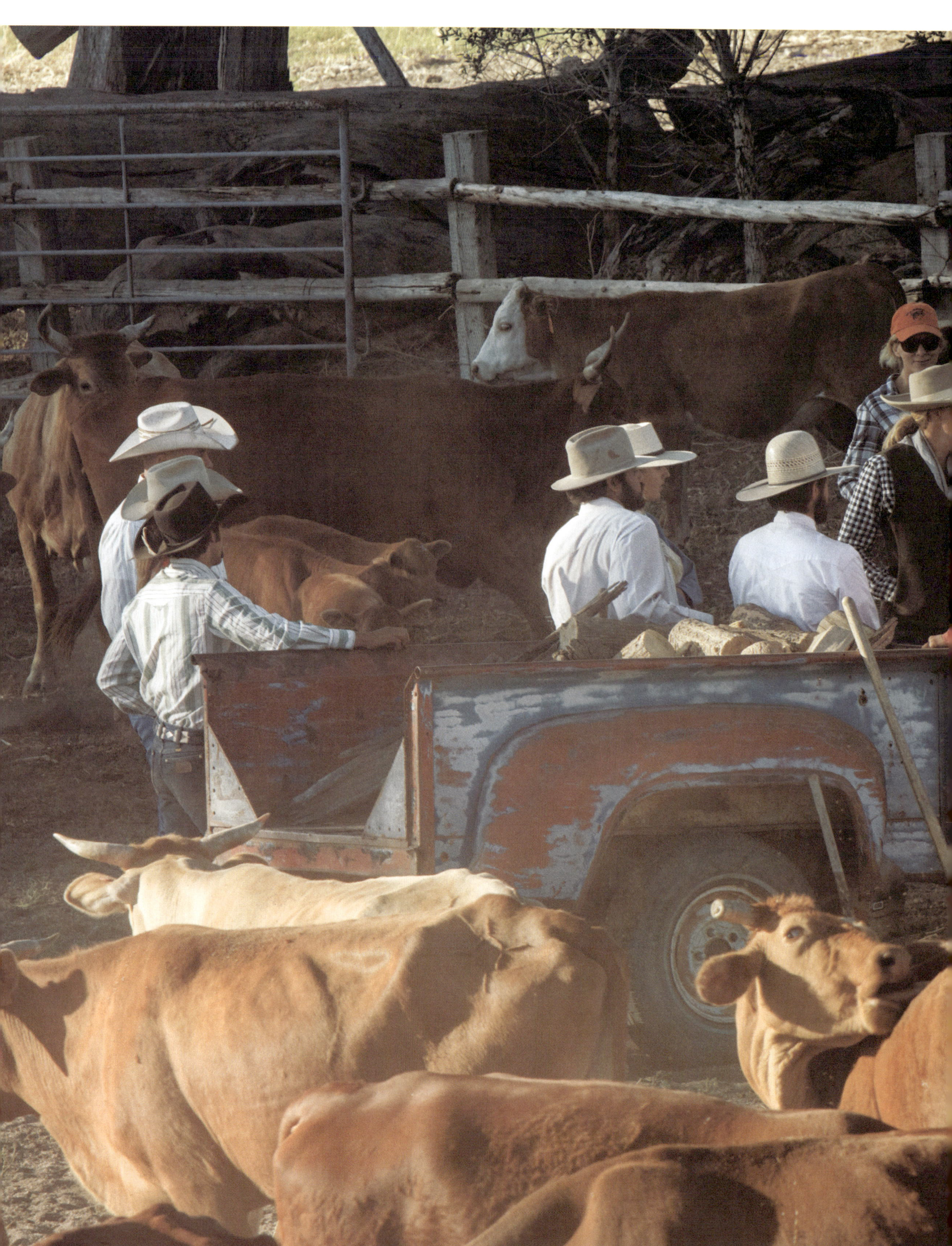

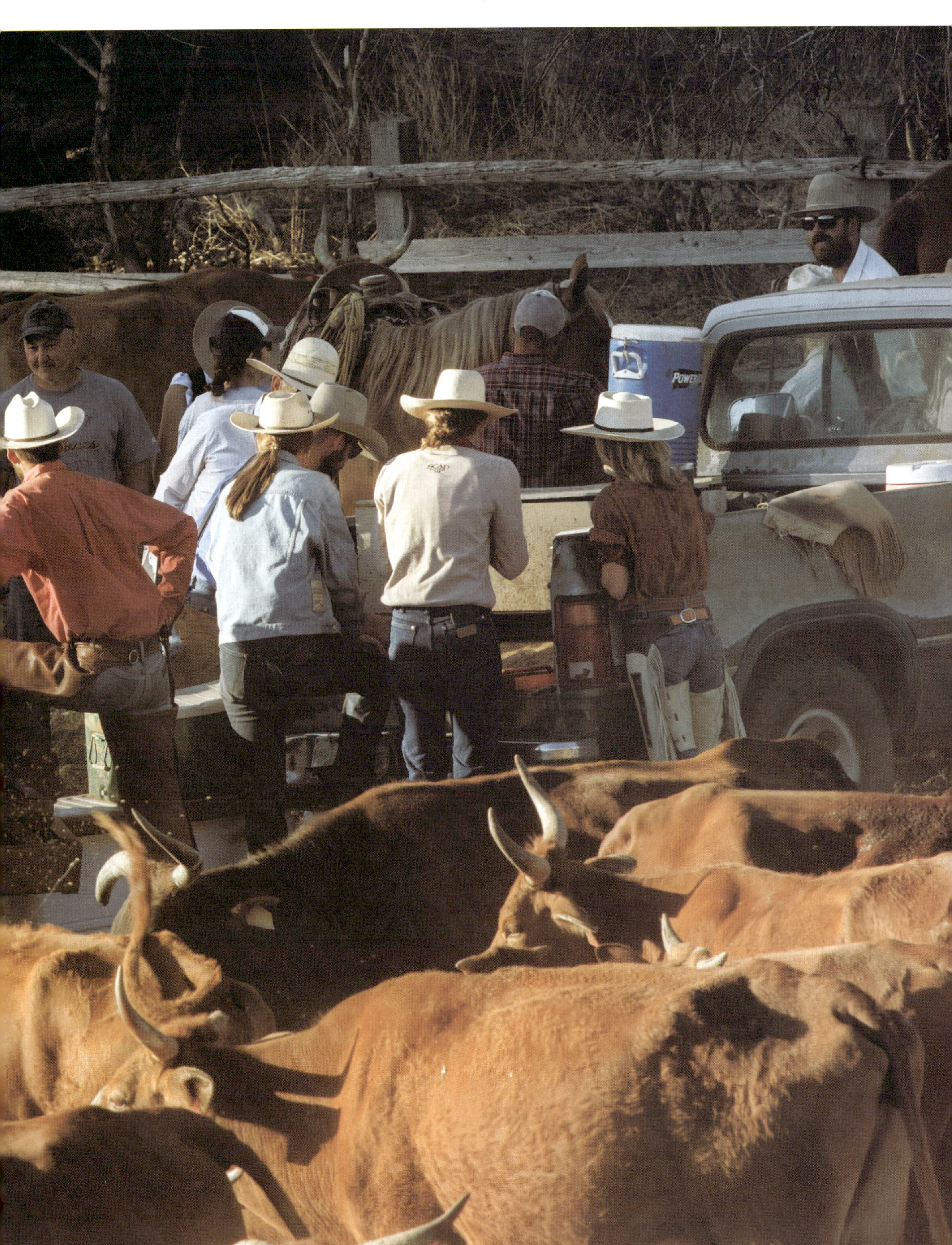

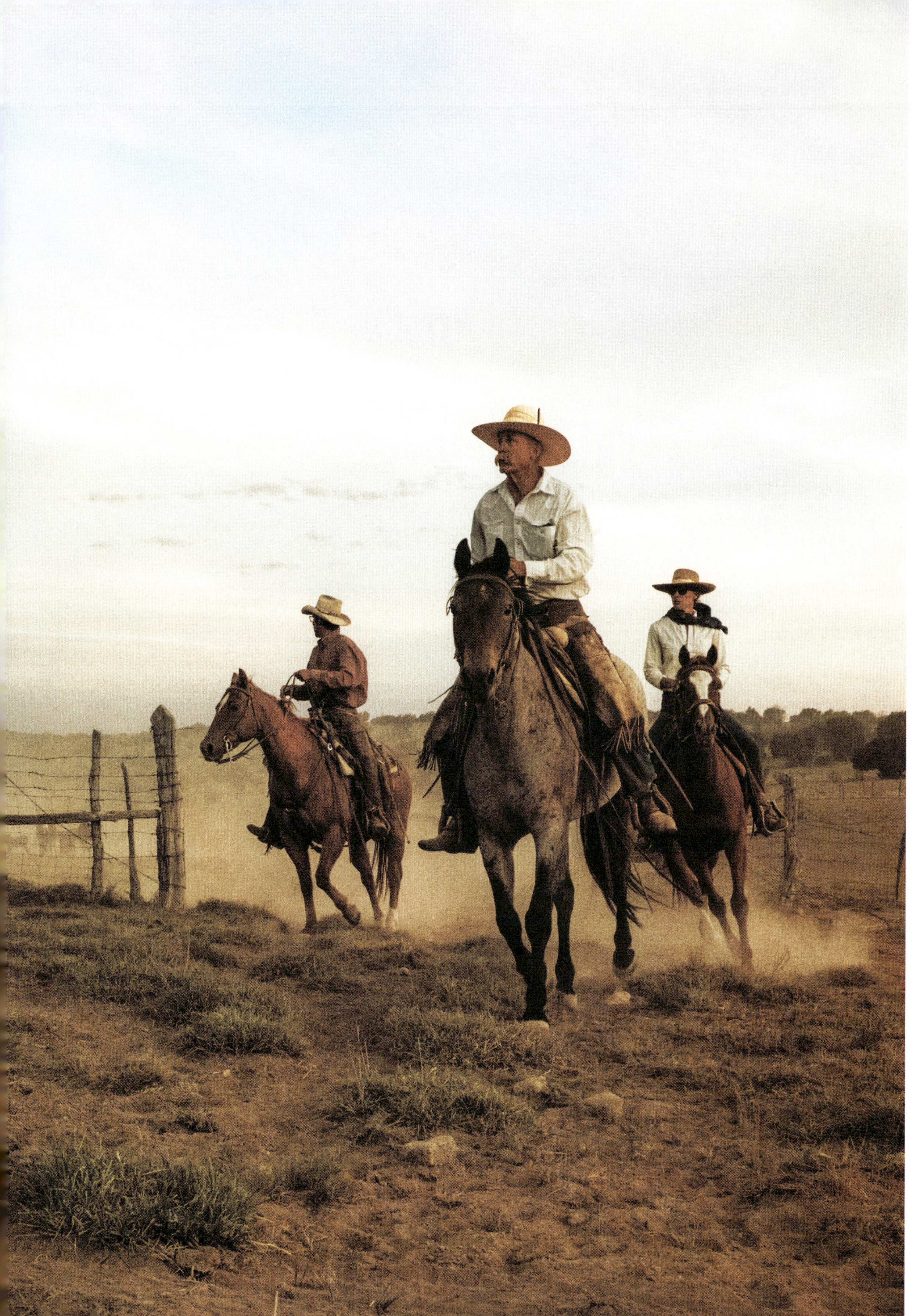

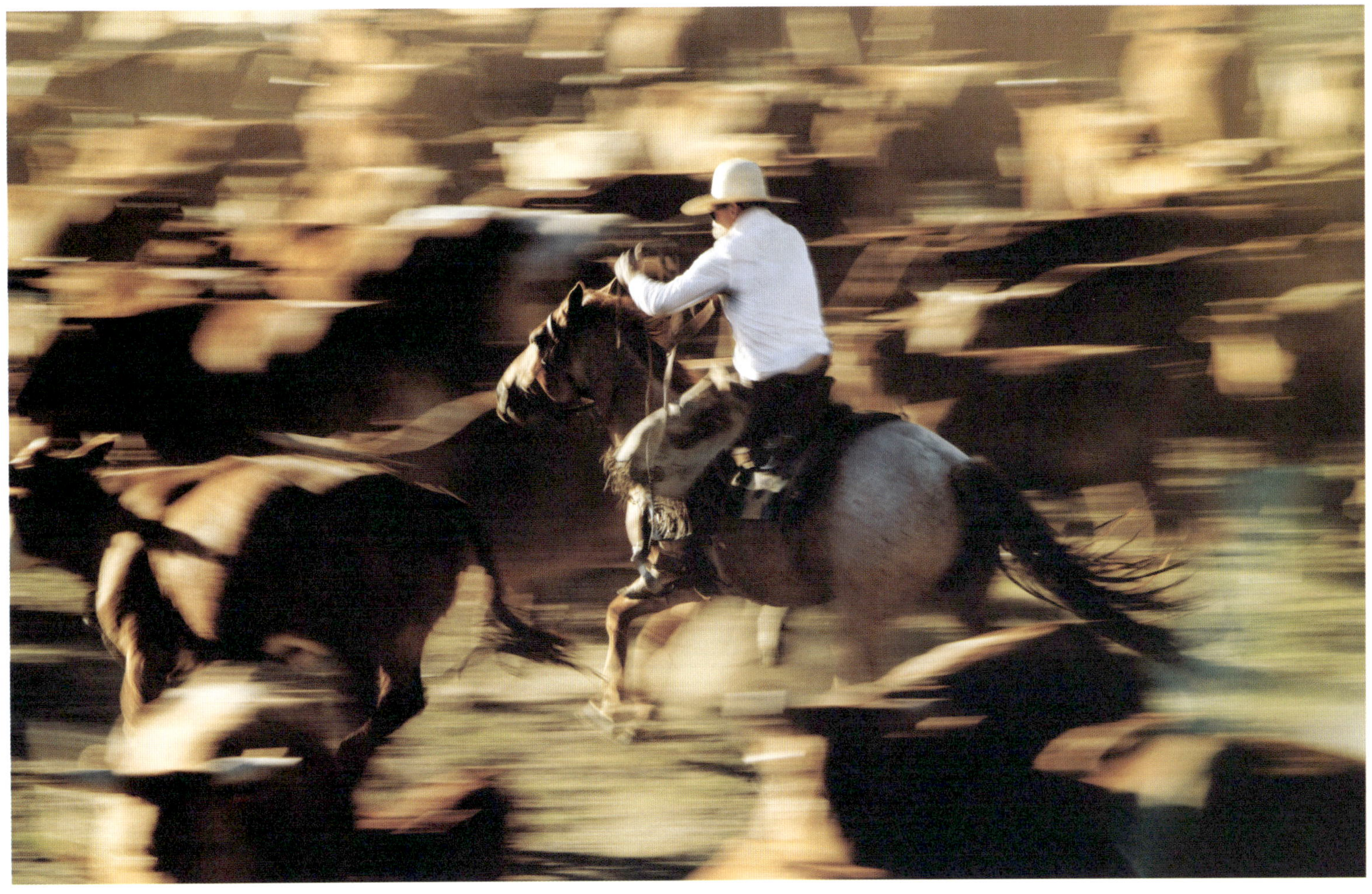

For someone who has not experienced the branding process before, it seems chaotic. There are calves jumping around on the end of long ropes and horses coming in one after the other.

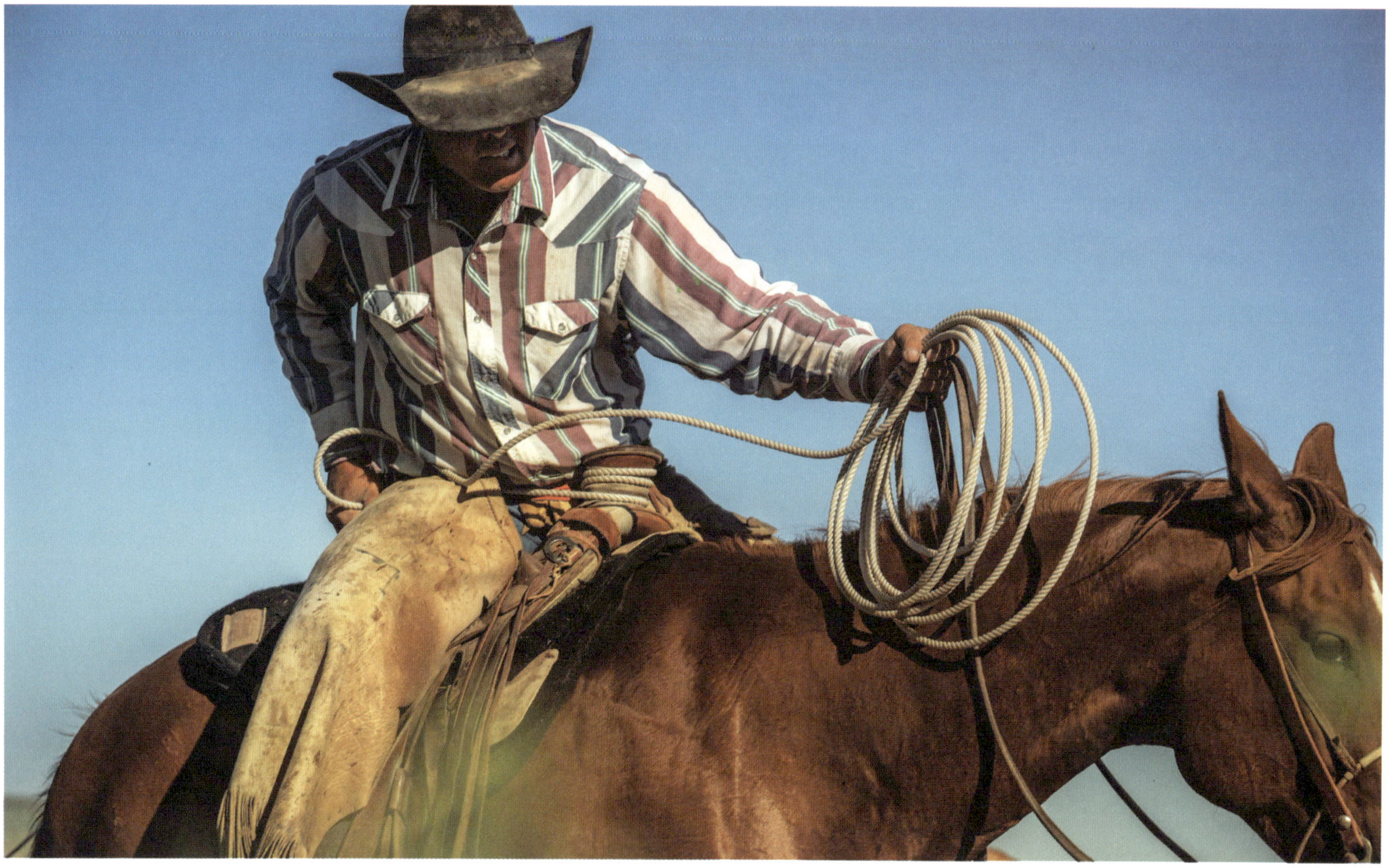

Ask anyone with ranch experience about their favorite time of year, and the majority will have the same answer: branding season. The sense of community, camaraderie, and coming together for a common goal embodies the branding season.

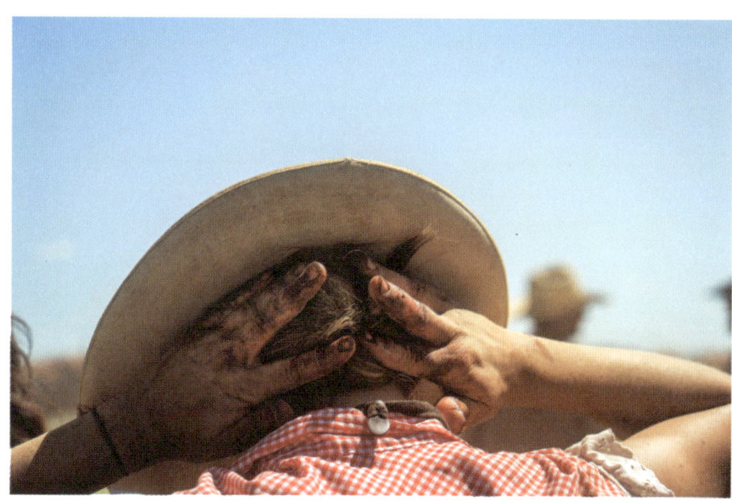

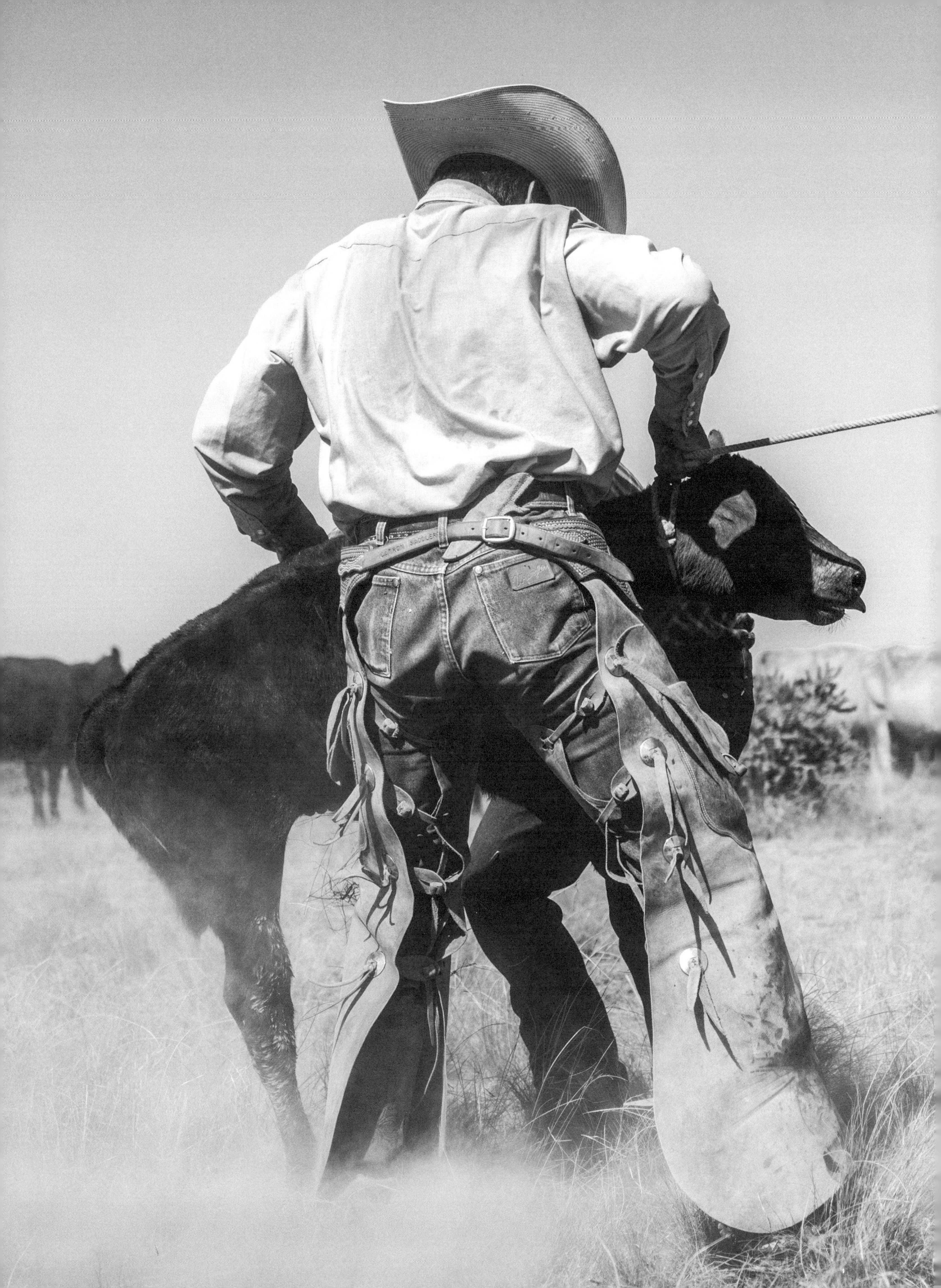

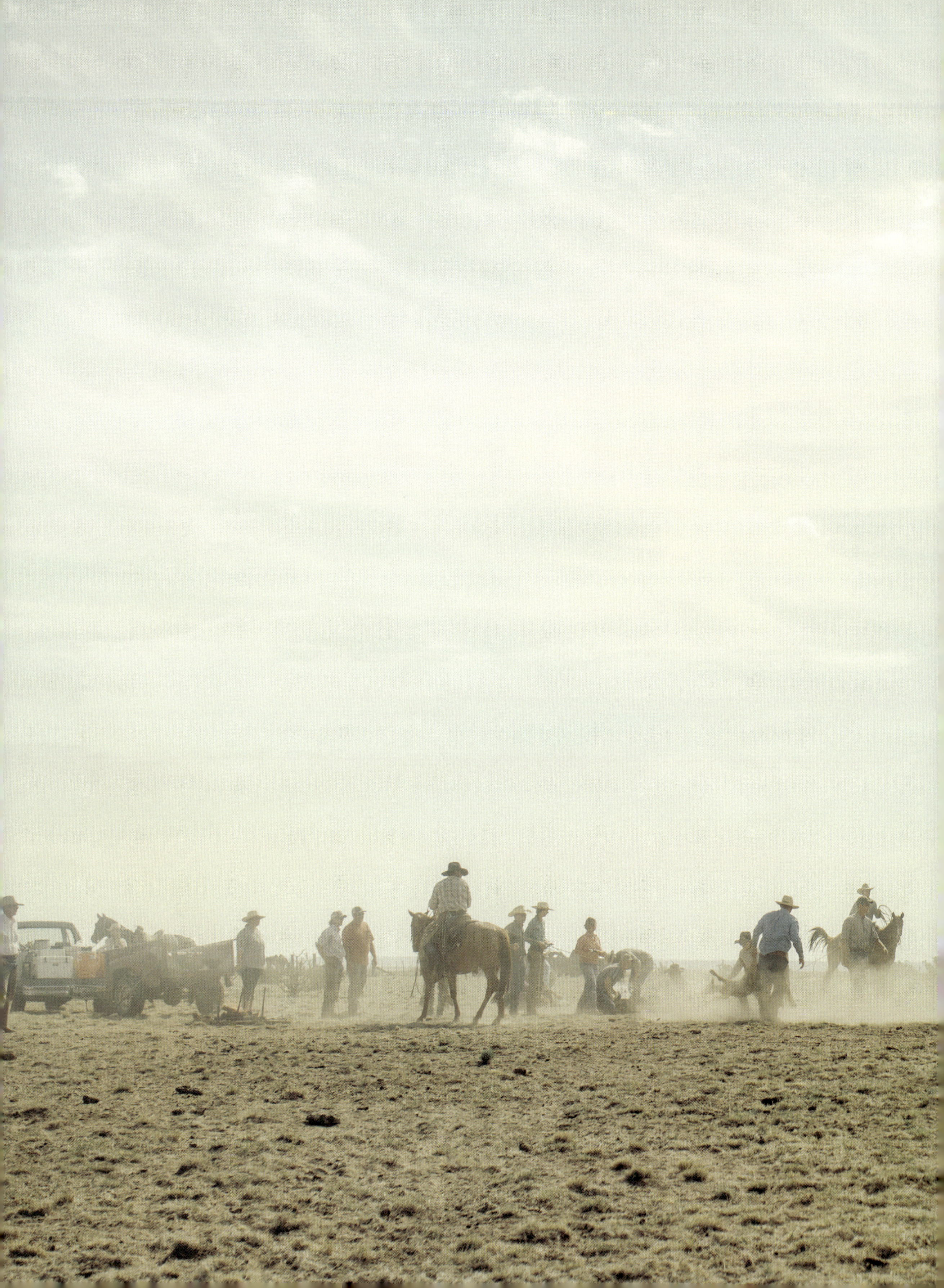

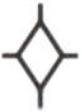

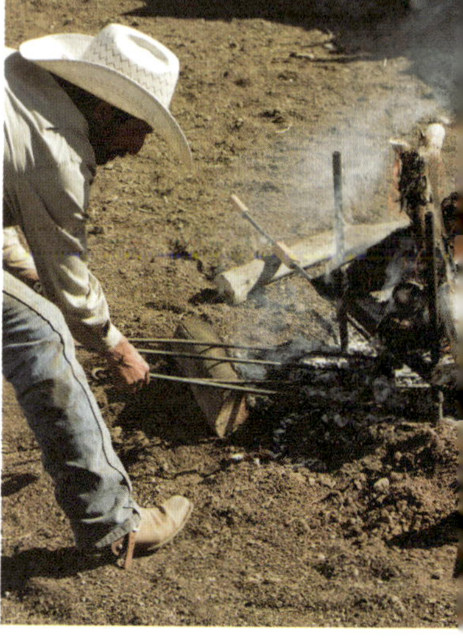

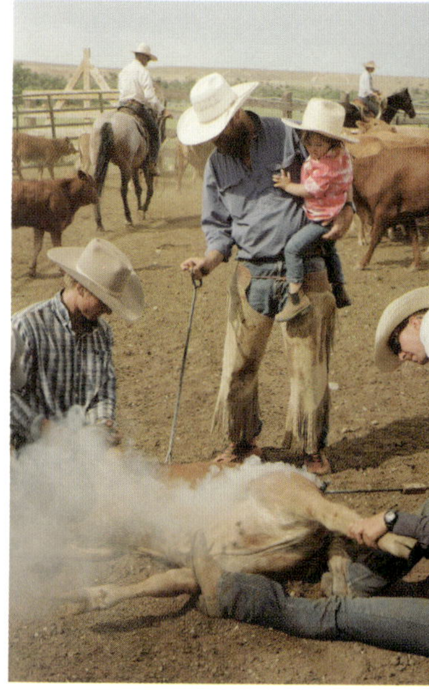

ORIGINS OF THE DIAMOND STAR

When I think back, I realize how fast times change. One morning in the early 1990s, at 6:00 a.m.—a time when there were no cell phones, no internet, and no email—I left the house and grabbed the *Arkansas Valley Journal* that had just come in as I went out the door. I was living on a ranch just outside of Pueblo, Colorado, and had been looking for a single-digit brand for a new cattle company I had started.

The *Arkansas Valley Journal* was the best periodical for finding brands for sale, but it was important to check the classifieds quickly because good brands went fast. As I headed down the driveway, I opened the paper and saw an amazing brand for sale. I couldn't believe it. I hadn't seen anything like it for a long, long time: a diamond that looked like a star, with rays coming out of each corner. Wow. I remember thinking how clean it would look on our cattle and feeling like it simply wasn't possible for something this nice to still be available.

I sped up in order to get to the Loaf 'N Jug store five miles ahead that had a pay phone. I put the coins into the slot of the phone and dialed. The woman who answered sounded sleepy and a bit annoyed, but I dove in and asked if the brand was still for sale. I remember the feeling of relief mixed with joy coming over me when she said yes. I sent her a check that afternoon, and the brand was ours.

Over the years, we have branded many cattle with it—cattle we have purchased and calves we've produced. When we built the iron, we split the brand in half so it could be applied in two rounds and the hide inside of the triangle wouldn't overheat and cause a scar. If a brand is applied right—meaning it's kept on for the appropriate amount of time, corresponding to the age of the animal and the heat level of the iron—it simply changes the direction of hair growth instead of creating a scar on the skin. Sitting on the hip, where we proudly place it, the diamond star looks like it's meant to be there.

Our brand has become a symbol of the pride we feel for breeding a herd of cattle that is the embodiment of our philosophy of ranching. It stands for who we are as people and ranchers who are preserving our traditions by living them out. It stands for how we care for the land and the wildlife that abounds here alongside our livestock. It stands for the pride we have in the group of people who we work with side by side, people who have become a part of the everyday life we are living on the ranches we manage.

Indeed, times have changed. Back then, we were just beginning our ranching operations and had few cattle of our own to brand. Today, we've been in the business for almost 30 years with an ever-growing herd of cattle wearing the diamond star. Beyond this, our brand has become a representation of our work and all the things we stand for. It's stamped on all the leather goods that we produce in our leather shop. It's even a bumper sticker that we place on our trucks and cars. And today, in a world run by cell phones, computers, and things that hardly existed when the brand first came to us, it's also a symbol of the things to come that we are working for.

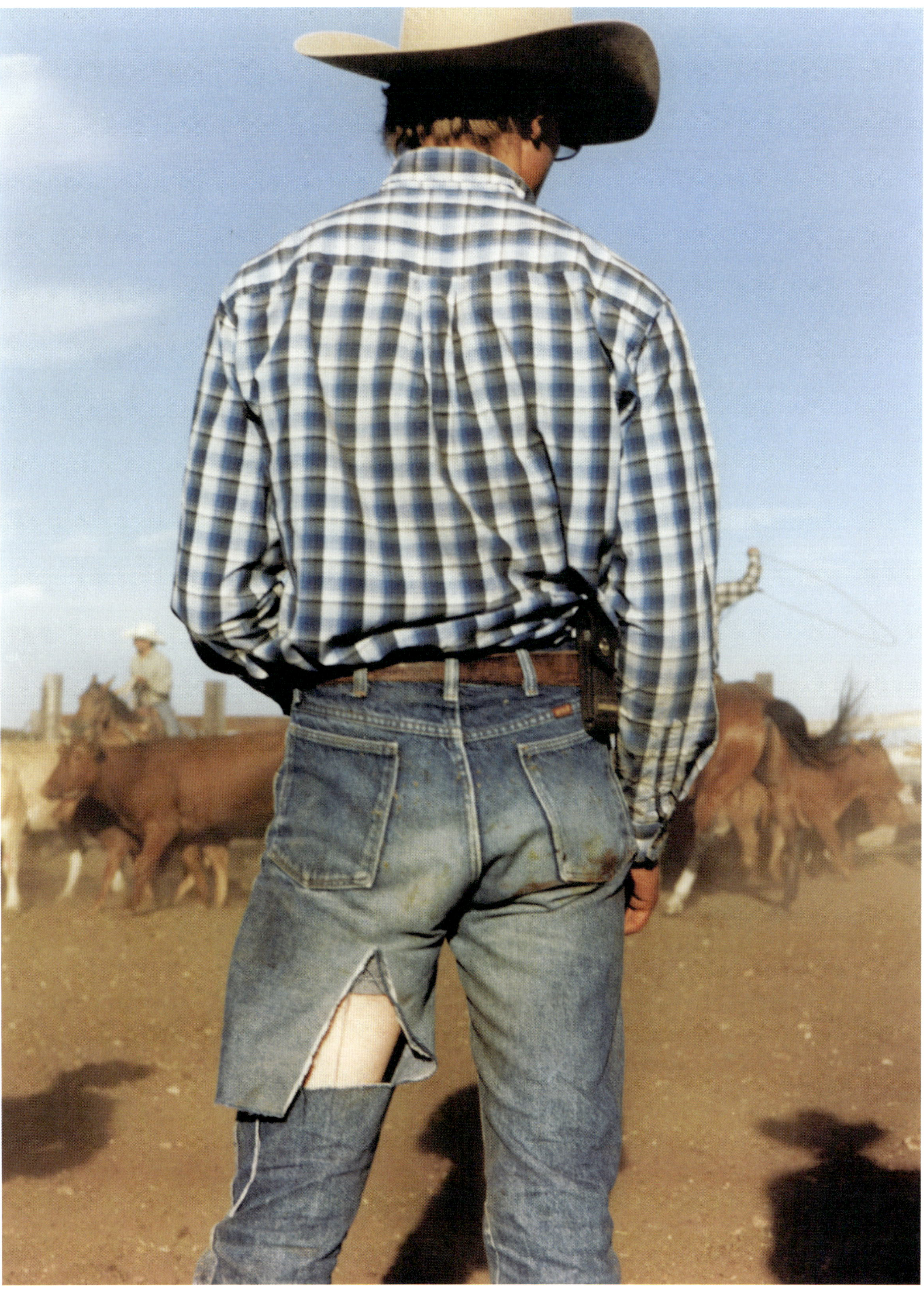

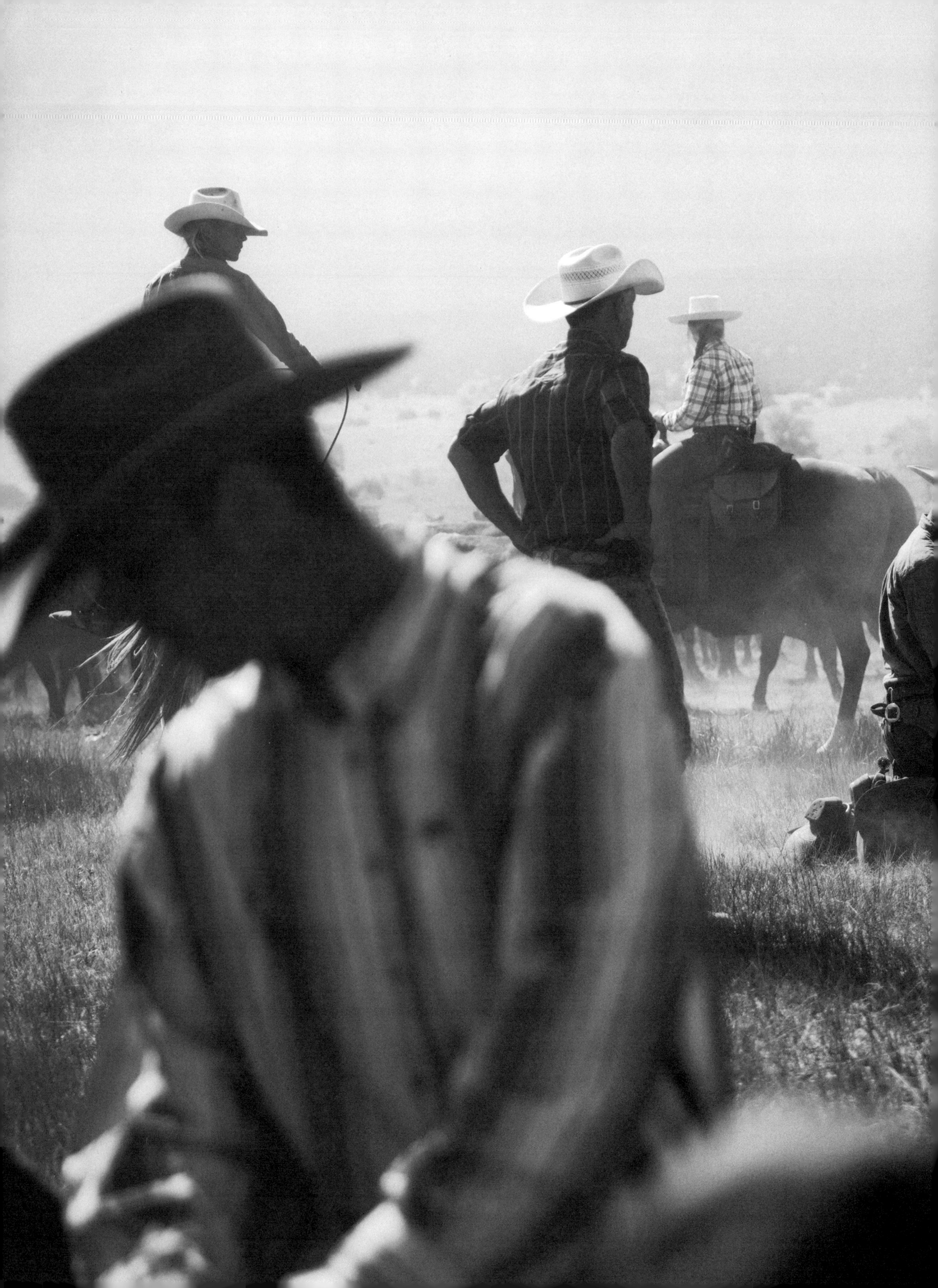

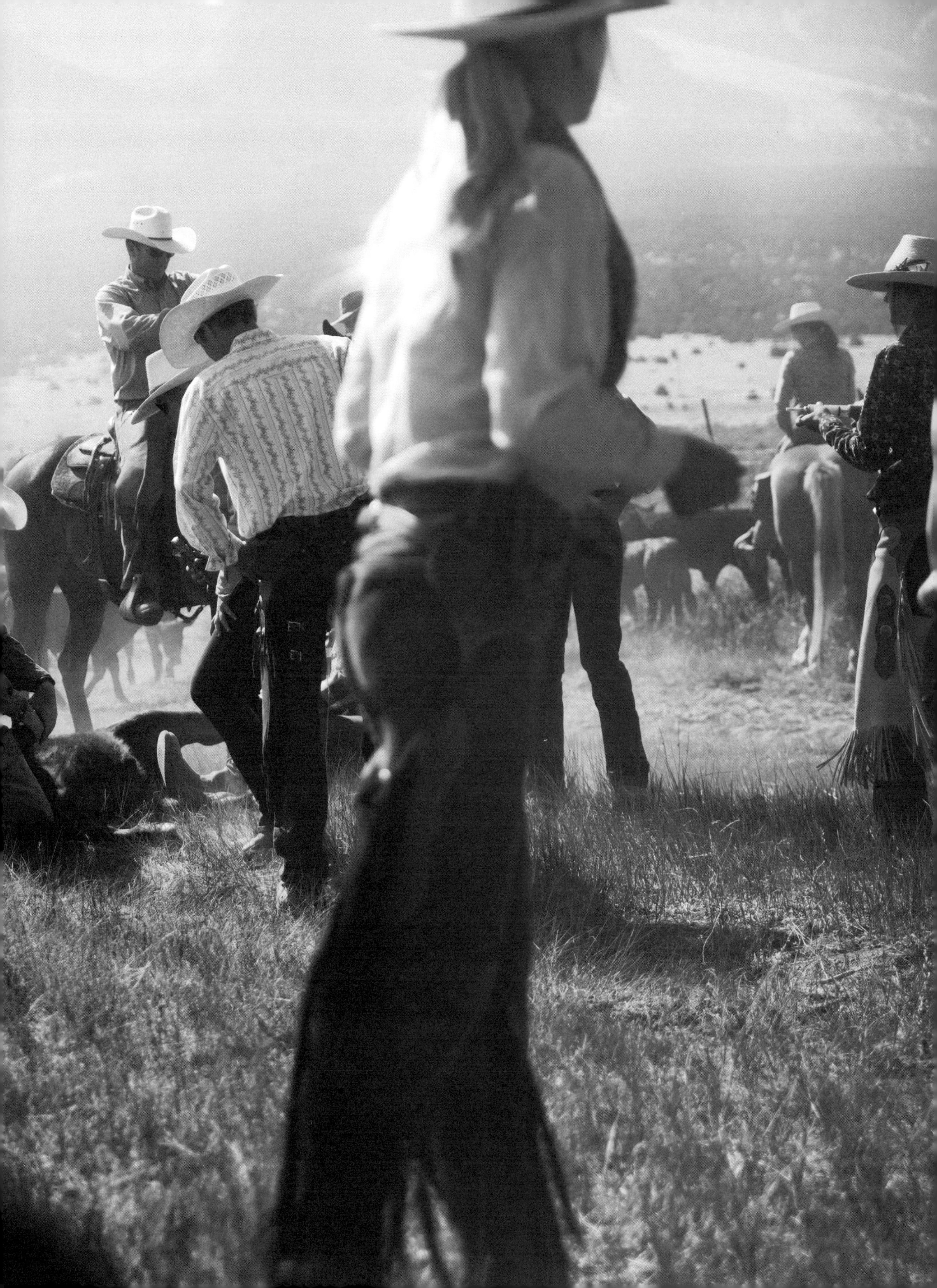

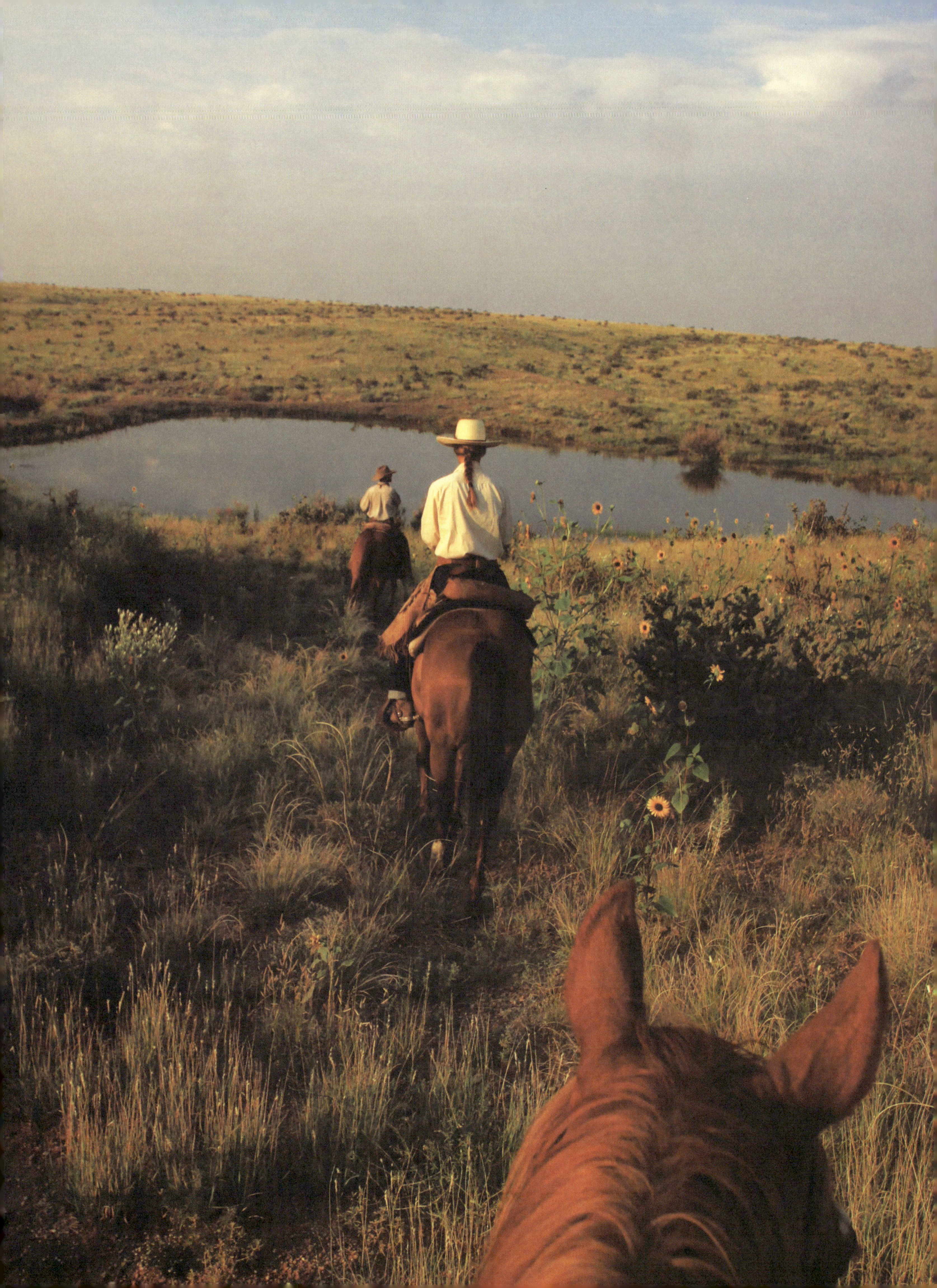

THE

FUTURE

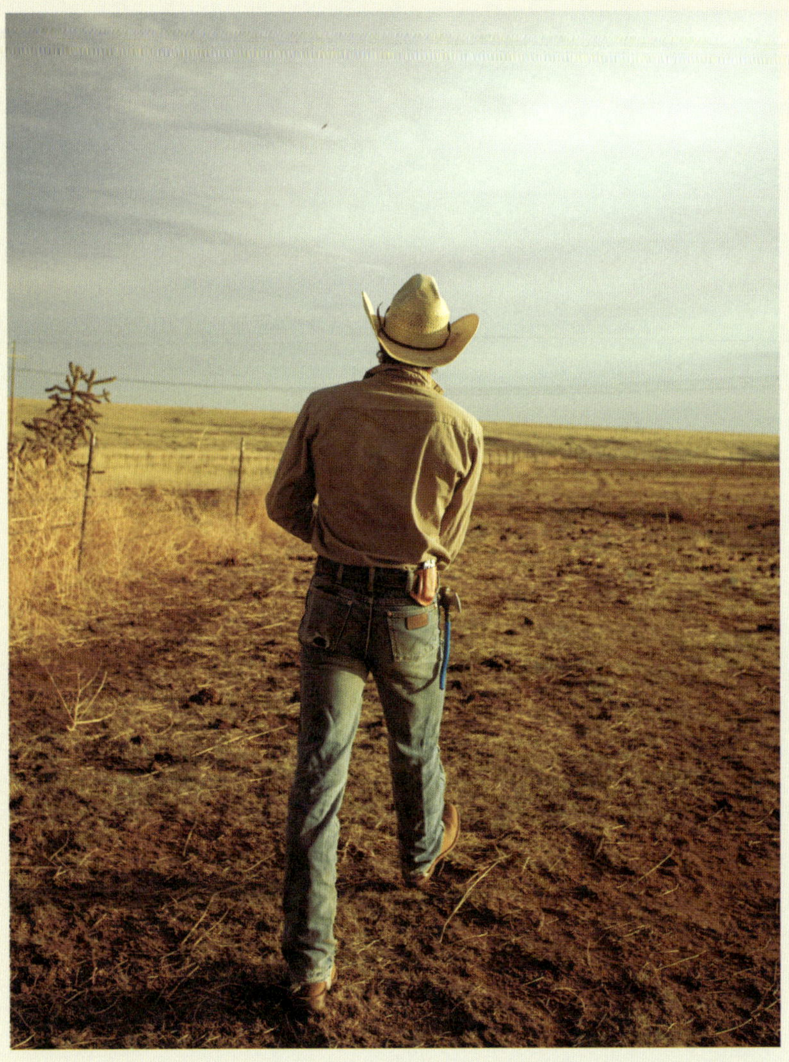

We've been working our cow herds the past couple of weeks, testing them for pregnancy, weaning the entire year's production of calves, and culling the cows that have not kept up. Filled to the brim with bawling animals, the corrals make us feel the Earth spinning, as the cattle trample the dirt into a fine dust that covers the faces of the crew of men and women, each a cog in the turning wheel that moves one cow at a time through the chute.

The outcome of this work tells us how we have done for the year. It is also the manifestation of decades of genetic mapping that involves selecting and culling cattle to create a herd that is fertile, productive, and adapted to its natural environment. The herd deeply represents Ranchlands' past and future.

For 24 years, we have followed the herds after letting them out of the corrals, and today, we're walking the animals back to the winter pasture. This year, it's the North Pasture. There's a thin string of cows stretching more than a mile ahead of us, to the west, as if they're pointing a long finger toward the future. They take each step forward with an eternal, in-the-moment presence that we humans cannot muster. Their steadfast calmness professes a profound courage that can only be found in the nature of animals, a courage that we must find within ourselves as we venture forward into the unknown. From these animals, we have learned that we must preserve our legacy by living it. They are the ones who show us the value of each blade of grass and kernel of soil, the depths that the water can reach, and what can happen when land is given time to rest and grow. They help us understand that community is simply the ability to communicate without speaking a single word. They are what tie us to the ground and intertwine all the dimensions that make Ranchlands what it is—our work with the land, our risks and opportunities, our financial viability, our joys and fears, and our people.

While this herd represents who we are at our core, Ranchlands has become much more. The connection is that we are stewards, working every day to preserve the future of the wild places that we call home. We are horsemen and women who partner with the animals to better understand a species besides our own. We are artists, reflecting the truths of nature through our craft and for the benefit of all. We are scientists, using our tools to better understand the natural world. We are travelers, exploring new places and immersing ourselves in different ways of life. We are old friends visiting from diverse backgrounds, sharing time together in nature on our ranches. We are entrepreneurs, using the power of our businesses in service of the greater good. We are craftspeople, keeping the highest quality of goods and traditional artisan techniques alive. We are educators, empowering the next generation to create a better future.

Because of this organic growth over the past 24 years, Ranchlands is now at a watershed moment. We see ranching as one of the most direct and powerful avenues for affecting large-scale changes on large-scale landscapes. By stewarding the land in a way that mimics natural ecosystems and relies on the inherent power of the land itself, we guide the regeneration of rangeland and protect the integrity of intact natural ecosystems in the American West. All of this is accomplished by working in partnership with the natural environment—not against it.

Ranchers are on the cusp of the greatest opportunity we have ever encountered, which is to show the world our role as naturalists. Our challenge as ranchers is to be aware of our own capacity and start changing the paradigm of viewing ourselves as strictly cattlemen, horsemen, or ranchers. In that list of names, we have to include environmentalists, naturalists, and conservationists. Maybe even weirdos. Although these labels carry deep emotional gravity within our cultural identity, in the end they are just words that bring to the surface our relevance in today's world and our place in its future.

We see land as a multidimensional resource and not just a home for ourselves, our cattle, and our horses. Our power, though, and the place where real change happens, is in the hearts and minds of the people who interact with us. The evolving model has ushered in a new community of people from outside our world who have come here to experience nature and ranch life and to support our mission to provide careful environmental stewardship to the land through ranching. This coalescence of people from all walks of life who care and who are actively doing something about the ecological problems we face—when combined with our deep consciousness for the land and our long legacy of tried-and-true ways—will fuel the momentum of a grassroots call to honor and protect the land.

This picture of future possibilities has lived beneath the surface of our vision, morphing day by day for more than two decades. Ranchlands is a culture, a growing, committed community of people. It is a way of life, a way of being, and a way of interacting—not only with people and our communities but also with animals, with nature, and with our planet. While all of this is manifested in the tangible form of ranching, Ranchlands is much bigger than just ranching.

Who would have thought that a bunch of cows grazing, having babies, and stomping on grass would become the power and driving force behind such an ambitious outlook? Who would have thought that this herd would be the current providing the energy and momentum and leading us to preserve a way of life—one that has the capacity to bring people together for the greater good of us all?

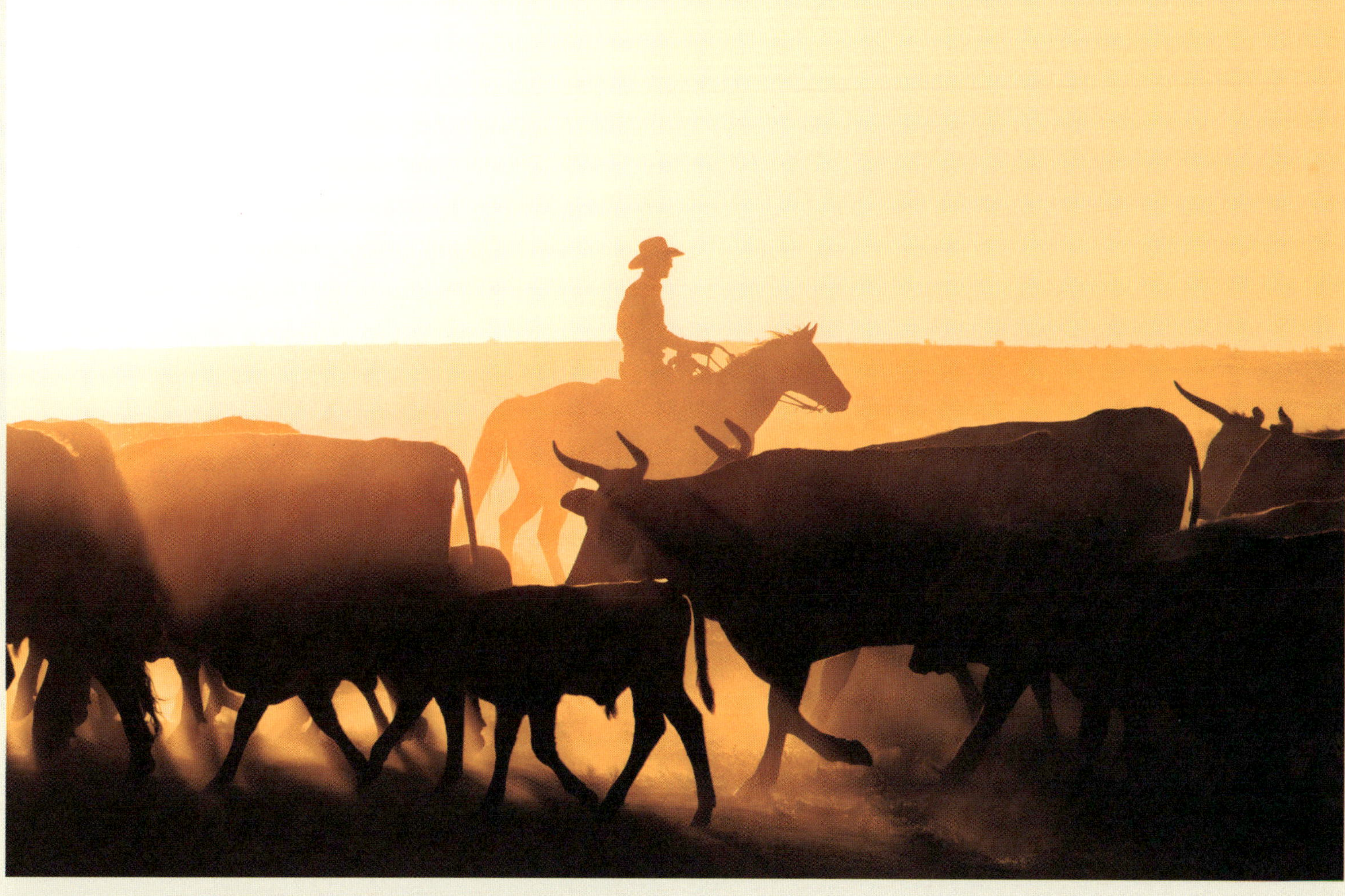

WALKING INTO TOMORROW

The year ahead is going to be interesting, to say the least. In most circumstances, what has happened would be catastrophic, taking our whole business under. We have just lost our two largest leased ranches that have been servicing the banknote on Paintrock Canyon Ranch in Wyoming, purchased three years ago. A tenant rancher, no matter how tenured or hardworking, is always living on borrowed time. However, as much as it impacts our business, it cuts deeper at the human level.

It means losing the home where we have lived the last 24 years, where I raised my kids and where my five grandchildren were born. We will miss the people we are leaving behind in our community and our business network, all the school kids who have been coming to the ranch over the years, and our neighbors. But most of all, it's the place—the land and the animals—that we will miss: the lakes, the endless prairie, the creek and its springs, and little things like owls dropping out of their perches in the creek bank and flying low in front of us as we are riding down the watercourse. We will miss the summer thunderheads, rising like massive anvils in the huge sky, that bring the monsoonal rains, giving life to this place of unbelievable beauty and resilience.

However, these feelings of sadness have been replaced by a deep gratitude for the life-changing opportunity to have lived there and for how it formed our lives, making us who we are as a family and an organization. It brought so many people into our lives, gathering around the ranching concept we call Ranchlands. This circle of people gives us the courage and confidence that we need as we contemplate navigating our pathway into the future.

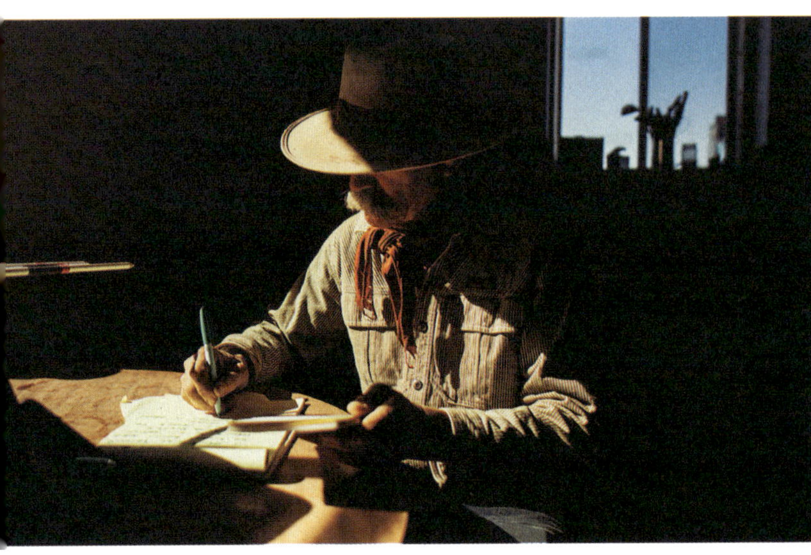

Twenty-four years ago, we started with a lot less, but through persistence and imagination, we rose each day to face the challenges and joys of working for ourselves on a bountiful piece of land. Today, we have a great deal more, and it runs deeper because of our community. We have learned that what is coming will be different but better than what we expect. A big swinging gate is opening before us into a green pasture of unforeseeable dimensions. We can't wait to see what we are going to find there!

DUKE PHILLIPS III
Paintrock Canyon Ranch
March 3, 2024

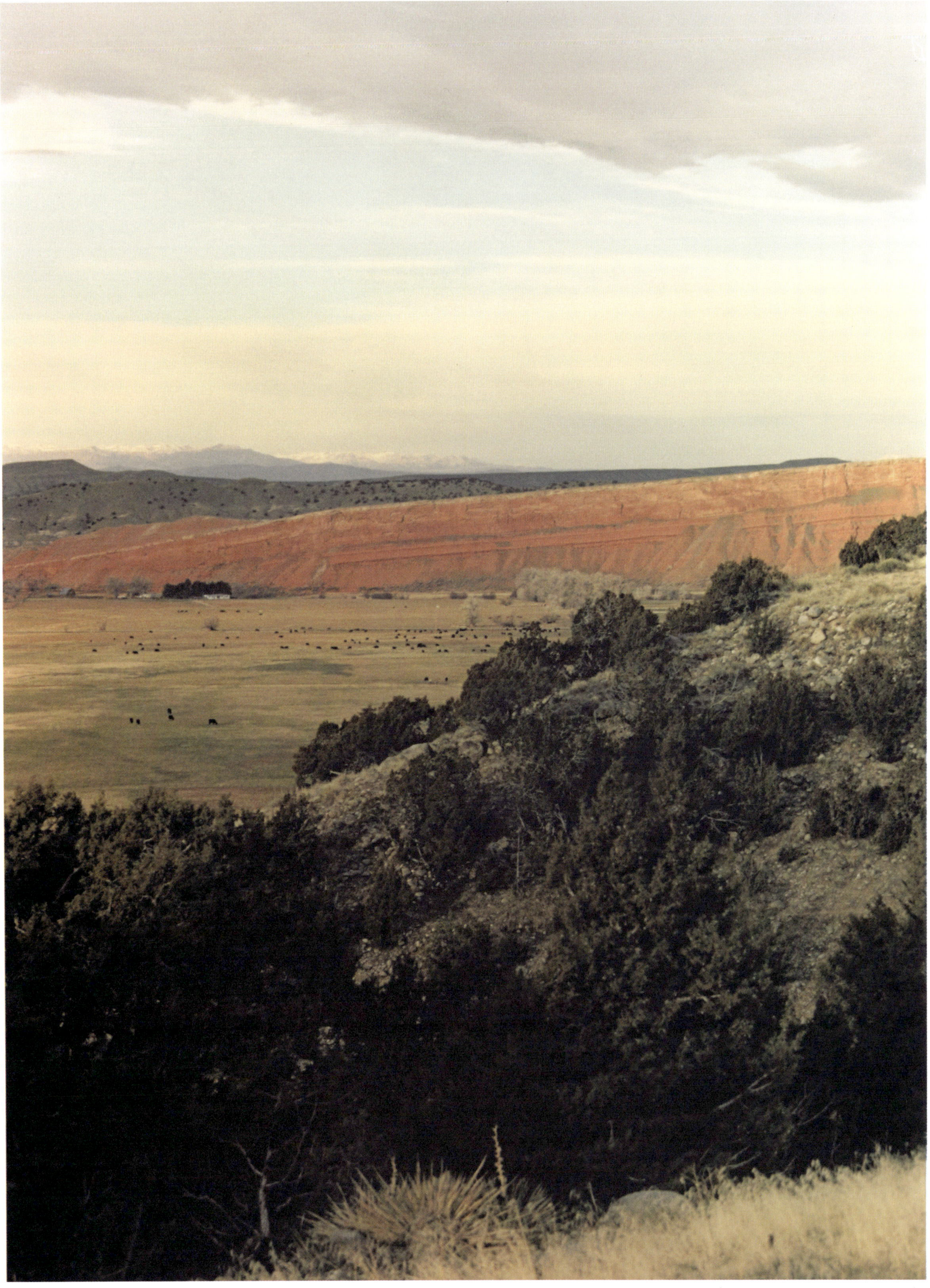

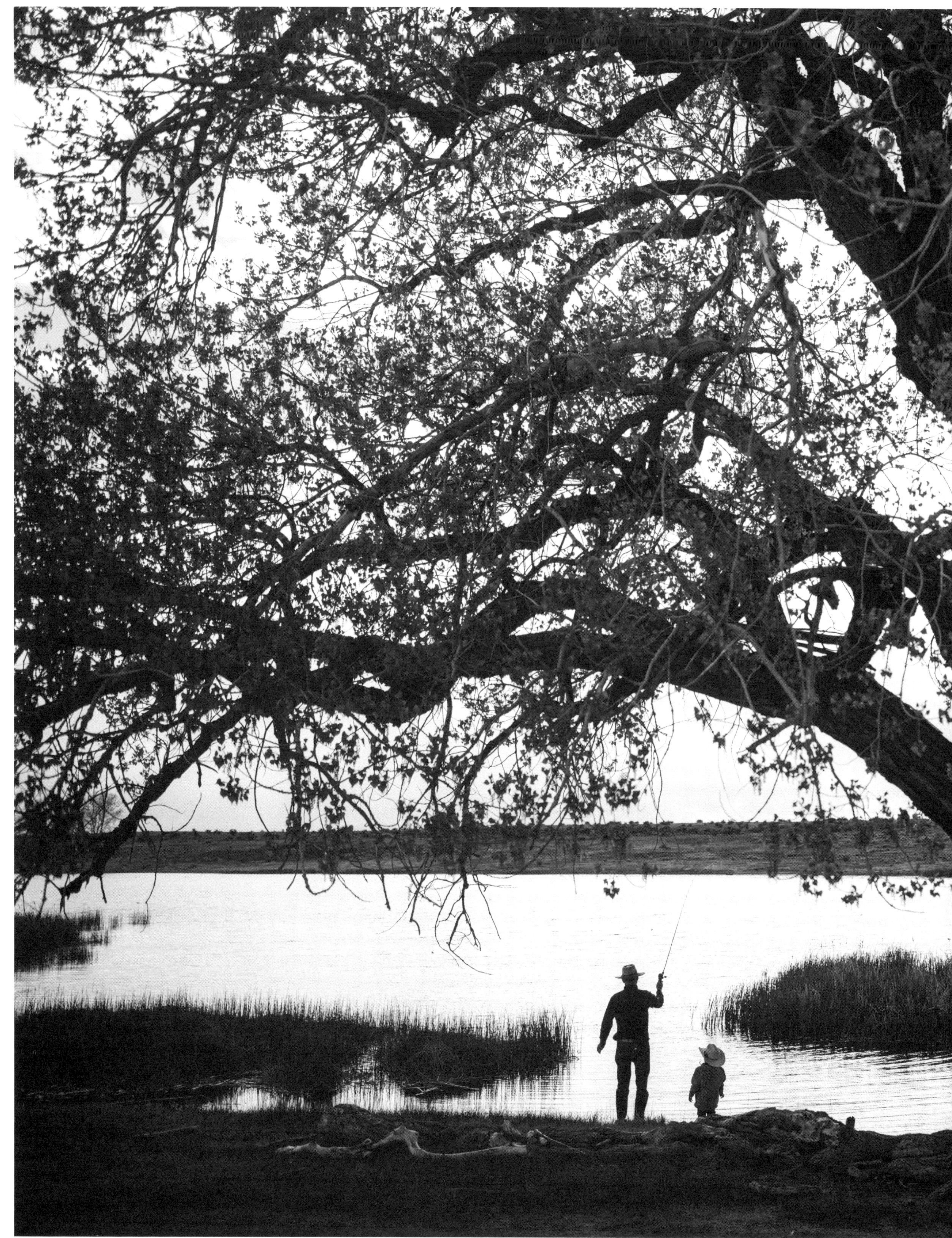

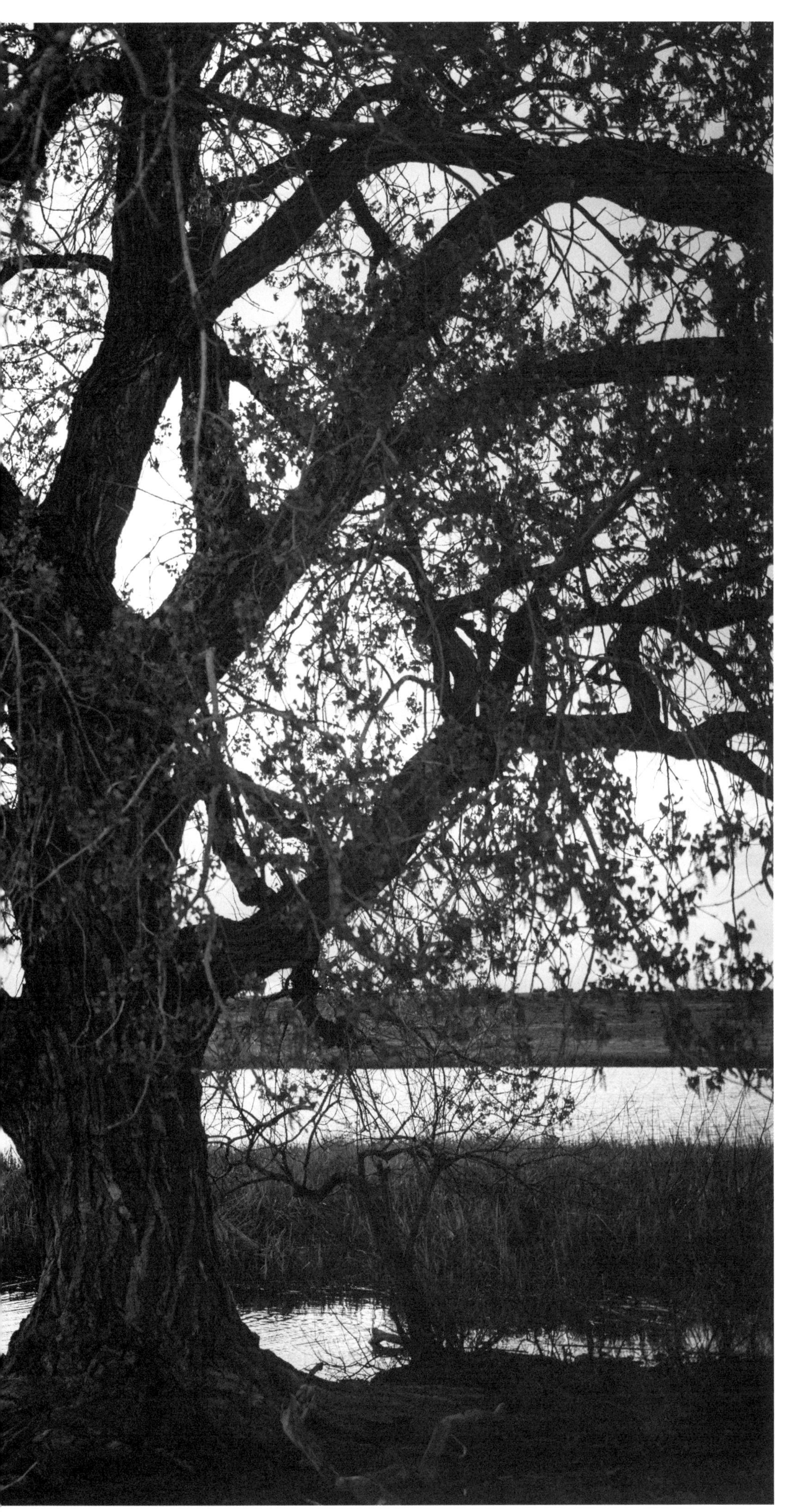

LEGACY AND TRADITION

We are an outgrowth of ranching, an operation that has held on to one of the oldest traditions in our country and pushed it into the future. It all began almost 62 years ago on a remote ranch set in an old-world horse culture. The old ways represent our past and where we came from. They are methods that have been tested by generations of men and women, working in conditions that generally have been more difficult and less reliant on technology than today. Now we have become an exemplary model of a forward-facing modern business that uses a mix of traditional ranching practices and new schools of thought—a highly unusual practicum of blending the old ways with the new.

In a nation where 98 percent of the population lives in towns and cities, ranchers still live on land, just as our ancestors did when our country was settled. We use horses to trot out to work instead of driving trucks. We are skilled with long ropes and know how to handle horses and cattle out on the range instead of bringing them into the corrals to work. Our free time is spent outdoors, often riding, fishing, or hiking. Our children grow up in the family business, working, learning our trade, and gaining a deep familiarity and connection with nature.

From this foundation, Ranchlands has evolved over the past 24 years into a deeply American brand by connecting its ranching legacy with conservation values that have become intrinsic to ranching, while also working to provide broad access to the constituency that it has created. This access is a means of communication that builds bridges between urban and rural communities as a means of fostering collaboration for the collective good. Combating the fast-growing chasm between nature and modern society has become a mainstay of Ranchlands' mission to develop a model that brings people together.

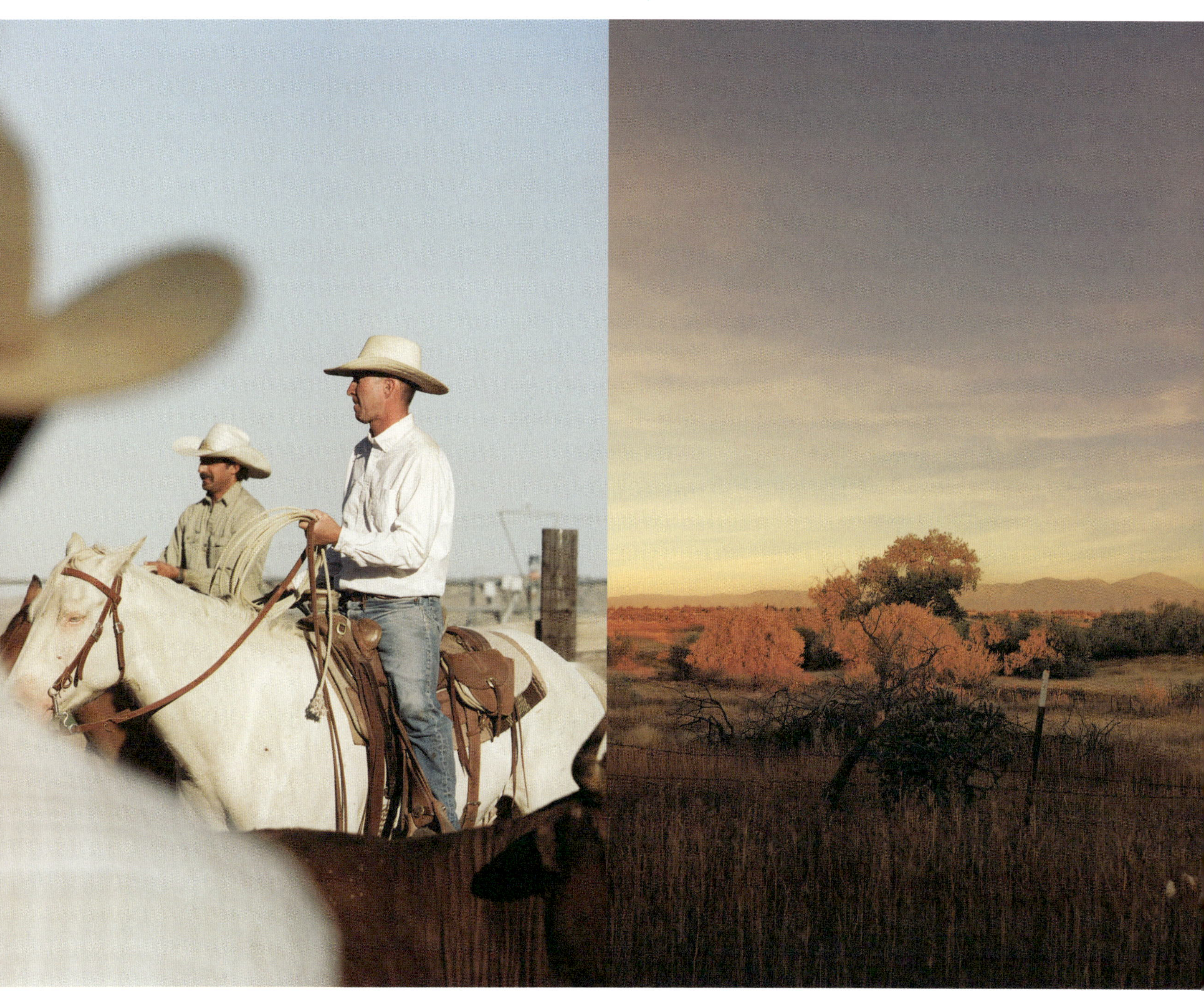

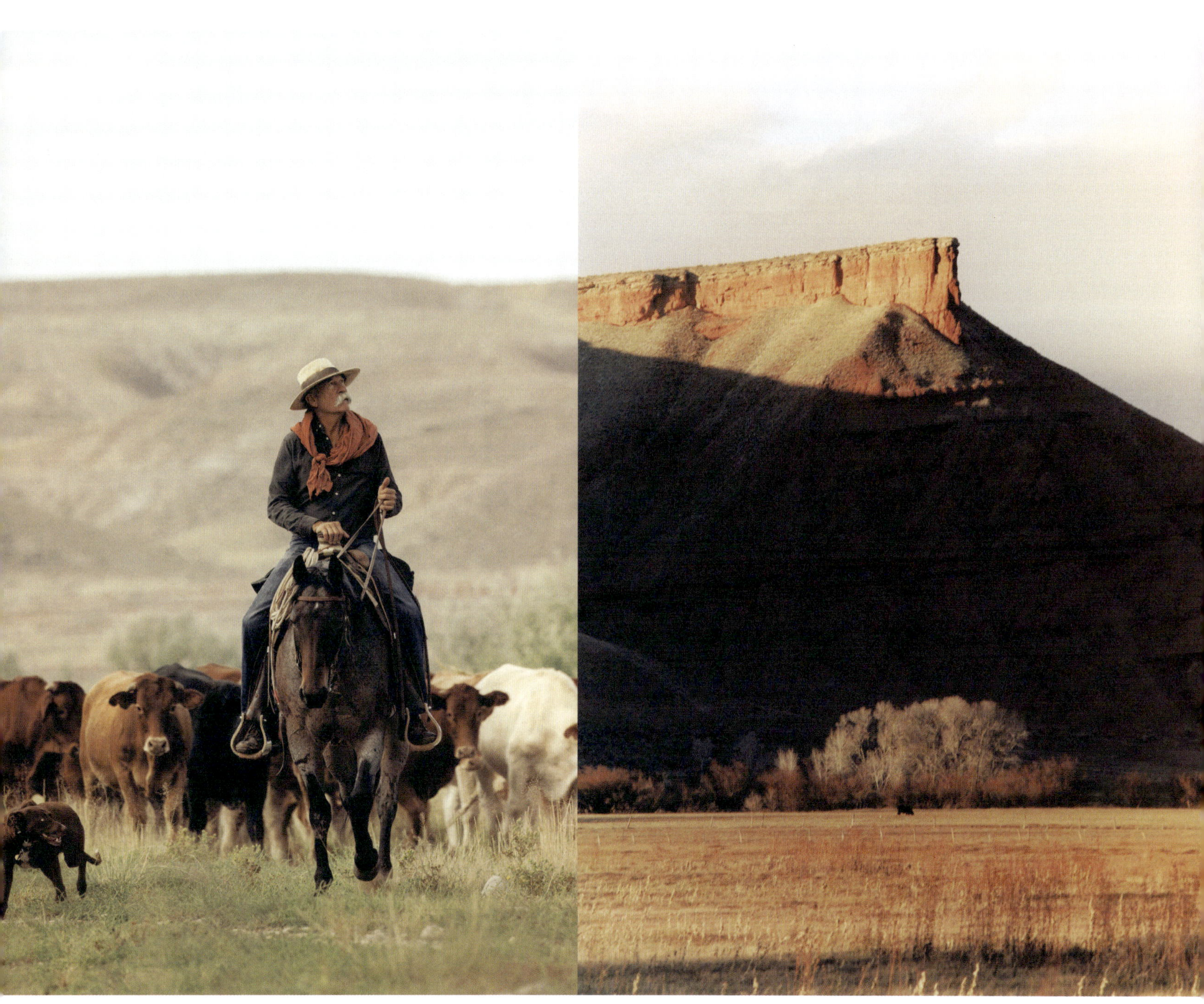

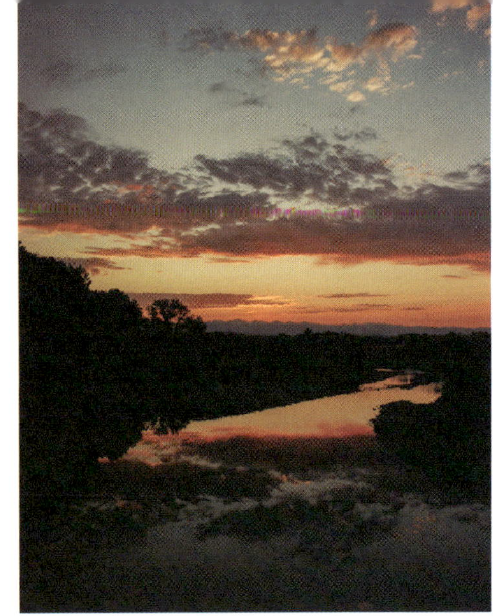

PAINTROCK CANYON RANCH

There is a rock wall close by that our creek is named after, scratched and painted upon by the people who lived here not so long ago. They roamed these canyons hunting, fishing, and gathering nuts and berries. When I am out on the Paintrock Canyon Ranch, I sense them sitting on the ridge, walking along the creek as I am, stalking a deer. Their connection to the land must have been similar to how I feel today. I think they felt protected by it, as I do. I have ridden and walked miles and miles across it, fished much of its waters, and hunted its canyons, yet I feel I have barely gotten to know it—an impossibility it seems. I can feel it gathering in me like the endless churn of water in the Paintrock Creek just outside my door. Even after I've been out on the land all day, when I come home and sit down to look out our glass doors, it pulls me back out for a walk up the creek.

This place is a refuge for me, this life with my wife, my children, and the people I work with. It is my identity. It is where I feel held together with them, as all ranching families I know also feel about their people and land. I suppose it's natural to feel this way when we are in it every day, but I also think it is because this specific parcel of land represents the culmination of the time we have spent building our business over the last couple of decades. All the other ranches we have leased; this is the first ranch we have owned.

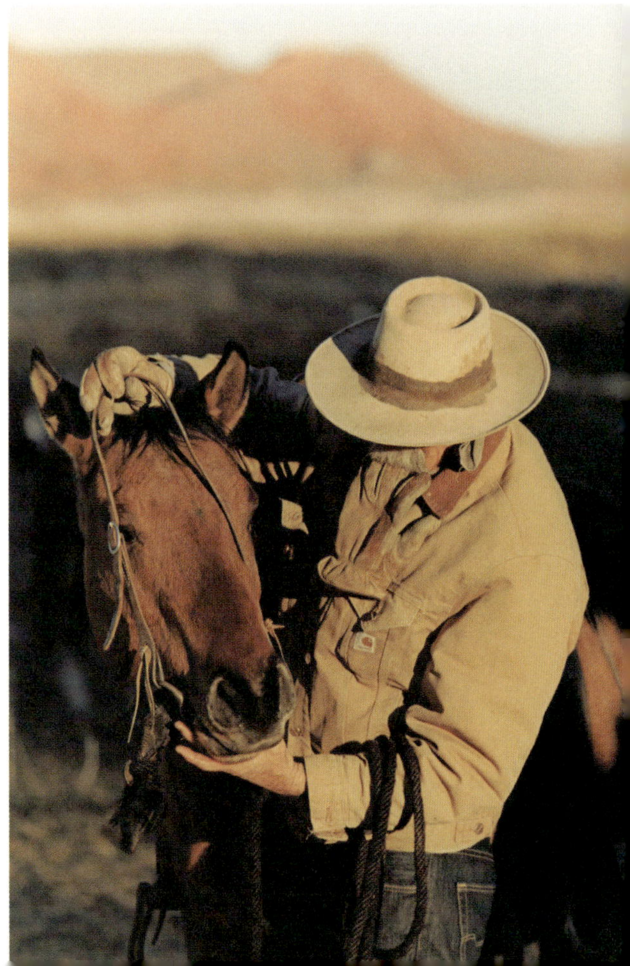

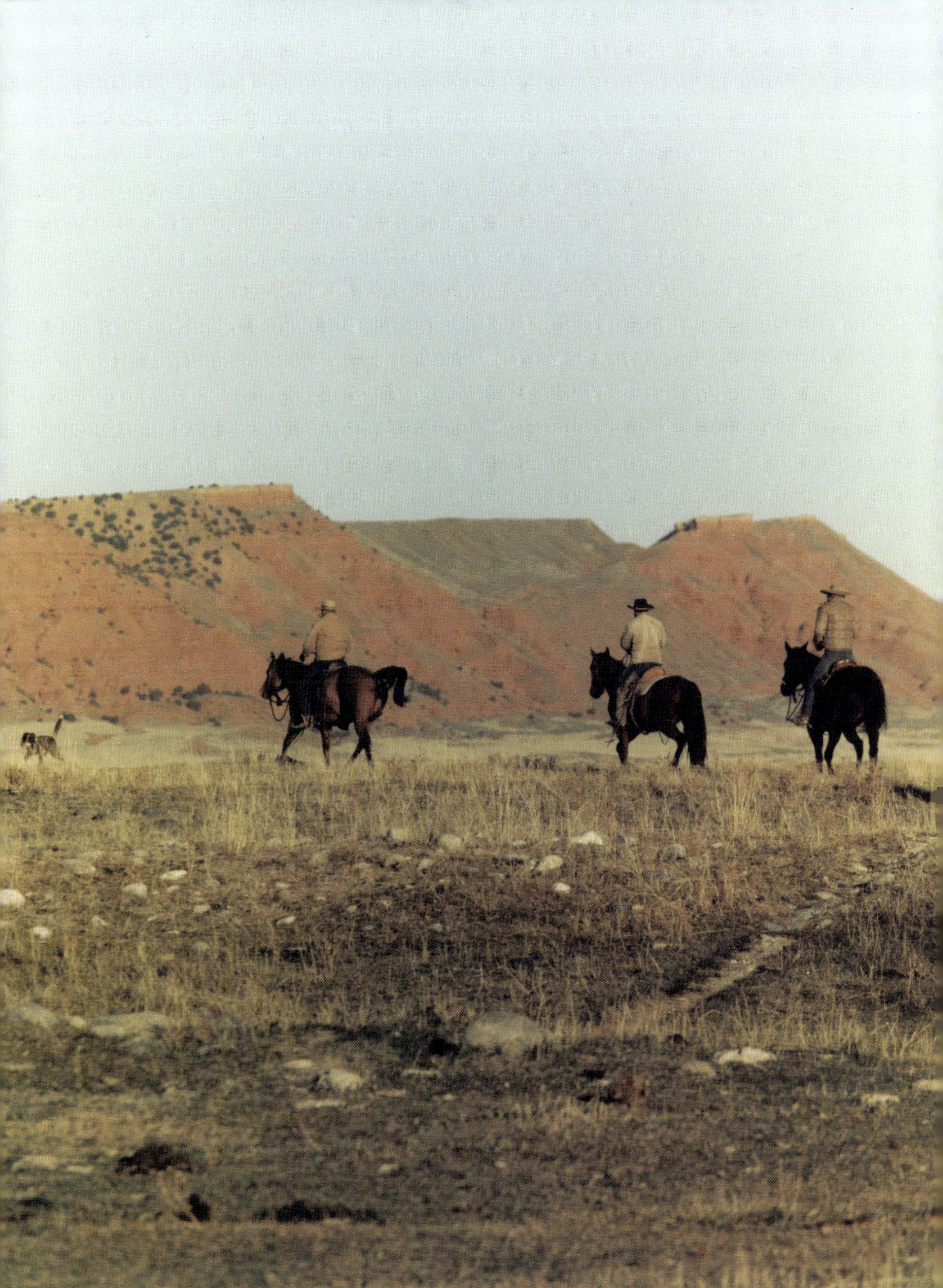

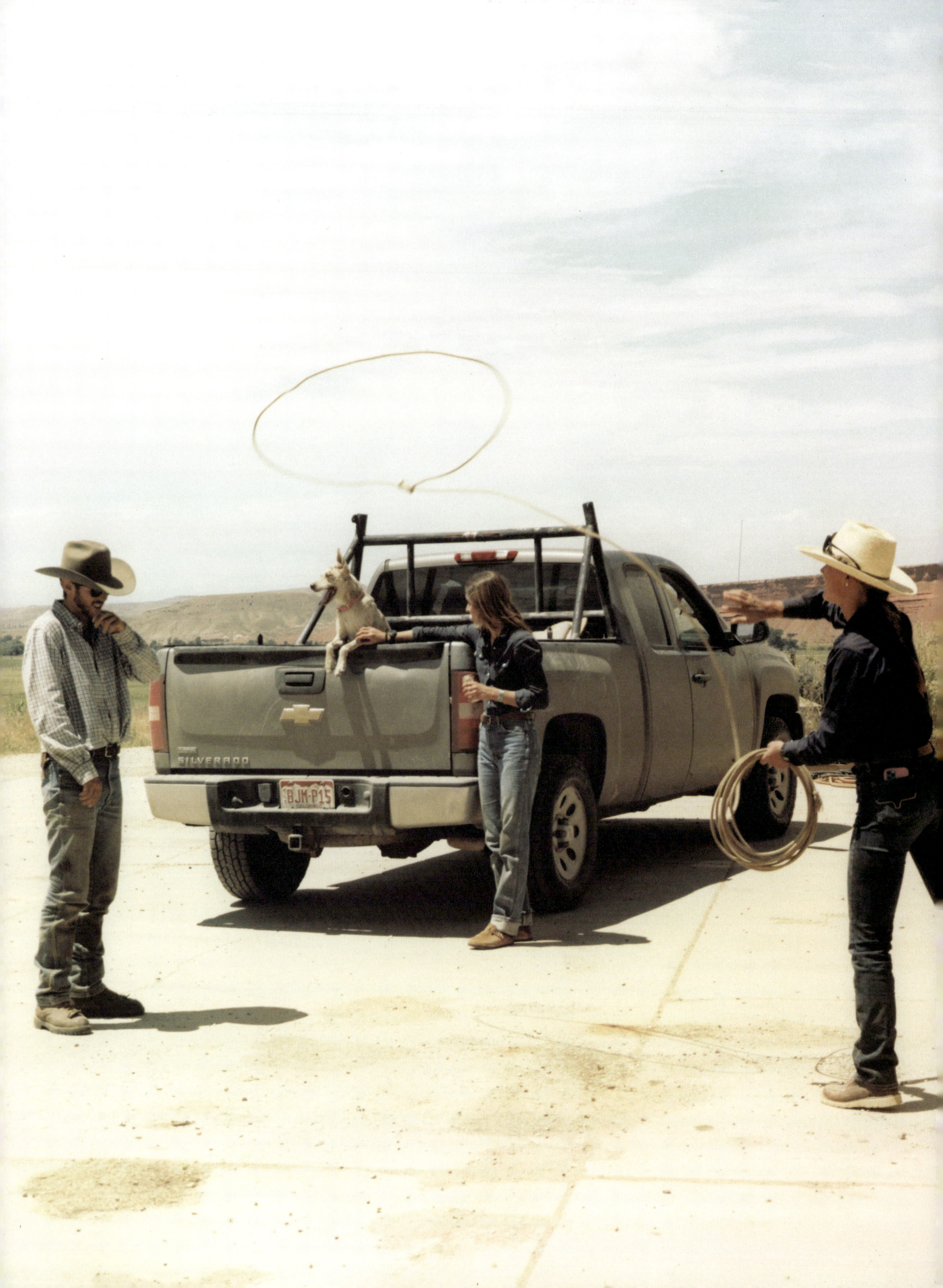
SILVERADO
BJM-P15

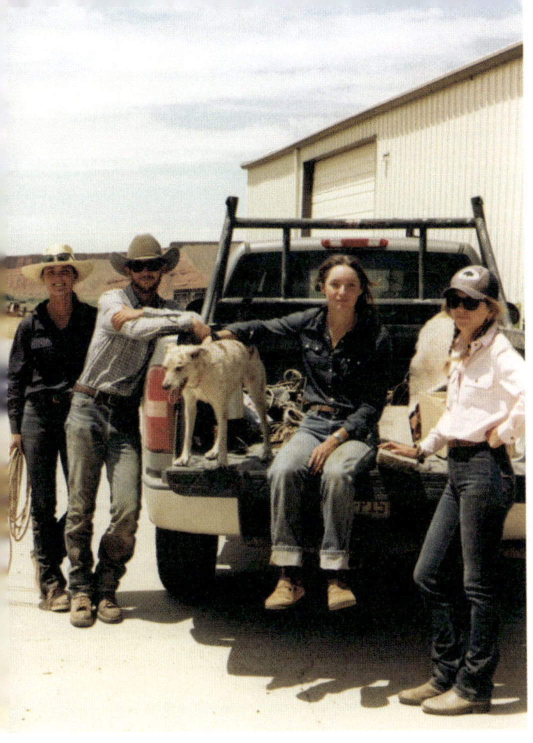

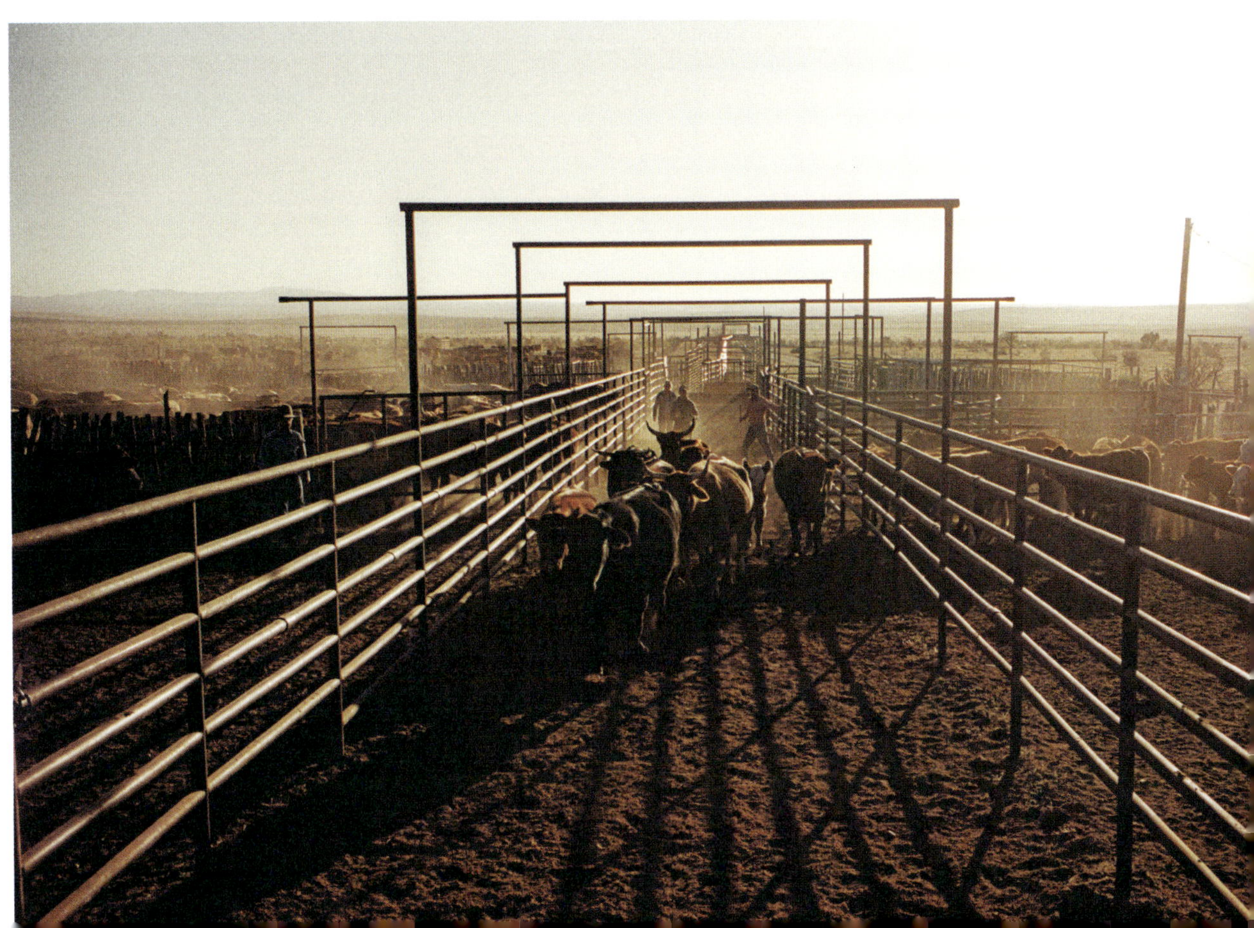

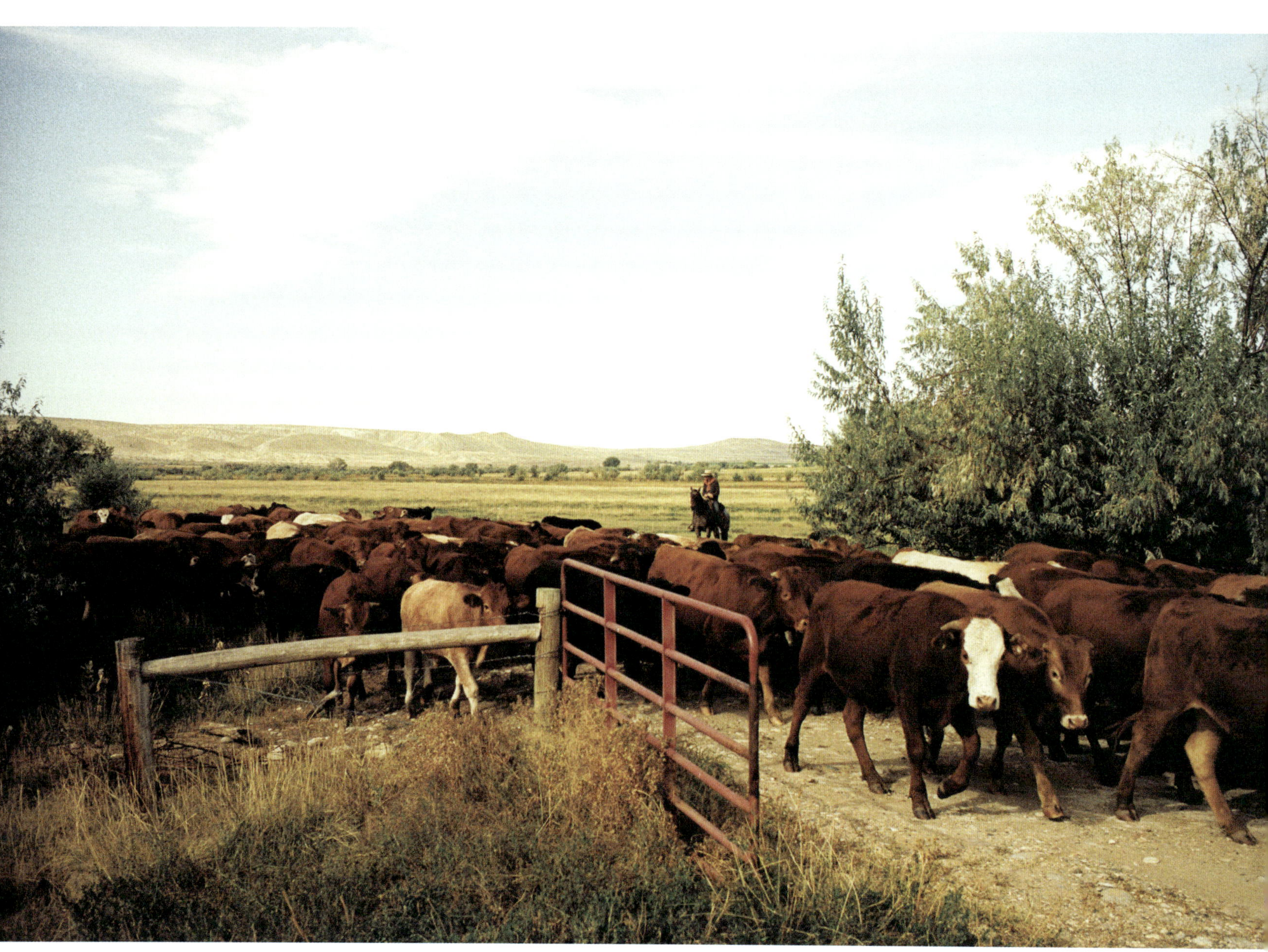

DUKE IV

"When I think about the idea of home, it's about the people. It's all about the memories and the experiences and what we've accomplished together. There are no borders across any of these ranches. In every way, we're all a family, intertwined and connected by this work, by these lands, and by one another."

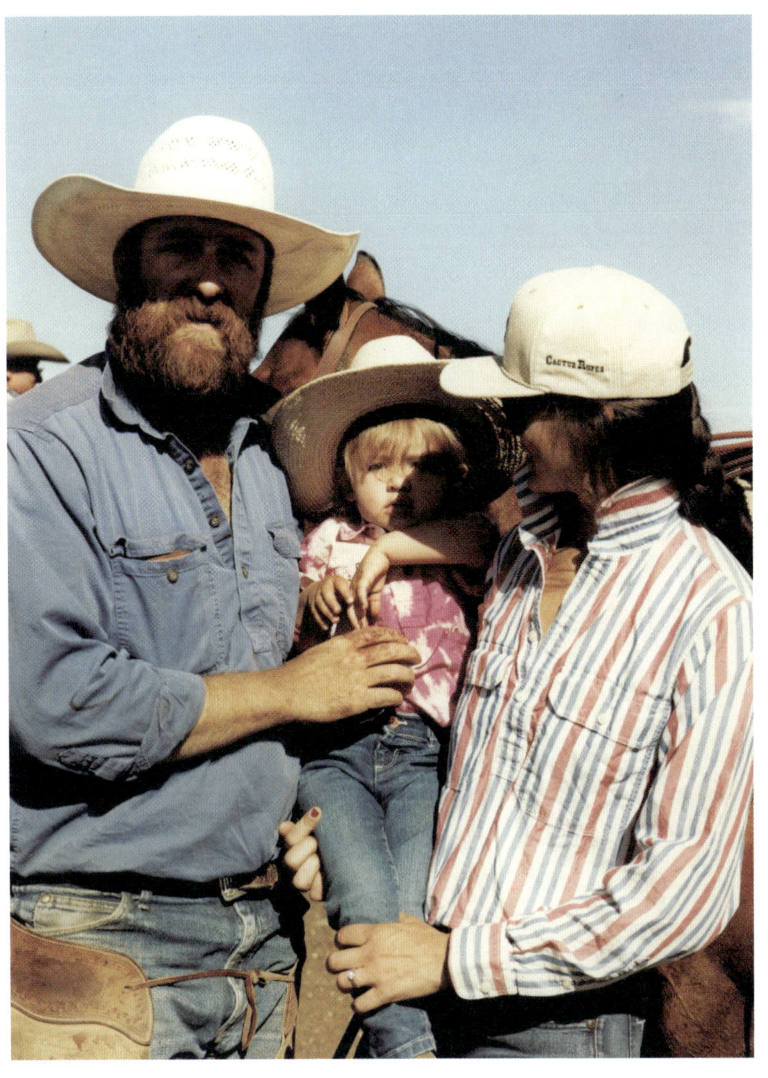

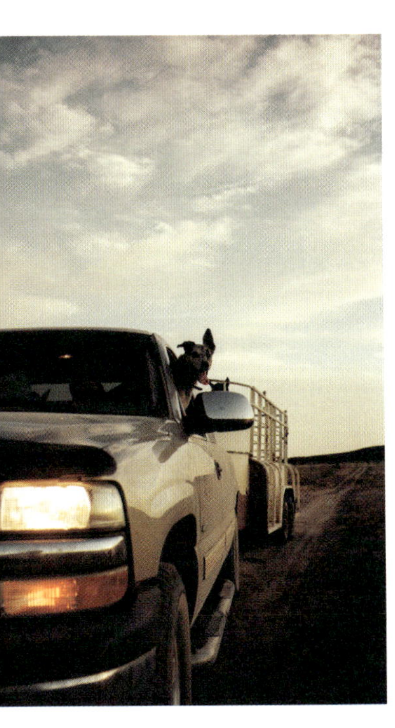

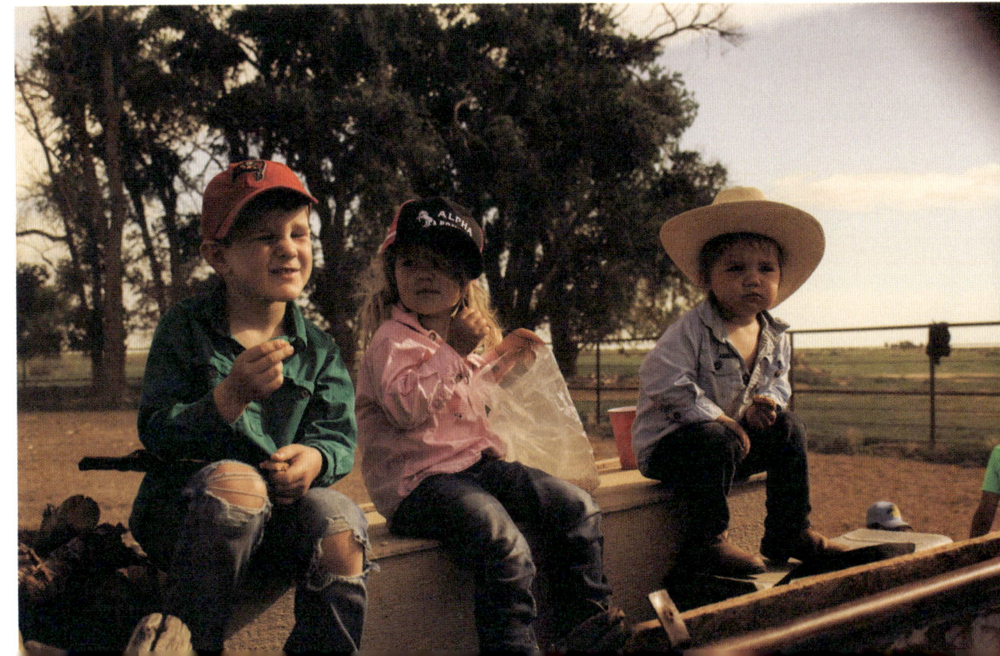

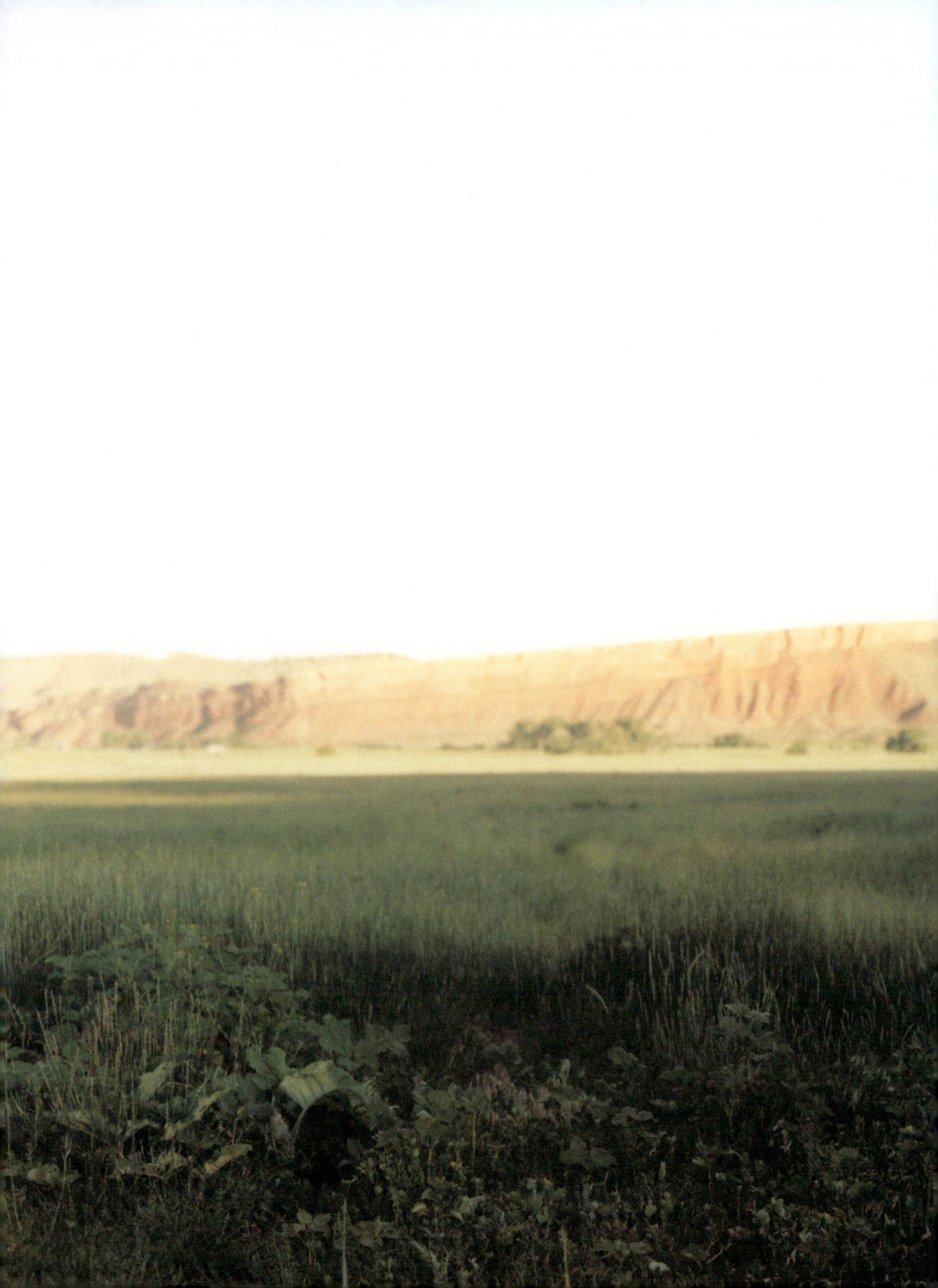

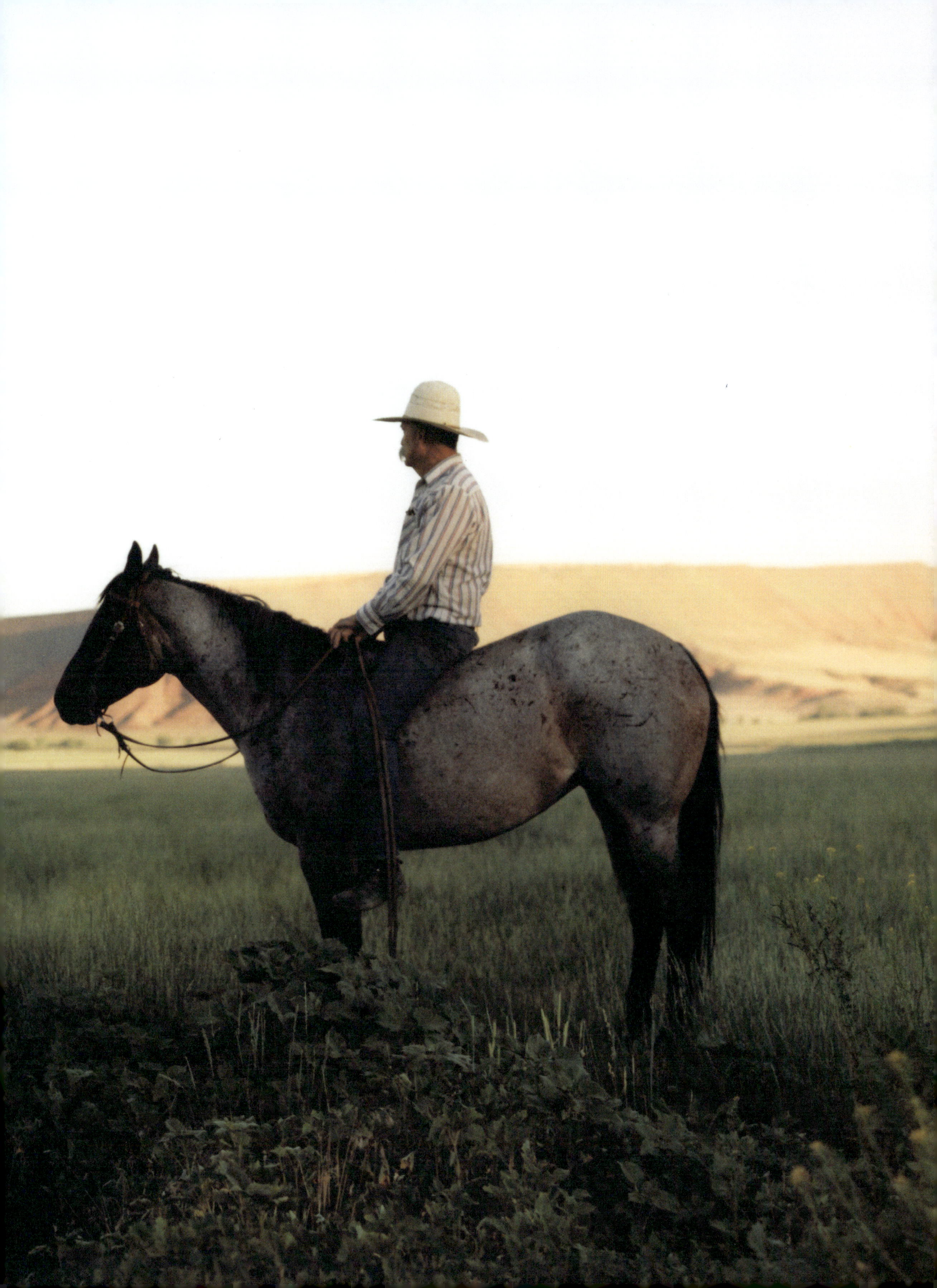

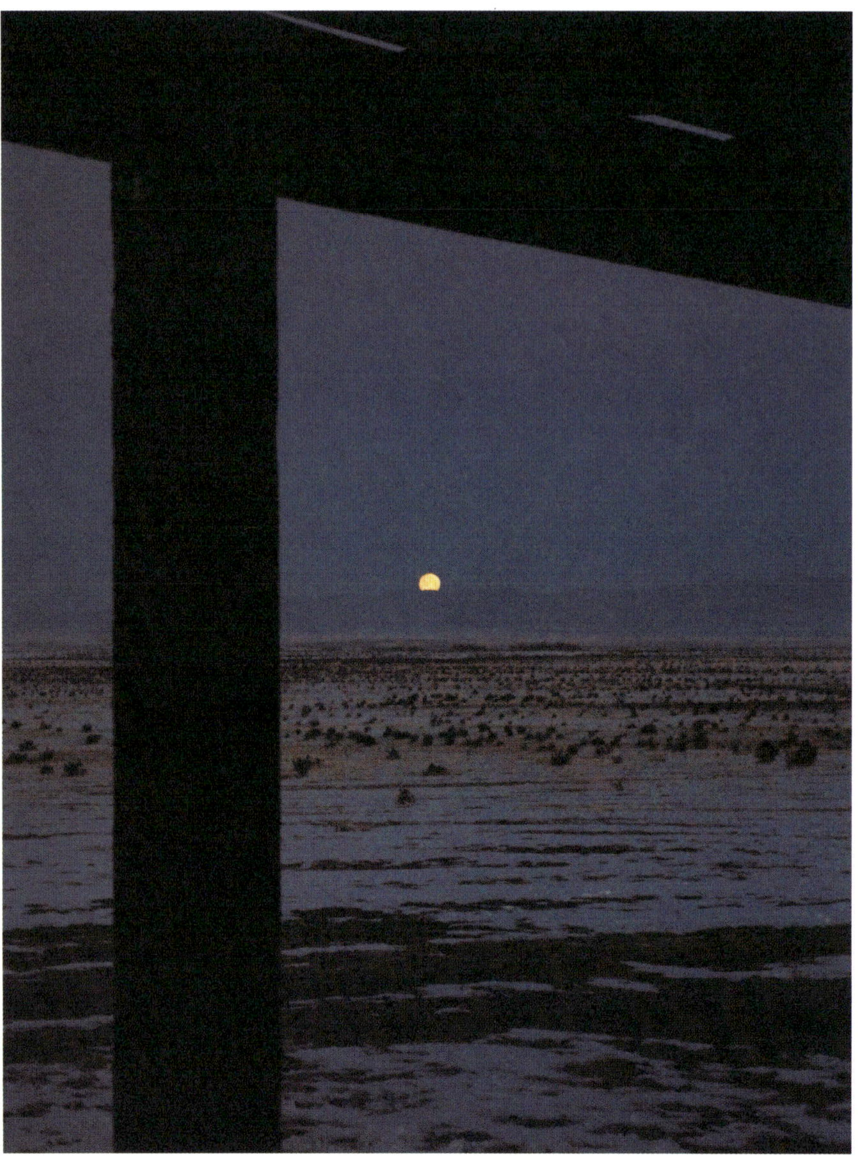

AFTERWORD

This book is a story about ranching in the American West. It is told through our family's journey of initiating and building our ranching business, which evolved into what is today Ranchlands. The images show what is typically seen on most large-scale, professionally run ranches in the United States. The words describe philosophies, circumstances, and values that exist in one form or another on most ranches and that guide us as a family, a business, and a collective of people who share a vision for ranching that reaches beyond our perimeter fences.

There were several purposes behind sharing our story. First, we wanted to show the beauty and romance that goes alongside living and working with the animals and lands that make up a ranch and exist as part of the natural world. Second, we wanted to describe what goes on behind the romance—things like why cattle genetics are important, the methodology around grazing techniques that build health and resilience within ecosystems, the everyday dimensions of ranching, and the closeness of the ranch community. Third, we wanted to communicate and inspire you, the reader, to understand the important role that ranching can play in preserving nature at a scale that nothing else can compare with. Last, we wanted to extend an invitation to you to join us, come visit us, be a part of our community, and become involved in a grassroots movement that has the potential to make real changes in the world.

We want our legacy to be that we have helped ranchers across the country perpetuate ranching into the future by preserving the very thing that sustains us—the land.

Thank you for picking up this book and reading our story. We hope you enjoyed it.

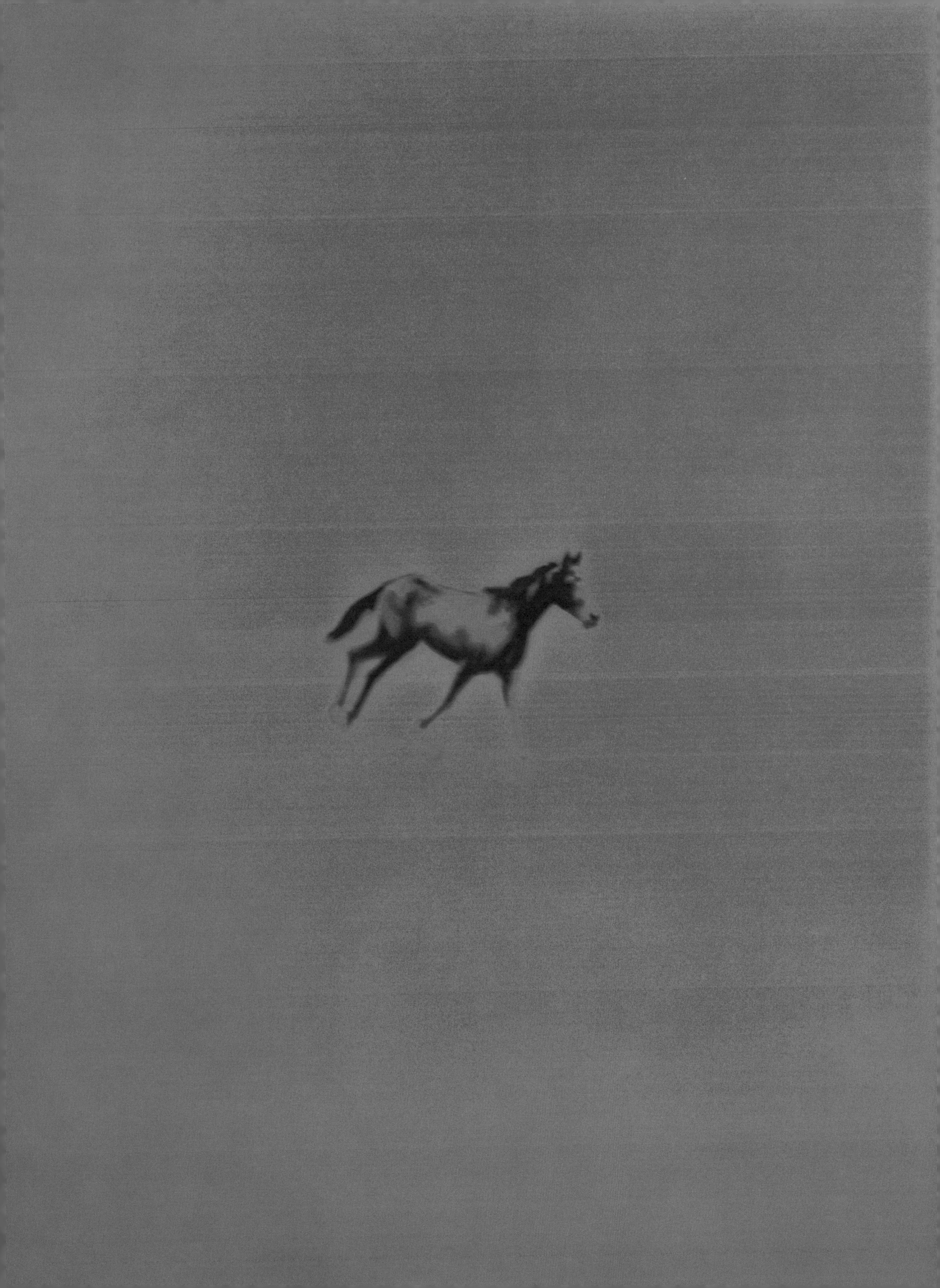

This book would not have been possible without the time and energy of Madeline Jorden and Molly Baldrige, nor the rest of the Ranchlands team. We extend a special thank-you to the landowners who entrust us with their ranches.

First published in the United States of America in 2024 by
Rizzoli International Publications, Inc.
300 Park Avenue South
New York, NY 10010
www.rizzoliusa.com

PUBLISHER: Charles Miers
ASSOCIATE PUBLISHER: James Muschett
MANAGING EDITOR: Lynn Scrabis
EDITOR: Candice Fehrman

PROJECT DIRECTOR: Parker Fitzgerald, RANSOM LTD.
PROJECT EDITOR: Sarah Rowland
DESIGN: David Blumberg, Jonathan Niega, CLARITY Studio
PROJECT COORDINATOR: Madeline Jorden, Ranchlands
TEXT: Duke Phillips III, Ranchlands
(except The Work on pages 176–181, which was written by Brennan Cira)

Printed in China

2024 2025 2026 2027 / 10 9 8 7 6 5 4 3 2 1

ISBN: 978-0-8478-3087-9

Library of Congress Control Number: 2024934626

Visit us online:
Facebook.com/RizzoliNewYork
Twitter: @Rizzoli_Books
Instagram.com/RizzoliBooks
Pinterest.com/RizzoliBooks
Youtube.com/user/RizzoliNY
Issuu.com/Rizzoli

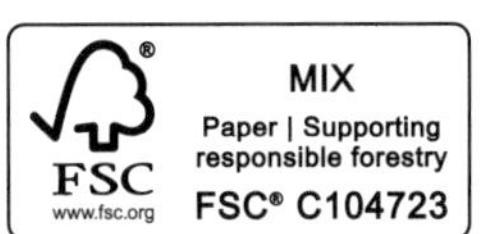

PHOTO CREDITS

FANNY LINDQVIST BJÖRK
12–13, 22–23, 34, 41, 128, 160–161, 184–185, 261 (left)

ISABEL BUTLER
250–251

JARED CHAMBERS
7, 29, 30, 46–47, 48, 67, 112 (both), 114–115, 170, 171, 172–173, 262 (top)

BRENNAN CIRA
35, 82, 88 (top), 89, 96 (bottom), 121 (bottom left), 163, 176, 240–241, 242, 243, 255

AVERY CLARK
9, 50–51, 60–61, 73, 98–99, 119 (bottom), 152–153, 204–205, 207, 211, 246–247

DAVEY JAMES CLARKE
83 (top), 86–87, 267 (bottom right)

MATTHEW DELORME
56–57, 66, 70–71, 74–75, 109 (bottom), 116, 121 (bottom right), 134–135, 146 (bottom), 147, 149 (bottom), 151 (both), 182–183, 186–187, 188–189, 190–191, 206, 212, 214–215, 216, 217, 244 (both)

JAMES FITZGERALD III
18–19

PARKER FITZGERALD
2–3, 8, 10, 20, 21, 26, 42, 43, 49, 65, 68, 69, 78, 84, 85, 92 (top), 93, 95, 97, 101, 102–105 (all), 120 (top), 121 (middle), 122, 138, 139, 140–143 (all), 144–145, 148, 149 (top), 166, 167, 177, 181 (bottom), 195, 196, 197, 198, 199, 200, 201, 202, 203, 208, 209, 218–219, 220 (left), 221(right), 223, 248 (both), 249, 257, 260 (left), 261 (right), 262 (bottom), 263, 264, 265 (top), 267 (top), 268–269

KATRINA FLYNN
76, 113, 120 (bottom)

MADELINE JORDEN
126–127, 222, 245

CLAUDIA LANDREVILLE
5, 14–15, 16, 17, 24–25, 32–33, 40, 44, 45, 54, 55, 58, 59, 63, 72, 83 (bottom), 90, 91, 96 (top), 108, 118, 124, 125, 129, 131, 133, 137, 146 (top), 150, 158, 162, 168–169, 174, 178, 179 (top), 180, 181 (top), 193, 210, 220 (right), 221 (left), 224, 225 (all), 226, 227, 228 (both), 229, 232 (both), 233 (both), 235, 236, 237, 238, 239, 252, 254, 270

SIERRA MACDONALD
38

DUKE PHILLIPS III
52–53, 92 (bottom), 106–107, 130, 159, 213

RANCHLANDS ARCHIVE
80, 81

DEREK SLAGLE
109 (top)

NIKLAS SÖDERLUND
1, 88 (bottom), 100, 121 (top), 154, 155, 179 (bottom), 230–231, 234 (both), 265 (bottom), 266, 267 (bottom left)

WES WALKER
36–37, 39, 62, 94, 110–111, 132, 136, 164–165

SAVANNAH WHITE
260 (right)

FOREST WOODWARD
119 (top), 123, 156–157, 256, 258–259

PAINTING CREDITS

DUKE BEARDSLEY
5 (series of his iconic lineup riders painted on Quonset at Chico Basin Ranch)

G. L. RICHARDSON
271 (Little Filly)